Practical Enzymology

Practical Enzymology

Austin Murphy

MURPHY & MOORE
www.murphy-moorepublishing.com

Published by Murphy & Moore Publishing,
1 Rockefeller Plaza,
New York City, NY 10020, USA
www.murphy-moorepublishing.com

Practical Enzymology
Austin Murphy

International Standard Book Number: 978-1-63987-445-3 (Hardback)

Cataloging-in-Publication Data

Practical enzymology / Austin Murphy.
p. cm.
Includes bibliographical references and index.
ISBN 978-1-63987-445-3
1. Enzymology. 2. Enzymes. 3. Biochemistry. I. Murphy, Austin.
QP601 .P73 2022
574.192 5--dc23

TABLE OF CONTENTS

PREFACE

Biological catalysts which assist in accelerating certain chemical reactions are called enzymes. These proteins are highly specific in nature and can only act under the optimal temperature and pressure. On exposure to excessive heat, enzymes get denatured and lose their catalytic properties. Enzymology is defined as the study of the kinetics, structures and function of these enzymes. The important biological applications of enzymes involve metabolism, transcription, muscle contraction, hormone release, etc. They also find use in the dairy industry, paper industry, brewing industry, biofuel industry and food processing. On the basis of functions performed, enzymes can be classified into oxidoreductases, transferases, hydrolases, lyases, isomerases and ligases. This book elucidates the concepts and innovative models around prospective developments with respect to enzymology. The various sub-fields of enzymology along with technological progress that have future implications are glanced at herein. Coherent flow of topics, student-friendly language and extensive use of examples make this book an invaluable source of knowledge.

A foreword of all chapters of the book is provided below:

Chapter 1 - Enzymes are the proteins that catalyze the biological reactions. The metabolic processes in the cells need enzymes to increase their rate of reaction. This is an introductory chapter which will briefly introduce all the significant aspects of enzymes such as enzyme technology, enzyme nomenclature and classification, enzyme structure, enzyme ionisable group, etc; **Chapter 2 -** The process of increase in the rate of a biological reaction by an enzyme is known as enzyme catalysis. A few of the significant topics studied in relation to enzyme catalysis and kinetics are reaction kinetics, Michaelis-Menten kinetics, multi-substrate systems, enzyme inhibition, etc. This chapter has been carefully written to provide an easy understanding of the varied facets of enzyme catalysis and kinetics; **Chapter 3 -** The choice of a source of enzymes is a crucial step in the process of production of enzymes. Large scale production of enzymes is achieved by fermentation. It is the metabolic process of breakdown of an organic substrate by the micro-organisms through the action of enzymes. This chapter covers in detail topics such as enzyme source, enzyme production, microbial enzymes and enzyme purification; **Chapter 4 -** The process of attaching an enzyme to an inert and insoluble material is known as enzyme immobilization. Some of the important topics associated with enzyme immobilization are support materials for enzyme immobilization, properties of immobilized enzymes, etc. These topics are crucial for a complete understanding of the subject; **Chapter 5 -** The process through which the sequence of amino acids is changed by using various techniques to design valuable and useful enzymes is known as enzyme engineering. Enzymes are used in numerous industries such as pharmaceutical industry, food industry, light industry and biotechnology. The diverse applications of enzyme engineering have been thoroughly discussed in this chapter.

At the end, I would like to thank all the people associated with this book devoting their precious time and providing their valuable contributions to this book. I would also like to express my gratitude to my fellow colleagues who encouraged me throughout the process.

Austin Murphy

Chapter 1

Enzyme: An Introduction

Enzymes are the proteins that catalyze the biological reactions. The metabolic processes in the cells need enzymes to increase their rate of reaction. This is an introductory chapter which will briefly introduce all the significant aspects of enzymes such as enzyme technology, enzyme nomenclature and classification, enzyme structure, enzyme ionisable group, etc.

Enzymes are synthesized by all living organisms including man. These life essential substances accelerate the numerous metabolic reactions upon which human life depends. Knowledge of enzyme activity in serum or plasma, blood cells. Homogenates and extracts of tissue and urine assists the physician in determining the origin of tissue damage and organ disease. Certain metabolites can be made to decompose to desired end products in a test tube. However, extremes in temperature, pH or salt are incompatible with life. The same metabolic processes occur in the body under physiologic conditions: a pH near neutrality and a temperature near 37E centigrade. This occurs through the action of enzymes–catalysts of biologic origin.

A chemical catalyst (enzyme) is a substance which increases the rate of a particular reaction without itself being consumed or permanently changed. A catalyst affects only the rate at which a reaction precedes, that is, the rate at which a reaction reaches equilibrium, it does not enable the reaction to proceed past its normal equilibrium point. Because catalysts are not altered during the reaction, they are capable of carrying on their catalytic effects repeatedly. Consequently, very small quantities of catalysts are capable of affecting relatively large quantities of the reactants.

Enzymes are proteins which catalyze specific metabolic reactions. They are synthesized in the cell and are important in its functions. Under normal conditions the activities of many enzymes in the blood are held at relatively constant levels by the balance between enzyme synthesis and breakdown. The blood serum levels of a particular enzyme may be increased or decreased by diseases that lead to increased rates of enzyme release; to increased amounts available for release; or to decreased rate of enzyme breakdown Changes in blood serum enzyme activity can thus be indicative of bodily disorders as well as in diagnosis, prognosis, and in assessing therapy effectiveness, enzyme determinations in tissue are generally still relegated to the research laboratory. However, Enzyme determinations in serum, and less often in other body fluids are now routine in many clinical laboratories.

As biologic catalysts, enzymes retain the characteristics of chemical catalysts: they increase the reaction rate, they remain unchanged after the reaction, and they are highly efficient. Because they are unaltered, a single enzyme molecule may interact with an astronomical number of reactant molecules. For example, one mole of enzyme may interact with 10,000 to 1,000.000 moles of reactant molecules per minute. Some enzymes are so specific that they catalyze a single type of reactant. For example, urease, which acts only on urea, others may act on many substrates. For example, alkaline phosphatase which may act on phenyl phosphate, beta glycine phosphate, etc.

As proteins, enzymes possess all the properties in common with other proteins, having molecular weights on the order of 10,000 to well over a million. The ribonuclease a molecule shown at the beginning of this presentation is relatively "small", having a molecular weight of "only" 12,700. Enzymes are labile: subtle changes in their structure, called denaturation, can cause them to lose their activity. Consequently, they must be handled quite carefully. Adverse conditions of temperature, pH, and salt concentration are only a few factors known to cause denaturation of an enzyme.

As proteins, enzymes are multivalent electrolytes containing ionizable groups. The ionization state, which affects the enzyme activity, depends on the hydrogen ion concentration, that is, pH. As electrolytes, enzymes migrate in an electrical field. Positive molecules migrating towards the cathode: negative molecules migrating towards the anode.

Naming of and Types of Enzyme Reactions

Enzymes are named according to the starting material Or chemical group they act upon by adding the suffix "ASE". Consequently, the enzyme hydrolyzing urea is called urease. That acting on phosphate esters is called phosphatase, and so on. Enzymes of diagnostic value may be classified as transferases: Oxidoreductases. Hydrolases, and lyases.

- Transferases catalyze the transfer of chemical groups, such as an amino or phosphate group, from one compound to another.
- Oxidoreductases, also referred to as dehydrogenases. Catalyze electron transfer.
- Hydrolases catalyze substrate breakdown with water fixation.
- Lyases are enzymes that split carbon-to-carbon bonds without the transfer of chemical groups.

Enzyme-Substrate Complex

Described simply, enzymatic reactions involve two steps, the first step is a combination of the enzyme (E) with the substrate (S) to form an enzyme-substrate complex (ES), the second step is the decomposition of the enzyme substrate complex to yield the reaction product (P) and the unchanged enzyme.

The velocity of an enzyme reaction will depend on the time required for the enzyme-substrate combination, and the time required for the enzyme to form the product. Obviously, the time required for the formation of the enzyme-substrate complex will depend on the enzyme concentration, and the substrate concentration.

Activity

The ability of an enzyme to catalyze a specific reaction is called its "activity". Activity is a measure of the rate at which the reaction proceeds. Enzymes are measured in terms of activity rather than concentration because of their chemical similarities, and their extremely low concentrations in biological fluids. The international unit of enzyme activity is defined as the amount of enzyme which will catalyze the transformation of one micro-mole of substrate per minute under certain specified conditions such as pH and temperature. A milli-unit is a thousandth of an international unit. Enzyme activity is normally expressed in international units per milliliter of enzyme-containing solution, or where this value is inconveniently small: as milli-units per millilitre.

Temperature

Enzyme activity depends on a number of factors such as temperature, pH, substrate, coenzymes and inhibitors. Increasing the temperature increases the rate of enzyme activity. At the same time, denaturation of the protein can result in the loss of enzyme activity, this thermal deactivation depends upon the length of time an enzyme is exposed to the particular temperature. Thermal denaturation is negligible for most enzymes below about 30°C. However, between 30° and 40° some denaturation occurs. At 60° centigrade, most enzymes are denatured after only a few minutes.

In 1961 the commission of enzymes of the international union of biochemistry recommended that all enzyme measurements be made at 30° centigrade. However, for historical reasons, convenience, or where low enzyme activity and small samples are involved, other temperatures such as ambient. 25°, 32E, and 37E continue to be used.

Generally speaking, for each 10E rise in temperature, many enzymes will exhibit a 10 two-fold increase in activity (Q_{10}). Temperature difference of only 1°C changes the activity of some enzymes by as much as 4-10%. Therefore, it is essential to carefully control the temperature of specimens, solutions, and spectrophotometer cells during enzyme assays.

A characteristic bell shaped curve is obtained when enzyme activity is plotted against pH. The pH optimum, the pH at which greatest activity occurs, corresponds to the vertex of the curve. This pH optimum depends on the environment in which. The reaction takes place. For example, the nature of the buffer, the nature and concentration of the substrate and the presence of activators or inhibitors can all affect the optimum pH.

Velocity

If all the other variables in an enzyme reaction remain constant, the reaction velocity depends on the substrate concentration (S). At relatively large substrate concentrations the enzyme becomes saturated with substrate and the reaction velocity reaches a maximum max (V_{max}). Any further addition of substrate will not significantly increase the reaction velocity.

During the period where the substrate concentration is low with respect to the enzyme concentration, the reaction is termed "first order". Under these conditions the reaction rate will be greatly influenced by the substrate concentration. At high substrate concentrations, the reaction rate reaches a maximum and the reaction is termed "zero order". Under these conditions, the reaction rate becomes independent of the substrate concentration.

When measuring enzyme activity only the zero order part of the curve should be used so that the assay is a reflection of enzyme activity only.

The Michaelis-Menton constant, k_m, can be determined from the substrate concentration curve examined earlier. K_m is numerically equal to the concentration of the substrate, in moles per liter, when the velocity of the enzyme reaction reaches one-half its maximum velocity, V_{max}. It is a number characteristic of each enzyme pair. In addition to having practical applications in the clinical laboratory, this constant can be useful in the development of new methods for measuring enzyme activity. For example, the affinity of a non-specific enzyme for different m's substrates, as defined by their $K_{m's}$, can help in selecting the appropriate substrate.

Coenzymes

Enzymes often require the presence of certain organic. But non-protein factors called "coenzymes". They are relatively low molecular weight with respect to enzyme proteins and are often required for the correct positioning of the enzyme and substrate so that a reaction can proceed. Many coenzymes are derived from dietary vitamins, especially the b-complex vitamins. For example, niacin is the precursor of two coenzymes, nicotinamide adenine denucleotide or nap, and nicotinamide adenine dinucleotide phosphate or NADP. NAD and NADP are essential for the activity of a group of dehydrogenases, inorganic ions such as magnesium. Manganese, zinc, and iron are also sometimes regarded as coenzymes since these metal ions have been found necessary to activate certain enzymes.

In either case, coenzymes must be added to many enzyme assay systems in sufficient amounts so as not to be limiting factors in the reaction.

Inhibitors

Inhibitors can cause enzyme reactions to proceed more slowly than expected for given conditions of pH, temperature, and substrate-coenzyme concentrations. Inhibitors may be substances which compete for the active sites on enzymes and generally have chemical structures very similar to those of the substrates. That is, the inhibitors compete with the substrates for the enzymes. For example, hexokinase, which acts upon d-glucose is inhibited by d-xylase or 6-deoxy d-glucose.

An inhibitor may also slow a reaction by binding to the enzyme at a point other than that at which the substrate is bound, in this case, the inhibitor alters the enzyme structure so as to decrease, or even stop the enzyme activity. An important example of this is fluoride ion which inactivates enolase. Enolase is involved in the metabolism of glucose. Consequently, fluoride ion is often used as a preservative in glucose determinations. Inhibitors have been used to gain information on complex enzyme systems, and to better understand single reaction mechanisms.

Naturally occurring inhibitors in blood serum do not have a significant influence on in vitro assays of enzymes. However, urine frequently contains inhibitors which dampen certain enzyme assays and the urine must be dialyzed prior to enzymatic analysis.

Iso-Enzymes

Enzymes which catalyze the same chemical reaction but exhibit differences in their biochemical, physical, and immunological properties are termed Iso-enzymes. Chief among these differences are differences in their rates of electrophoretic migration. For example, under electrophoresis the LDH enzyme can be separated into five distinct fractions.

The various bodily organs and tissues contain characteristic proportions of the different Iso-enzymes. When a specific organ or tissue is damaged, the cellular enzymes are released into the blood plasma where they superimpose their Iso-enzyme patterns on those normally present: case in point - cardiac muscle release Iso-enzymes following a heart attack and, thus, are used to clinically confirm such a medical event. Consequently, a measure of the serum Iso-enzymes patterns obtained by electrophoresis can often reveal the damaged tissue.

Modern Equipment

Modern ultraviolet visible spectrophotometers now permit enzyme activities to be measured conveniently, quickly, and accurately. If the substrate acted upon or the product formed are colored, or can be coupled to secondary reactions which produce colored products, then the reaction can be monitored with a colorimeter, that is, a spectrophotometer measuring changes in visible light. Quantitative ultraviolet measurements of enzyme activity are becoming the more accepted approach, however. This preference arises for a number of reasons, but mainly because ultraviolet methods avoid the additional steps, reagents, and undesirable colored side reactions usually required producing colored reactants and products. For example, a very common ultraviolet approach involves NAD and NADH. As shown in ultraviolet spectrum, NAHD has a characteristic and relatively high absorption at 340 nanometers. NAD has virtually no absorption at the same wavelength. Consequently, any changes in the NADH concentration, either increasing or decreasing, can easily be monitored with little or no interference due to corresponding changes in NAD concentrations.

A specific example of the usefulness of these ultra violet absorption characteristics is in the assay of LDH enzyme. In one case, pyruvate and NADH are catalyzed by LDH-P to produce lactate and NAD. The NADH concentration would decrease and produce decreasing ultraviolet absorption. In the other case, the opposite reaction is catalyzed by LDH-L to produce increasing concentrations of NADH.

Enzyme reactions which do not involve NADH or NAD may often be coupled with other reactions which do involve these compounds; for example, in GOT assays. Aspartate and alpha-ketoglutarate may be catalyzed by GOT to produce glutamate and oxalacetate. Coupled to the first reaction, involves the generated oxalacetate, NADH, and a second enzyme, malic dehydrogenases to produce malate. Consequently, for every mole of oxalacetate produced by GOT, a mole of NADH is consumed and the absorbance change is directly proportional to GOT activity. There are several ultra-violet visible spectrophotometric methods which are commonly used to determine enzyme activity levels. The two-point procedure involves measuring the amount of product formed, or the decrease in substrate, over a given interval over time. For example, blood serum samples are added to substrate solutions of known concentration and, after a period of incubation, the substrate or product is again measured. The difference then represents the amount of substrate transformed during the incubation period and provides an indirect measure of enzyme activity.

Enzyme Technology

Enzyme technology is primarily engaged in the production, isolation, purification and use of enzymes either in the soluble or immobilized form, for the benefit of humankind. With the advancements in, recombinant DNA technology, enzyme engineering produces more effective and diverse group of enzymes with useful applications in microbiology, biochemistry, diagnostics, therapeutics, biocatalysis, structural biology etc. The overall objective of this emerging technology is to produce unique sustainable products with specific function to fulfil the need of growing population.

Enzyme Production Technology

The basic steps taken for enzyme production include:

- Organism selection: Selection of organism depends on the ability of microorganism to yield increased concentration of desired enzyme in a short time period with decreased amounts of secondary metabolites. Microbial inoculums introduced in sterilized liquid medium for fermentation, providing optimal growth conditions such as pH, temperature, O2 supply, nutrient.
- Isolation and purification of enzyme: Desired enzyme produced may be excreted into the culture medium or may be present within the microbial cells is recovered and purified by cell disruption techniques and downstream processing respectively ensuring minimal loss of enzyme activity.
- Immobilization of enzyme: Immobilization of enzyme is achieved on cellulose, polyacrylamide, starch and beads etc., by the following methods:
    - Adsorption: It involves the attachment of enzyme molecule on an inorganic or organic inert solid support such as silica gel, beads or glass, starch, cellulose etc., by forming weak van dar Waals forces or hydrogen bonding. Later enzymes are removed by slight alterations in pH, ionic strength or temperature.
    - Cross linking: Enzyme can be immobilized by forming cross-linkage to another chemical such as cellulose or glyceraldehydes but it might cause denaturation of enzyme.
    - Entrapment: Enzymes can be immobilized by physical entrapment inside a mesh, capsule or a gel matrix of an inert material such as gelatin, polyacrylamide gel, starch, collagen, silicone, cellulose and rubber.
    - Covalent binding: Immobilization of the enzymes by means of covalent bonding with the support providing the strongest enzyme-support interaction.

Applications of Enzyme Technology

- Textile Industry:
    - Amylases enzyme is used as softening agents for starched clothes.
    - Cellulases is useful for fabric finishing and often used for better dye uptake in cotton.
- Leather Industry:
    - Proteolytic enzymes are used in dehairing of the skin and for softening /plumping of dehaired skin.
- Detergent Industry:
    - Proteases are efficient in removing protein stains such as grass, blood, egg and human sweat.
    - Lipases can easily break down lipid and protein based stains.

- Amylases is efficient in removing starch-based residues.
- Cellulase used in detergent causing restoring fibre smoothness, its original color and softening.

- Pharmaceutical Industry:
  - Pegadamase bovine is employed for SCID treatment.
  - Pancreatic enzymes is effective against fat malabsorption in HIV patients.
  - Spinal injuries could be treated with chondroitinases.
  - Hyaluronidase aids in the rebuilding of damaged nerve tissues.
  - Lysozyme is used as an antibacterial agent and also possess activity against HIV.
  - Human melanoma and hepatocellular carcinomas could be inhibited by PEGylated arginine deaminase.
- Food Industry:
  - Amylase is added to the dough to ensure its quality and increased shelf life.
  - Enzymes are also employed in cheese making by milk coagulation.
  - Proteolytic enzyme help to increase the shelf life of meat and sea food.
  - Lipases used to control the lipid content of food products.
  - Glucose oxidase and catalase are used as food preservative agents.
  - Renin is used in the manufacture of cheese.
  - Lactase enzymes is applied in the making of ice cream, yoghurt and frozen and invertase in the making of chocolate covered berries and candies.
  - Glucose isomerase is used for the production of fructose and high fructose syrups.

Enzyme Nomenclature and Classification

The nomenclature of enzymes is derived from their substrates or the catalyzed chemical reactions, and "ase" is usually added as a suffix. Enzymes can be indexed with letters and numbers: the letter EC plus four numbers representing four elements. The first number represents enzymes that are classified according to the mechanism of enzymatic reaction.

Classification

According to the type of reactions that the enzymes catalyze, enzymes are classified into seven categories, which are oxidoreductases, transferases, hydrolases, lyases, isomerases, ligases, and translocases. Oxidoreductases, transferases and hydrolases are the most abundant forms of enzymes. Individual enzyme classes are further classified systematically based on the chemical name of the substrate and its reaction mechanism.

Enzyme class	Reaction type	Description
EC 1 Oxidoreductases	$A_{red} + B_{ox} \rightleftharpoons A_{ox} + B_{red}$	Catalyze redox reaction and can be categorized into oxidase and reductase.
EC 2 Transferases	$A\text{-}B + C \rightarrow A + B\text{-}C$	Catalyze the transfer or exchange of certain groups among some substrates.
EC 3 Hydrolases	$A\text{-}B + H_2O \rightarrow A\text{-}H + B\text{-}OH$	Accelerate the hydrolysis of substrates.
EC 4 Lyases	$A\text{-}B \rightleftharpoons A + B$ (reverse reaction: synthase)	Promote the removal of a group from the substrate to leave a double bond reaction or catalyze its reverse reaction.
EC 5 Isomerases	$A\text{-}B\text{-}C \rightleftharpoons A\text{-}C\text{-}B$	Facilitate the conversion of isoisomers, geometric isomers or optical isomers.
EC 6 Ligases	$A + B + ATP \rightarrow A\text{-}B + ADP + P_i$	Catalyze the synthesis of two molecular substrates into one molecular compound with the release energy.
EC 7 Translocases		Catalyze the movement of ions or molecules across membranes or their separation within membranes.

Each group of enzymes in the above seven categories can be further divided into several subgroups according to the characteristics of the functional groups or bonds in the substrates. In order to show the properties of substrates or reactants more accurately, each subclass is further divided into subclasses and directly contains a quantity of enzymes.

Moreover, on the basis of the molecular composition, enzymes can be divided into pure enzymes and binding enzymes. Enzymes containing only protein are called pure enzymes. Binding enzymes are composed of proteins and cofactors. Only when the two components are combined, can the enzyme have catalytic activity.

Oxidoreductases

Oxidoreductases consist of a large class of enzymes catalyzing the transfer of electrons from an electron donor (reductant) to an electron acceptor (oxidant) molecule, generally taking nicotinamide adenine dinucleotide phosphate (NADP) or nicotinamide adenine dinucleotide (NAD) as cofactors. Since so many chemical and biochemical transformations comprise oxidation/reduction processes, it has long been an important goal in biotechnology to develop practical biocatalytic applications of oxidoreductases. During the past few years, significant breakthrough has been made in the development of oxidoreductase-based diagnostic tests and improved biosensors, and the design of innovative systems for the regeneration of essential coenzymes. Research on the construction of bioreactors for pollutants biodegradation and biomass processing, and the development of oxidoreductase-based approaches for synthesis of polymers and functionalized organic substrates have made great progress. Proper names of oxidoreductases are in a form of "donor:acceptor oxidoreductase"; while in most cases "donor dehydrogenase" is much more common. Common names also sometimes appeared as "acceptor reductase", such as NAD^+ reductase. "Donor oxidase" is a special case when O_2 serves as the acceptor.

Classification

Oxidorecuctases can be either oxidases or dehydrogenases. Oxidases are generally involved when molecular oxygen functions as an acceptor of hydrogen or electrons. However, dehydrogenases work by oxidizing a substrate through transferring hydrogen to an acceptor that is either NAD/NADP or a flavin enzyme. Peroxidases, hydroxylases, oxygenases, and reductases also belong to oxidoreductases. Peroxidases are placed in peroxisomes, and could catalyze the reduction of hydrogen peroxide. Hydroxylases give hydroxyl groups to its substrates. Oxygenases could incorporate oxygen from molecular oxygen into organic substrates. In most cases, reductases can act like oxidases, but catalyzing reductions.

Oxidoreductases are sorted as EC 1 in the EC number classification of enzymes and can be further classified into 22 subclasses.

EC number	Description
EC 1.1	Act on the CH-OH group of donors
EC 1.2	Act on the aldehyde or oxo group of donors
EC 1.3	Function on the CH-CH group of donors
EC 1.4	Functioning on the CH-NH_2 group of donors
EC 1.5	Act on CH-NH group of donors
EC 1.6	Act on NADH or NADPH
EC 1.7	Take other nitrogenous compounds as donors
EC 1.8	Act on a sulfur group of donors
EC 1.9	Act on a heme group of donors, respectively
EC 1.10	Treating diphenols and related substances as donors
EC 1.11	Act on peroxide as an acceptor (peroxidases)
EC 1.12	Act on hydrogen as donors
EC 1.13	Act on single donors with incorporation of molecular oxygen
EC 1.14	Function on paired donors with incorporation of oxygen
EC 1.15	Act on and act on superoxide radicals as acceptors
EC 1.16	Oxidize metal ions
EC 1.17	Take effect on CH or CH_2 groups
EC 1.18	Apply iron-sulfur proteins as donors
EC 1.19	Take reduced flavodoxin as donors
EC 1.20	Dispose phosphorus or arsenic in donors
EC 1.21	Form a X-Y bond from X-H and Y-H bond
EC 1.97	Some other oxidoreductases

Reactions

The catalyzed reactions are similar to the following reaction in Figure, where A is the reductant and B is the oxidant. In biochemical reactions, the redox reactions are sometimes more difficult to observe, such as this reaction from glycolysis: Pi + glyceraldehyde-3-phosphate + $NAD^+ \rightarrow$ NADH + H^+ + 1, 3-bisphosphoglycerate, where NAD^+ is the oxidant (electron acceptor), and glyceraldehyde-3-phosphate functions as reductant (electron donor).

Redox reaction:

$$A_{red} + B_{ox} \rightarrow A_{ox} + B_{red}$$

Functions

Oxidoreductase enzymes play significant roles in both aerobic and anaerobic metabolism. They can be found in biological procedures like glycolysis, TCA cycle, oxidative phosphorylation, and amino acid metabolism. In glycolysis, the enzyme glyceraldehydes-3-phosphate dehydrogenase accelerates the reduction of NAD^+ to NADH. However, the re-oxidization of the generated NADH to NAD^+ occurs in the oxidative phosphorylation pathway in order to maintain the redox state of the cell. Additional NADH molecules are produced in the TCA cycle. The glycolysis product pyruvate takes part in the TCA cycle in a form of acetyl-CoA. During anaerobic glycolysis, the oxidation of NADH is accomplished through the reduction of pyruvate to lactate, which is then oxidized to pyruvate in muscle and liver cells. Moreover, the pyruvate is further oxidized in the TCA cycle. All twenty of the amino acids, except for leucine and lysine, can be degraded to intermediates in TCA cycle, which allows the carbon skeletons of the amino acids to be converted into oxaloacetate and subsequently into pyruvate. The gluconeogenic pathway can then exploit the formed pyruvate.

Transferases

Transferases are a class of enzymes that transfer specific functional groups from one molecule (donor) to another (acceptor). Transferases are implicated in hundreds of different biochemical pathways, and are essential to some of most important processes in lives. Transferases participate in a myriad of cell reactions and are also utilized during translation. Mechanistically, an enzyme catalyzing the following reaction would be considered as a transferase:

Redox reaction:

$$A_{red} + B_{ox} \rightarrow A_{ox} + B_{red}$$

Where X is the donor that is often a coenzyme, and Y is the acceptor. Group would be the functional group that is transferred on account of transferase activity.

Nomenclature

Systematic names of transferases are based on the form of "donor:acceptor grouptransferase." For example, methylamine:L-glutamate N-methyltransferase is the normative name for the transferase methylamine-glutamate N-methyltransferase, where methyltransferase is the EC category, methylamine is the donor, and L-glutamate is the acceptor. Nonetheless, the more frequently used nomenclature for transferases are often in a form of "acceptor grouptransferase" or "donor grouptransferase." Practically, many molecules are not mentioned by taking advantage of this terminology because of the application of more prevalent common names.

It occurs as early as the 1930s that some of the most important transferases were discovered. Transamination that means the transfer of a NH2 group from an amino acid to a keto acid through an aminotransferase is noticed for the first time in 1930 after the disappearance of glutamic acid appended to pigeon breast muscle, which is subsequently confirmed by the discovery of its reaction mechanism in 1937. This reversible reaction could be also applied to other tissues, which lays the basis for the possibility that there similar transfers act as a main choice of producing amino acids via amino transfer. Later in 1953, enzyme UDP-glucose pyrophosphorylase is revealed to be a transferase, since it is found to be capable of reversibly generating UTP and G1P from UDP-glucose and an organic pyrophosphate. Another historical significant discovery in transferase is illumination of the breakdown mechanism of catecholamine by catechol-O-methyltransferase, which accounts a large part for Julius Axelrod's 1970 Nobel Prize in Physiology or Medicine.

Classification

Up to now, the classification of transferases is still under way for new ones discovered frequently. The category of transferases is described primarily according to the type of biochemical group transferred, and can be divided into ten groups based on the EC Number classification, which comprises more than 450 different unique enzymes and have been assigned a number of EC 2 in the EC numbering system. Hydrogen is not recognized as a functional group when it refers to transferase targets. On the contrary, hydrogen transfer is divided into oxidoreductases in consideration of electron transfer.

EC number	Description
EC 2.1	Single carbon transferases under EC 2.1 are enzymes that transfer single-carbon groups, which contain functional groups of hydroxymethyl, methyl, carboxy, carbamoyl, formyl, and amido substituents.
EC 2.2	EC 2.2 includes aldehyde and ketone transferases transferring aldehyde or ketone groups, mainly comprising a variety of transketolases and transaldolases that play important role in pentose phosphate pathway and catalyze the transfer of dihydroxyacetone functional group to glyceraldehyde 3-phosphate.
EC 2.3	Acyl transferases as key aspects of EC 2.3 could transfer acyl groups or acyl groups that are converted into alkyl groups during the process of being transferred. Furthermore, this category also distinguishes amino-acyl from non-amino-acyl groups.
EC 2.4	Enzymes divided into EC 2.4 could transfer glycosyl, hexosyl and pentosyl groups. Glycosyltransferase under the subcategory of EC 2.4 takes participate in the biosynthesis of disaccharides and polysaccharides by transferring monosaccharides to other molecules.
EC 2.5	Currently, EC 2.5 only possesses enzymes that are involved to transferring alkyl or aryl groups, which yet do not include methyl group. This is different from functional groups that are transformed into alkyl groups when transferred.
EC 2.6	EC 2.6 is a group of enzymes that are consistent with transfer of nitrogenous groups, including transaminase, oximinotransferases and other nitrogen group transferring enzymes. Amidinotransferase is previously grouped into EC 2.6, while it has recently been reclassified as a subcategory of EC 2.1.
EC 2.7	EC 2.7 consists of not only enzymes that transfer phosphorus-containing groups, but also nuclotidyl transferases. Subcategory of phosphotransferase is further divided in accordance with the type of group experiencing the transfer. Phosphate acceptors mainly include alcohols, carboxy groups, nitrogenous groups, and phosphate groups. Various kinases are also constituents of this subclass of transferases.
EC 2.8	Sulfur transferases transferring sulfur-containing groups are covered by EC 2.8 and are further subdivided into the subcategories of sulfurtransferases, sulfotransferases, and CoA-transferases, as well as alkylthio groups transferring enzymes. Some specific sulfotransferases could employ PAPS as a sulfate group donor, within which alcohol sulfotransferase has a broad targeting capacity. Therefore, alcohol sulfotransferase is also acknowledged as "steroid sulfokinase," "hydroxysteroid sulfotransferase," and "estrogen sulfotransferase." Decrease in the activity has been concerned with human liver disease.
EC 2.9	Selenium transferases belong to EC 2.9 and only contain two transferases, which therefore is one of the smallest categories of transferase.
EC 2.10	The class of EC 2.10 covers enzymes transferring molybdenum or tungsten-containing groups. However, only one enzyme molybdopterin molybdotransferase, a component of MoCo biosynthesis in Escherichia coli, has been added until 2011.

Applications in Biotechnology

Terminal transferase is one of the few DNA polymerases functioning without an RNA primer, and could label DNA or produce plasmid vectors by adding deoxynucleotides in the form of a template to the downstream end or 3' end of an existing DNA molecule.

Glutathione transferases with high diversity can be applied by plants to segregate toxic metals from the rest of the cell, which thus can be processed as biosensors to detect contaminants such as herbicides and insecticides. Glutathione transferases could also increase resistance to both biotic and abiotic stress in transgenic plants, and now they are being explored as targets for anti-cancer medications owing to their functionality in drug resistance. Currently the only available commercial source of natural rubber is the Hevea plant.

Hydrolases

Hydrolase is a class of hydrolytic enzymes that are commonly used as biochemical catalysts utilizing water to break a chemical bond in order to divide a large molecule into two smaller ones. Hydrolases are pivotal for the body since they digest large molecules into fragments for synthesis, excrete waste materials, and provide carbon sources for the production of energy, during which many biopolymers are converted to monomers. Some hydrolases could release energy when they take effect. A number of hydrolases, especially proteases, are connected with biological membranes as peripheral membrane proteins or affixed through a single transmembrane helix. Some others are multi-span transmembrane ones. The names of hydrolases are systematically formed as "substrate hydrolase, while the commonly adopted names are typically in a form of "substratease."

In biochemistry, a hydrolase catalyzes the hydrolysis of a chemical bond like the following reaction:

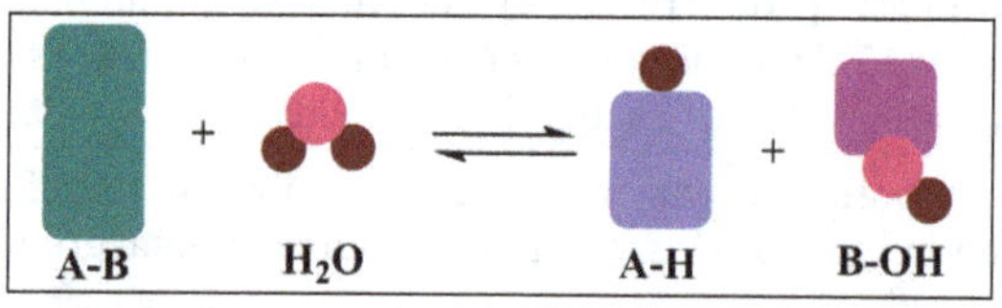

Hydrolysis reaction catalyzed by hydrolase.

Classification

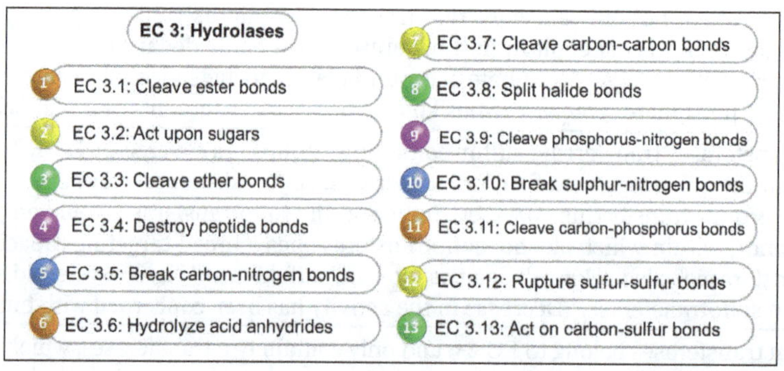

Hydrolases belong to EC 3 in the EC classification system and can be further grouped into thirteen subclasses on the basis of the bonds they act upon. EC 3.1 represents a kind of enzymes rupturing ester bonds, which are called esterases. Some common esterases include nucleosidases, phosphatases, proteases, and lipases, among which phosphatases cut phosphate groups off molecules. Acetylcholine esterase is a potent neurotransmitter for voluntary muscle and it as one of the most crucial esterases contributes to the transform of the neuron impulse into acetic acid after it degrades acetylcholine into choline and acetic acid. Some dangerous toxins such as the exotoxin and saxitoxin could impede with the action of cholinesterase, and many nerve agents react by hindering the hydrolytic efficacy of cholinesterase. Nucleosidases are capable of hydrolyzing the bonds of nucleotides. Glycerides could be hydrolyzed by lipases, which also make contribution to the breakdown of fats, lipoproteins and other larger molecules into smaller molecules like fatty acids that are used for synthesis and as a source of energy. Hydrolases in EC 3.2 mainly act upon sugars such as DNA glycosylases and glycoside hydrolase. Acetic acid has become a nice intermediate for glycolysis catalyzed by glycosidases that chop sugar molecules into carbohydrates and peptidases hydrolyze peptide bonds. EC 3.3 includes ether bonds destroying enzymes. EC 3.4 covers hydrolases that act upon peptide bonds like proteases and peptidases. For example, acylpeptide hydrolase as a member of the peptidase family could deacetylate the acetylated N-terminus of polypeptides. Some other type of hydrolases comprise enzymes breaking carbon-nitrogen bonds, not peptide bonds, acid anhydrides (acid anhydride hydrolases, including helicases and GTPase), carbon-carbon bonds, halide bonds, phosphorus-nitrogen bonds, sulphur-nitrogen bonds, carbon-phosphorus bonds, sulfur-sulfur bonds, and carbon-sulfur bonds, with EC number sequentially ranging from 3.5 to 3.13.

Various Functions of Various Hydrolases

Hydrolases could participate in a variety of biological procedures due to their diversification of acting position. Hydrolase expressed by *Lactobacillus jensenii* in the human gut could stimulate liver to secrete bile salts which facilitate the digestion of food. It has been suggested that the activity of bile salt hydrolase commonly found among intestinal microbiota could increase hydrogel-forming potencies of certain bile salts, whose occurrence in physiological conditions of human gut is thought to be able to enhance bacterial potnetiality to colonize the gastrointestinal tract and their survival rate in this specific ecological niche. Furthermore, new information about the activity of bile salt hydrolase in bacteria can be beneficial to more consciously make use of living microorganisms as food additives and also serves as guidance in the development of new medicines for the prevention and treatment of gastrointestinal disorders. Acylpeptide hydrolase found in human erythrocytes could be potentially treated as a biomarker for low dose exposure to organophosphorus in humans. Glycoside hydrolases play unique roles in various biological processes like the metabolism of cell wall, the biosynthesis of glycans, signalling, plant defence, and the mobilization of storage reserves. It has been found that a single residue in plant GH32, equivalent to Asp239 in AtcwINV1, seems to be important for sucrose stabilization in the active site and indispensable in determining the specificity of sucrose donor. The leukotriene A_4 (LTA_4) hydrolase could act upon epoxide in the final step in the biosynthesis of leukotriene B_4 that occupies a position in a variety of acute and chronic inflammatory diseases. LTA_4 hydrolase as a bifunctional zinc metalloenzyme is a significant enzyme in the 5-lioxygenase pathway and possesses a peptide-cleaving activity. Nucleoside hydrolase plays a central role in the purine salvage pathway and also functions as a primary target for the explorations of anti-parasitic drugs.

Lyases

In physiology, lyase is an enzyme catalytically aiding in breaking various chemical bonds by means of an "elimination" reaction, other than hydrolysis and oxidation. This reaction often results in the formation of a new cyclic structure or a new double bond, and a reverse reaction called a "Michael addition" might also possibly happen under the catalysis of lyase. To obtain either a double bond or a new ring, lyase acts upon the single substrate and a molecule is eliminated. Lyases are different from other enzymes for only one substrate is required for the reaction in one direction, but two substrates are essential for the reverse reaction. Lyases can be commonly observed in the reactions of the Citric Acid Cycle (Krebs cycle) and in glycolysis. In glycolysis, aldolase could readily and reversibly degrade fructose 1,6-bisphosphate into the products glyceraldehyde 3-phosphate and dihydroxyacetone phosphate, which is an example of a lyase cleaving carbon-carbon bonds. Lyase works without the necessary requirements for cofactor recycling and gives an absolute stereospecificity with a theoretical yield of 100%, being much more efficient compared with enantiomeric resolutions of only 50% productive rate. Therefore, considerable researches have been addicted to the exploration of lyases as biocatalysts to synthesize optically active compounds, which have also been already found application in a few large commercial processes. Lyases are systematically named as "substrate group-lyase", such as decarboxylase, dehydratase, aldolase, etc. When a product is more significant, synthase may be expressed in a name like phosphosulfolactate synthase.

Classification

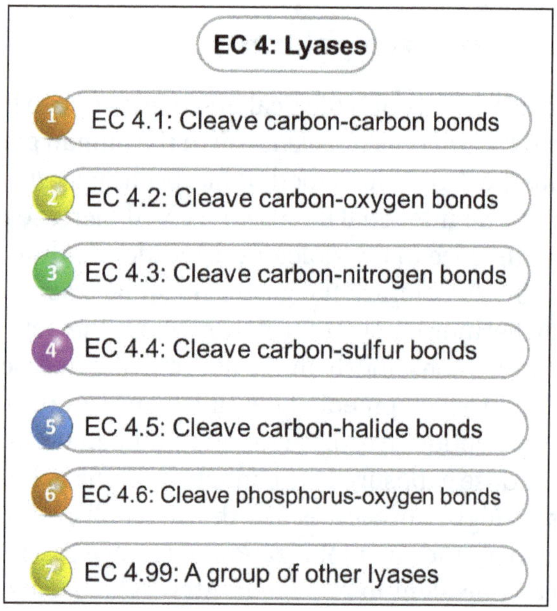

In the EC number classification of enzymes, EC 4 could represent lyases, which can be further classified into seven subclasses. Lyases in EC 4.1 cleave carbon-carbon bonds, and include decarboxylases (EC 4.1.1), aldehyde lyases (EC 4.1.2) facilitating the reverse reaction of aldol condensations, oxo acid lyases (EC 4.1.3) that catalyzes the cleavage of many 3-hydroxy acids, and others (EC 4.1.99). EC 4.2 contains a group of lyases that break carbon-oxygen bonds, such as dehydratases. Hydro-lyases being a part of carbon-oxygen lyases could facilitate the cleavage of C-O bonds by the elimination of water. Some other carbon-oxygen lyases promote the elimination of a phosphate or the removal of an alcohol from a polysaccharide. Lyases cleaving carbon-nitrogen bonds are sorted into EC 4.3. They could release ammonia with powerful cleaving ability and simultaneously form a double bond or ring. Some of these enzymes can also help to eliminate an amine or amide group. EC 4.4 represents lyases that split carbon-sulfur bonds, which could eliminate or substitute dihydrogen sulfide (H2S) from a reaction. Carbon-halide bonds cleaving enzymes are lyases in EC 4.5 and that utilize an action mode that removes hydrochloric acid from a synthetic pesticide dichloro-diphenyl-trichloroethane (DDT). EC 4.6 comprises lyases fracturing phosphorus-oxygen bonds, like adenylyl cyclase and guanylyl cyclase, and they eliminate diphosphate from nucleotide triphosphates. EC 4.99 is a group of other lyases.

Substrate Specificity

Narrow substrate specificity is usually considered to be a drawback for the commercialization of an enzyme in that it greatly restricts the flexibility of an enzyme as an assistant in the production of related compounds. Lyases are generally, but not always, found with narrow substrate specificity. Most hydratases and ammonia-lyases indeed possess quite narrow substrate specificity, while the substrate specificity for aldolases, decarboxylases and oxynitrilases is much broader. It is noteworthy here that the substrate specificity of a specific lyase varies depending on its source. However, it is not an absolute prerequisite for enzymes to own unrestricted substrate specificity for their commercial exploitation. In fact, there are several of the lyases in commercial use bearing a rather narrow substrate spectrum.

Cofactor Requirements

The commercial potential of enzymes can be severely limited by the requirement for expensive cofactors. Since the addition catalyzed by lyase does not implicate a mere oxidation or reduction, it is not an essential requirement for cofactors. However, up to now, most of the lyases identified do require cofactors, which are involved in stabilization of reaction intermediates, polarization of the substrate, substrate binding, temporary binding of the nucleophile, and so on. The majority of these cofactors are not very expensive, and covalently bound to the enzyme. Thereby, the cofactors of lyases do not establish a barrier to their commercialization. The requirements for cofactors of lyases are varied according to their different sources.

Lyase Deficiency Disorder

Lyase deficiency, also mentioned as HMG-CoA lyase deficiency, is a rare inherited disorder that causes a mess in the metabolism of amino acid leucine and prevents the body from synthesizing ketones that are used for energy production during periods of lacking food in the body. The condition could be inherited in an autosomal recessive pattern, meaning that both copies of the gene in each cell undergo mutations. The symptoms of lyase deficiency usually express within the first year of life and mainly include diarrhea, vomiting, lethargy, dehydration, and weak muscular development. During an episode of lyase deficiency, blood sugar concentrations can be hypoglycemic or extremely low, and some harmful compounds would be accumulated to cause the blood to become quite acidic. If left untreated, the disorder could even initiate coma, convulsions, breathing problems, and even death. Strenuous exercise, infection, and other physical stresses can probably give rise to bouts with the lyase deficiency symptoms.

Relieve of Lyase Deficiency Disorder

The symptoms of lyase deficiency would be amplified after going long periods without ingestion of food. Therefore, people attacked with this disease should be ensured regular ingestion and good balance. Intaking foods rich in carbohydrates and low in protein and fat may be beneficial to prevent low blood sugar levels. The application of L-Carnitine may help the body produce cellular energy, which would be extremely advantageous when the body feels lethargic and dehydrated.

Isomerases

The enzyme isomerase is a general class of enzymes that can change a molecule from one isomer to others. They typically make intramolecular rearrangements easier. Intramolecular rearrangements are where bonds are broken and formed. The usual form of this reaction is:

$$A - B \rightarrow B - A$$

There is a singular substrate which also produces just one product. Even though the product and the substrate have the same molecular formula, they have different bond connectivity and spatial arrangement. The isomerases are known to catalyse many different biological reactions. Glycolysis and carbohydrate metabolism are a couple of examples for the same.

The Process of Isomerization

The isomerases function by catalysing changes within just one molecule. Since they change one isomer to another, the end product has the same molecular formula but an alternate physical structure. Even though there are many different existing varieties of isomers they can be classified into two different groups-structural isomers and stereoisomers. In structural isomers, the bonds are ordered differently or they differ in their bond connectivity from one other. Structural isomers can be catalysed by intramolecular lyases, oxidoreductases and transferases. On the other hand, stereoisomers have the same order of individual bonds and even the same connectivity but there is a difference in three-dimensional arrangement of the bonds. The stereoisomers are catalysed using racemases, epimerases and cis-trans isomers.

Classification of Isomerases

Enzyme catalysed reactions are known to have their own unique classification number. The enzyme classification category for isomerase-catalysed reactions is EC 5. The isomerase can then be further categorised into five sub-classes:

- Racemases/Epimerases: The racemases and epimerases function by inverting stereochemistry at the target chiral carbon. While racemases act on molecules with one chiral carbon, the epimerases act on molecules with multiple chiral carbons but act only one of them. This category is then further broken down depending on what the enzyme acts upon, for example, action on amino acids or carbohydrates.
- Cis-trans Isomerases: This class is for the isomerases that catalyse the isomerization of the cis-trans isomers. Certain alkenes and cycloalkanes could have cis-trans stereoisomers. These isomers are distinguished through the positioning of the substituent groups relative to that of the plane of reference instead of absolute configuration. The difference between

cis-trans isomers is that cis isomers have their substituent groups on the same side while the trans isomers have their groups on opposite sides. This class is not broken down further into sub-classes.

- Intramolecular Oxidoreductase: These isomerases function by catalysing the transfer of electrons from one molecule to the other. Simply put, they catalyse the reaction that oxidises one part of the molecule while reducing the other part. Depending on their processes intramolecular oxidoreductases can be divided into further sib-classes.
- Intramolecular Transferases: Intramolecular transferases (mutases) are used to catalyse the movement of functional groups from one part of the molecule to another. The intramolecular transferases can be sub-classified down depending on which functional group the enzyme moves.
- Intramolecular Lyases: Intramolecular lyases function in reactions where a group is considered to be removed from one part of the molecule that forms a double bond while still being covalently attached to the molecule. Some of the reactions catalysed by intramolecular lyases involve the breaking of the ring structure. This class cannot be further classified.

Mechanisms of Isomerases

- Ring Expansion and Contraction via Tautomers: An example of this kind of mechanism, i.e. ring-opening and contraction of the ring is the isomerisation of glucose to fructose. The overall reaction causes the ring to form an aldose through acid/base catalysis and then subsequently forms cis-ethanol intermediate. The ring is closed after the formation of a ketose.
- Epimerisation: A classic example of epimerization is the Calvin cycle when D-ribulose-5-phosphate is converted into D-xylulose-5-phosphate by ribulose-phosphate3-epimerase. The difference between the substrate and the product is in the stereochemistry at the third carbon in the chain. The process involves the deprotonation of the third carbon to make a reactive enolate intermediate.
- Intramolecular Transfer: An example of an intramolecular transferase is chorismate mutase. Chorismate mutase catalyses the change of chorismate to prephenate. The latter is used as a precursor for L-tyrosine and L-phenylalanine in some plants and bacteria. This reaction is a Claisen modification that can continue with or without the isomerase, however, the rate increments 10 6 fold due to the chorismate mutase. The process experiences a chair transition state with the substrate in a trans-diaxial position.
- Intramolecular Oxidoreduction: A good example of this reaction mechanism is the activity of Isopentenyl diphosphate delta isomerase type I (aka IPP isomerase) in the synthesis of cholesterol. It typically catalyzes the conversion of isopentenyl diphosphate (IPP) to dimethylallyl diphosphate (DMAPP). In this isomerization process, a steady carbon-carbon double bond is repositioned to make a profoundly electrophilic allylic isomer. IPP isomerase catalyzes this process by the stereoselective antarafacial transposition of a solitary proton.

The Role that Isomerases Play in Human Diseases

Isomerase enzyme has a major role in human diseases. Lack of this biocatalyst can be the cause of many diseases in people. Both of the following diseases are very rare:

Phosphohexose Isomerase Deficiency (PHI)

It is also known as the glucose-6-phosphate isomerase deficiency or the phosphoglucose isomerase deficiency. It is a hereditary enzyme deficiency. PHI is the second most frequent erthoenzyopathy in glycolysis other than pyruvate kinase deficiency and is related with the non-spherocytic haemolytic anaemia of variable seriousness. This infection is fixated on the glucose-6-phosphate protein. This protein can be found in the discharge of some cancer cells. PHI is the aftereffect of a dimeric enzyme that catalyzes the reversible interconversion of fructose-6-phosphate and glucose-6-phosphate. The insufficiency of phosphohexose isomerase can result in a condition alluded to as a hemolytic disorder. In humans, the hemolytic syndrome, which is described by a lessened erythrocyte number, lower hematocrit, lower haemoglobin, higher number of reticulocytes and plasma bilirubin concentration, and higher liver-and spleen-somatic indices, was solely showed in homozygous mutants.

Triosephosphate Isomerase Deficiency

The illness alluded to as triosephosphate isomerase inadequacy (TPI), is an extreme autosomal recessive acquired multisystem issue of glycolytic digestion. It is described by hemolytic anaemia and neurodegeneration and is brought about by anaerobic metabolic dysfunction. This dysfunction is caused by a missense mutation that influences the encoded TPI protein. The most widely recognized mutation is the substitution of gene, Glu104Asp, which creates the most serious phenotype, and is answerable for around 80% of clinical TPI deficiency. TPI insufficiency. TPI has a 25% reoccurrence chance due to heterozygous parents. It is a congenital disease that regularly happens with hemolytic anaemia and shows with jaundice.

Industrial Applications

- Sugar Manufacturing: The most widely recognized utilization of isomerases in industrial applications is in sugar manufacturing. Glucose isomerase (otherwise called xylose isomerase) catalyzes the transformation of D-xylose and D-glucose to D-xylulose and D-fructose. Similar to most sugar isomerases, glucose isomerase catalyzes the interconversion of aldoses and ketoses.

- High-Fructose Corn Syrup: The transformation of glucose to fructose is a key segment in the production of high-fructose corn syrup. Isomerization is more particular than older chemical methods for fructose creation and hence results in a better yield of fructose without any side products. The fructose that is produced from this isomerization process is cleaner with no lingering flavours from contaminants. High-fructose corn syrup is favoured by numerous confectioneries and soft drink producers due to the higher sweetening power of fructose compared to sucrose (twice that of sucrose), it's moderately decreased price and it's the inability to crystallise. Fructose is likewise utilized as a sweetener by diabetics.

- Production of Ethanol: The isomerization of xylose to xylulose has its own business applications as researchers are becoming more interested in for biofuels. This process is typically observed in microscopic organisms that feed on rotting plant matter. Its most regular industrial use is in the generation of ethanol, accomplished by the fermentation of xylulose. The utilization of hemicellulose as source material is exceptionally normal. Hemicellulose contains xylan, which itself is made out of xylose in β(1,4) linkages. The utilization of glucose isomerase effectively changes xylose to xylulose, which would then be able to be acted upon by fermenting yeast.

Ligases

An enzyme is a catalyst or a chemical produced by cells to speed up a biochemical reaction. It is usually a protein molecule with a characteristic sequence of amino acids that fold to produce a specific three-dimensional structure, which gives the molecule unique properties. Other molecule with catalytic activity is ribozyme, an enzyme made of RNA rather than protein. Enzymes may be classified and named according to the reaction they catalyze:

- Oxidoreductases
- Transferases
- Lyases
- Isomerases
- Ligases

Ligases are a group of enzymes that catalyze the binding of two molecules. An example is a DNA lygase that link two fragments of DNA by forming a phosphodiester bond. Ligases are classified into six subclasses:

- EC 6.1 (ligases forming carbon-oxygen bonds)
- EC 6.2 (ligases forming carbon-sulfur bonds)
- EC 6.3 (ligases forming carbon-nitrogen bonds)
- EC 6.4 (ligases forming carbon-carbon bonds)
- EC 6.5 (ligases forming phosphoric ester bonds)
- EC 6.6 (ligases forming nitrogen-metal bonds)

Translocases

Translocase is a general term for a protein that assists in moving another molecule, usually across a cell membrane. These enzymes catalyze the movement of ions or molecules across membranes or their separation within membranes. The reaction is designated as a transfer from "side 1" to "side 2" because the designations "in" and "out", which had previously been used, can be ambiguous. Translocases are the most common secretion system in Gram positive bacteria.

Classification

The subclasses designate the types of ion or molecule translocated:

EC number	Description
EC 7.1	Catalyzing the translocation of hydrons
EC 7.2	Catalyzing the translocation of inorganic cations
EC 7.3	Catalyzing the translocation of inorganic anions and their chelates
EC 7.4	Catalyzing the translocation of amino acids and peptides
EC 7.5	Catalyzing the translocation of carbohydrates and their derivatives
EC 7.6	Catalyzing the translocation of other compounds

The further classification concerns the reaction that provided the driving force for the translocation:

EC number	Description
EC 7.x.1	linked to oxidoreductase reactions
EC 7.x.2	linked to the hydrolysis of a nucleoside triphosphate
EC 7.x.3	linked to the hydrolysis of a diphosphate
EC 7.x.4	linked to a decarboxylation reaction

Functions of Several Common Translocases

Translocases play an important role in biological activities because of their critical functions. For example, most of the proteins needed for mitochondrial function are encoded by the nucleus of the cell. Translocase of the outer membrane (TOM) can work in conjunction with translocase of the inner membrane (TIM) to transport proteins into the mitochondrion. In the case of preproteins, transportation into the mitochondrial matrix requires the presequence translocase of the inner membrane (TIM23 complex) and the presequence translocase-associated motor (PAM). Carnitine-acylcarnitine translocase is responsible for transporting both carnitine-fatty acid complexes and carnitine across the inner mitochondrial membrane, which is required since fatty acids cannot cross the mitochondrial membranes without assistance. ADP/ATP translocases are transporter proteins that enable the exchange of cytosolic adenosine diphosphate (ADP) and mitochondrial adenosine triphosphate (ATP) across the inner mitochondrial membrane. Free ADP is transported from the cytoplasm to the mitochondrial matrix, while ATP produced from oxidative phosphorylation is transported from the mitochondrial matrix to the cytoplasm, thus providing the cells with its main energy currency. DNA translocases are motor proteins that can convert the chemical energy of ATP hydrolysis into directional movement along DNA. Different transport enzymes play different functions in organisms, and maintain the stability of the internal environment, which is of great significance to the life activities.

Enzyme Structure

Enzymes have four levels of structures as shown in figure. These are:

- Primary structure
- Secondary structure

- Tertiary structure
- Quaternary structure

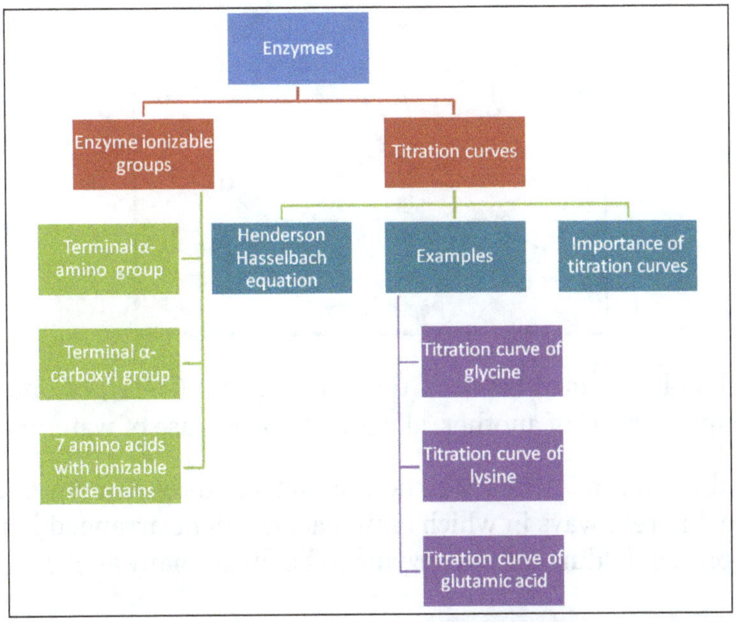

The enzyme structure ranges from a basic amino acid sequence to a three dimensional (3D) structure in a folded protein. The amino acid sequence in polypeptide chains in each enzyme is distinct and determines the three-dimensional shape. Further, it is the 3D structure of an enzyme that determines the enzyme activities.

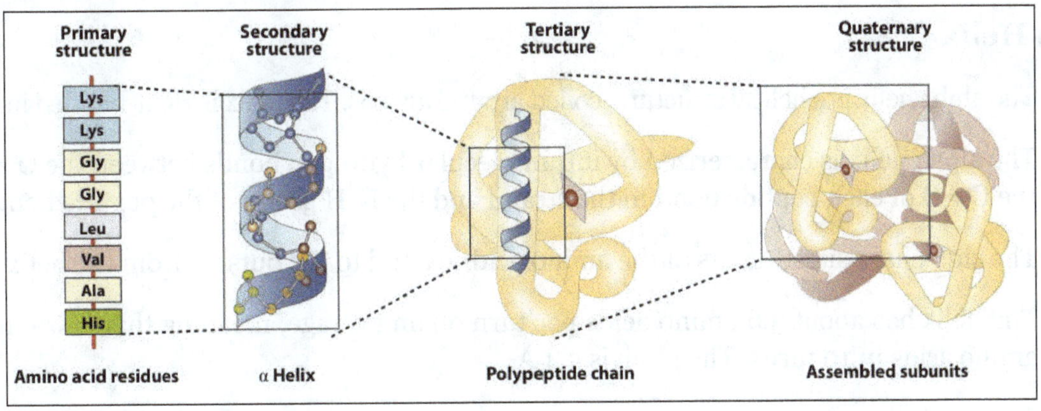

Primary Structure

- The sequence of amino acids in an enzyme is the primary structure.

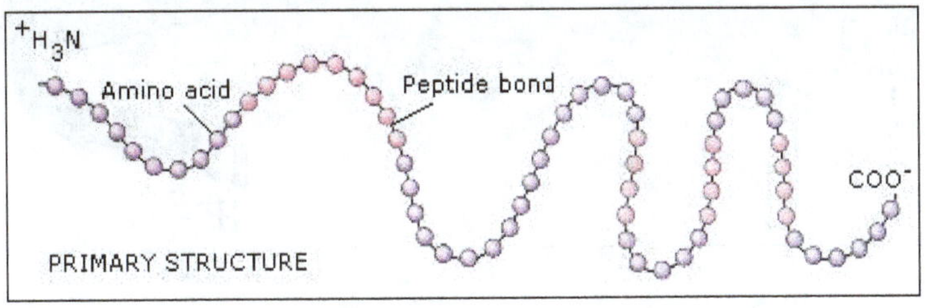

- In the primary structure, the constituent amino acids are linked by peptide bonds (-CONH-) bonds.

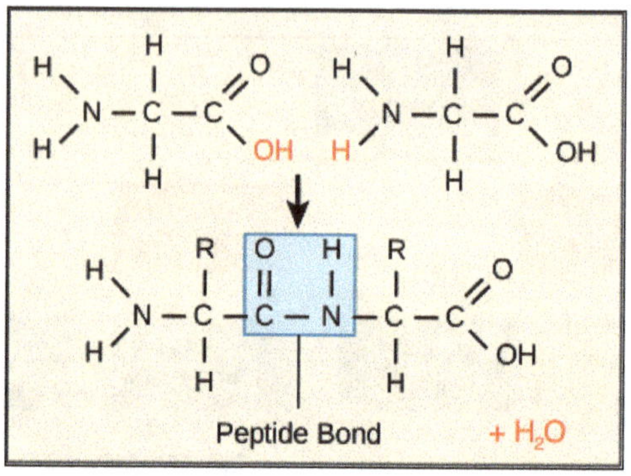

- The peptide bond is formed between the amino group ($-NH_2$) of one amino acid and the carboxyl group (-COOH) of another, along with the release of water molecule.
- As mentioned above, the primary structure dictates three dimensional structures of the proteins. The different ways in which amino acids will be arranged in the chain will influence proper protein folding for the enzyme to be functionally active.

Secondary Structure

The secondary structure in enzymes refers to the interaction of amino acids in a chain (primary structure) which are closely located. There are two types of secondary structures: helical (called α helices) and pleated sheets (called β pleated sheets).

Alpha Helix

- The alpha helix is a helical structure, coiled around an axis. The helix is right-handed in nature.
- The alpha helix is characterized by intramolecular hydrogen bonds between the O atom of the C = O of each peptide bond in the strand and the N-H group of the peptide bond.
- The side-chain substituents of the amino acids extend to the outside from the helix.
- The helix has about 3.6 amino acids per turn on an average, meaning that it will have 36 amino acids in 10 turns. The pitch is 5.4 Å.

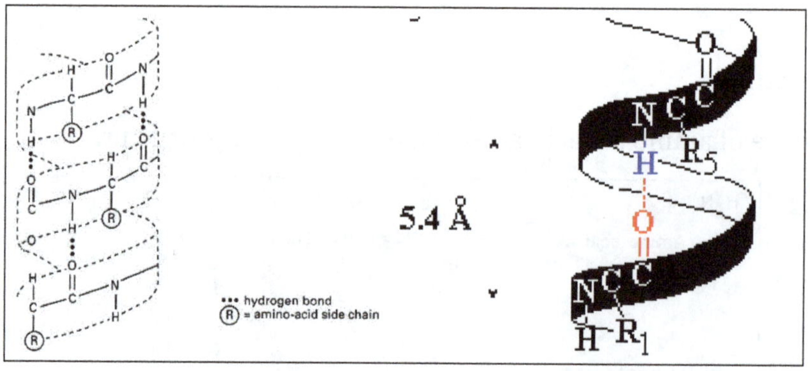

Alpha helices form more readily in enzymes than any other possible conformations owing to the optimal use of internal hydrogen bonds is made in these arrangements for attaining stability.

Beta Pleated Sheet

The second form of secondary structure in enzymes is the beta pleated sheet. This structure is formed by intermolecular hydrogen bonding between two or more straight chains. The O atom of the C = O of peptide bond in one strand hydrogen bonds with the N-H group of the peptide bond in an adjacent strand.

Again, the two strands involved in the formation of beta pleated sheets can run either parallel to each other or anti-parallel to each other. If the amino groups of both chains are on the same side, the sheet are said to be parallel to each other. On the other hand, if the amino groups of both chains are on the opposite side, the chains are said to run in the opposite direction. In this case, the sheet is termed antiparallel. The anti-parallel ß-sheet is more stable than parallel sheet owing greater alignment in the hydrogen bonds.

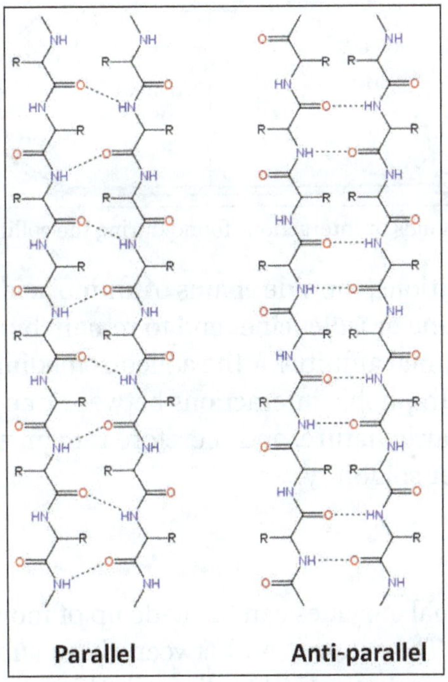

Table: Difference between alpha helix and beta pleated sheet structure.

Properties	Alpha helix	Beta pleated sheet
Structure	Right handed, helical	Sheet like
Formation	Formed by H bonds within the polypeptide chain	Formed by H bonds between two adjacent polypeptide chains
Side chains	Side Chains (R groups) orient out of the helix	Side chains (R groups) are directed both inside and outside the pleated structure
Types	Only one type	Can be of two types: parallel and antiparalle
Preferred amino acids	Ala, Leu, Met, Phe, Glu, Gln, His, Lys, Arg	Tyr, Trp, Phe, Met, Ile, Val, Thr, Cys

Tertiary Structure

- The arrangement of amino acids in the three dimensional space defines the tertiary structure of enzymes.
- The protein molecule arranges itself three dimensionally in such a way as to achieve low energy and maximum stability.
- The various interactions involved in the formation/stabilization of a tertiary structure are Hydrogen bonds, polar-polar interaction, hydrophobic interaction, ionic interaction, formation of disulphide bonds, Van der Waals forces.

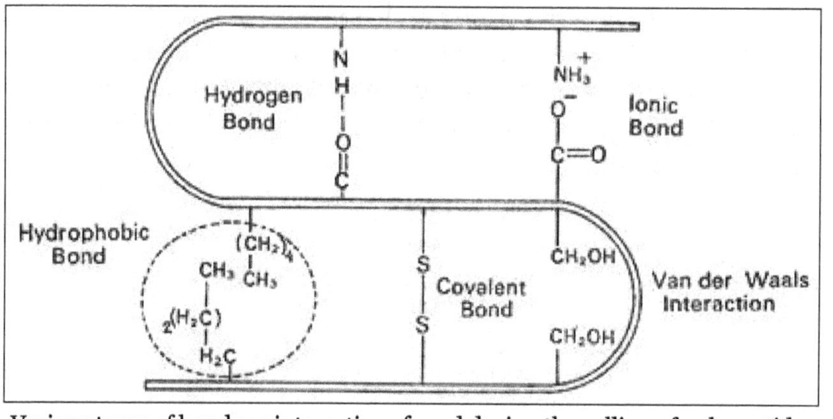

Various types of bonds or interactions found during the colling of polypeptide.

- Under physiologic conditions, the side chains of amino acids which are hydrophobic in nature such as phenylalanine or isoleucine tend to remain buried within the protein/enzyme core, owing to their minimal affinity for the aqueous medium. The alkyl groups of Ala, Val, Leu, Ileu often form hydrophobic interactions between one-another. Acidic or basic amino acid side-chains are polar in nature, and therefore remain exposed on the enzyme surface, to allow for greater water solubility.

Quaternary Structure

Sometimes, proteins or functional enzymes can be made up of more than one polypeptide chains, which are known as subunits. The interaction between these subunits is called the quaternary structure. Various interactions, including H-bonding, disulfide-bridges and salt bridges are also involved in stabilizing the overall complex.

Enzyme Structure Analysis

Analysis of enzyme/protein structure can be done with the help of following advanced analytical techniques/equipments.

- Determination of amino acids which are present in an enzyme and the molar ratios of each can be analyzed/ determined by an amino acid analyzer.
- The sequence of amino acids in the enzyme can be analyzed by peptide mapping, Edman degradation or mass spectroscopy.

- The secondary structure of an enzyme can be determined by circular dichroism spectroscopy (CD).
- The tertiary structure of an enzyme can be determined by fluorescence spectroscopy.
- X-ray crystallography or nuclear magnetic resonance (NMR) analysis can be used to obtain a highresolution analysis of the 3D structure of enzyme.

Enzyme Ionizable Group

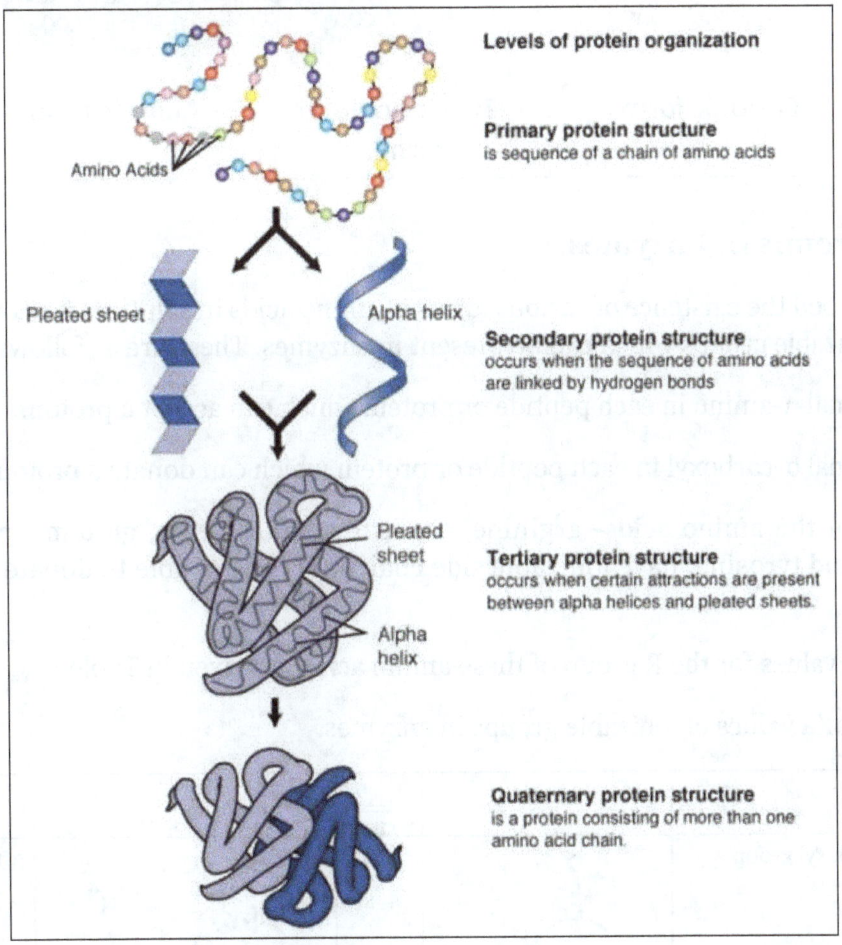

Enzymes are protein molecules which contain a large number of acid and basic groups. Depending on the type of amino acid present within the enzyme and their ionization with the pH of their environment, the charges on these groups vary. The overall charge on the enzyme, along with its activity, stability and solubility is therefore affected.

Existence of Amino Acids in Solution

The functional groups in amino acids in the enzyme have the ability to readily ionize. Each amino acid has at least two acid-base groups (an amino group and a carboxyl group). Certain amino acids within enzyme may also have ionizable side chains. At physiological pH (pH = ~7.4) the $-NH_2$ group remains basic and protonated while the –COOH group is acidic and remains deprotonated.

Amino acids can act both as acid or a base, i.e., they can exist both in their cationic form and anionic form. Additionally, amino acids can also exist in a dipolar state, which is known as the zwitterionic form. In this dipolar state, amino group is protonated and carboxyl group is deprotonated and these dipolar ions can act as either an acid (proton donor) or a base (proton acceptor). Such substances have dual character and are called as ampholytes.

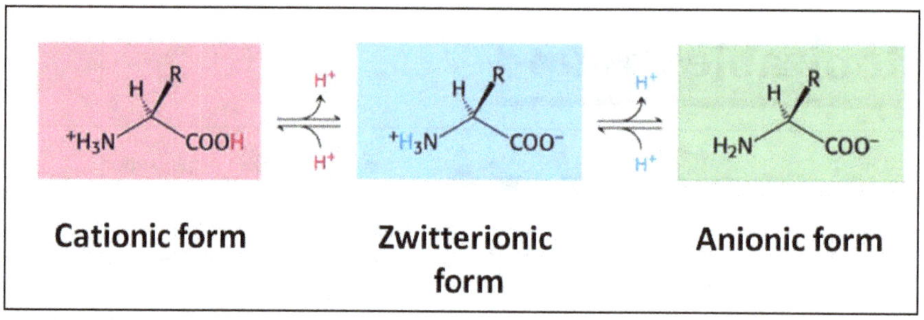

Ionizable Groups in Enzymes

Having understood the existence of various forms of amino acids in solution, we now take a look at the various ionizable groups which can be present in enzymes. These are as follows:

- A terminal α-amino in each peptide or protein which can accept a proton.
- A terminal α-carboxyl in each peptide or protein which can donate a proton.
- Seven of the amino acids- arginine, aspartic acid, cysteine, glutamic acid, histidine, lysine and tyrosine; have ionizable side chains. These are able to donate or accept protons.

The typical pK_a values for the R group of these amino acids are given in Table.

Table: Typical pKa values of ionizable groups in enzymes.

Group	Acid	⇌	Base	Typical pK_a*
Terminal α-carboxyl group	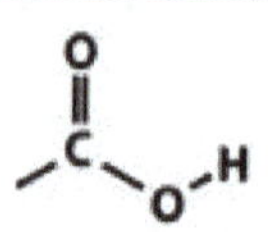	⇌	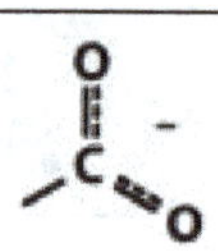 + H^+	3.1
Aspartic acid Glutamic acid	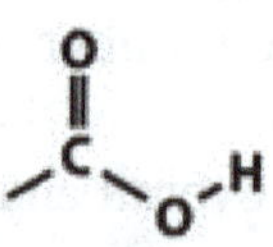	⇌	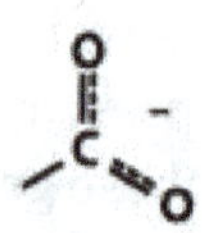 + H^+	4.1
Histidine	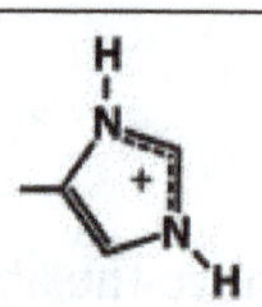	⇌	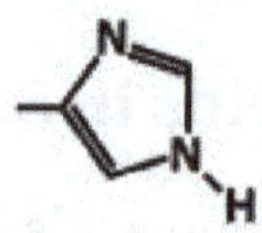 + H^+	6.0

Terminal α-amino group	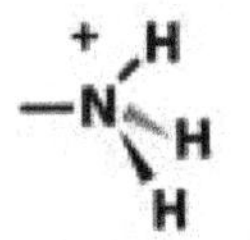	⇌	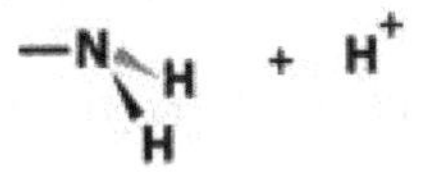	8.0
Cysteine	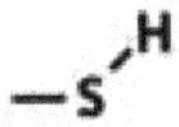	⇌		8.3
Tyrosine	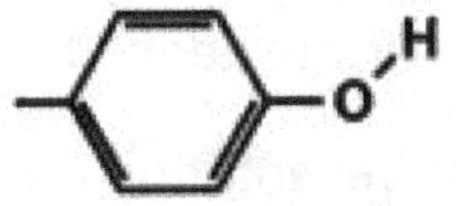	⇌	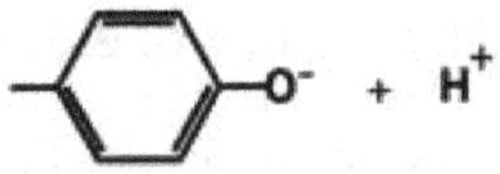	10.9
Lysine	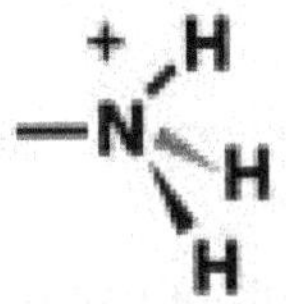	⇌	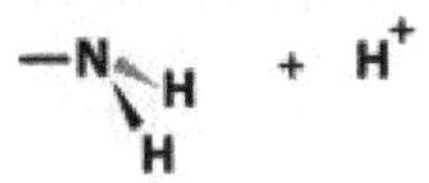	10.8
Arginine	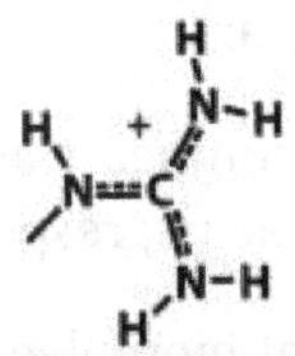	⇌	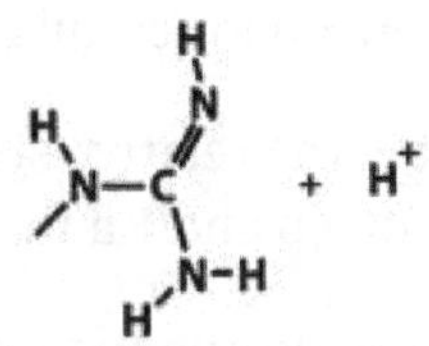	12.5

*pK_a values depend on temperature, ionic strength, and the microenvironment of the ionisable group.

Henderson-Hasselbalch Equation

$$pH = pK_a + \log\frac{[\text{conjugate base}]}{[\text{weak acid}]} (\text{for weak acid})$$

$$pOH = pK_b + \log\frac{[\text{conjugate acid}]}{[\text{weak base}]} (\text{for weak base})$$

The Henderson-Hasselbalch equation is the equation commonly used in chemistry and biology to determine the pH of a solution.

- This equation shows a relationship between the pH or pOH of the solution, the pKa or pKb, and the concentration of the chemical species involved.
- This equation was developed independently by the American biological chemist L. J. Henderson and the Swedish physiologist K. A. Hasselbalch to determine the pH of the bicarbonate buffer system in blood.
- This type of kinetic analysis has enabled us for nearly a century to relate theoretically the changes of the acidic intensity of dilute solutions to a quantity of acid or base added or subtracted.
- This equation can be considered as the backbone of acid-base physiology.
- This equation is commonly used to determine the amount of acid and conjugate base required to prepare a buffer of the desired pH.

Objectives of Henderson Hasselbalch Equation

The principal objectives of the Henderson Hasselbalch equation include the following:

- To calculate the pH, pOH, [H3O+]tot, [OH-]tot, [H3O+]water, and [OH-]water in a solution containing a strong acid (base) given the initial concentration of the acid (base).
- To describe how a buffer solution (either acidic or basic) can resist significant changes in pH when small amounts of either acid or base are added to the buffer solution.
- To describe how either an acidic or basic buffer solution is prepared.
- To describe a “buffer solution”.
- To describe “buffer capacity”.
- To determine whether an aqueous solution of salt will be acidic, basic, or neutral given values of Ka and Kb for conjugate acid-base pairs.
- To describe how the relative strengths of the conjugate acids or bases can be evaluated using the values of Kb and Ka for the bases and acids, respectively.
- To determine the protonation state of different biomolecule functional groups in a pH seven buffer.

Principle/Theory/Derivation of Henderson Hasselbalch Equation

- According to the Bronsted-Lowry theory of acids and bases, an acid (HA) can donate a proton (H+) while a base (B) can accept a proton.
- An acid after losing a proton forms a conjugate base (A–), and the protonated base exists as conjugate acid (BH+).
- The dissociation of acid is expressed in terms of the equilibrium equation as:

  $$HA \leftrightarrow H^+ + A^-$$

- This relationship can be described in terms of the equilibrium constant as:

  $$Ka = \frac{[H+][A-]}{[HA]}$$

Now, taking negative log on both sided of the equation gives:

$$-\log K_a = -\log \frac{[H+][A-]}{[HA]}$$

Or

$$-\log K_a = -\log\left[H^+\right] + \left(-\log \frac{[A-]}{[HA]}\right)$$

By definition,

$$-\log K_a = pK_a \text{ and } -\log\left[H^+\right] = pH$$

Thus,

$$pK_a = pH - \log\frac{[A-]}{[HA]}$$

This equation is then arranged to form the Henderson Hasselbalch equation as:

$$pH = pK_a + \log\frac{[A-]}{[HA]}$$

Equation

- The Henderson Hasselbalch equation can be expressed in two ways:

$$pH = pK_a + \log\frac{[\text{conjugate base}]}{[\text{weak acid}]}(\text{for weak acid})$$

$$pOH = pK_b + \log\frac{[\text{conjugate acid}]}{[\text{weak base}]}(\text{for weak base})$$

Where,

pK_a= dissociation constant of acid,

pK_b= dissociation constant of base.

Applications of Henderson Hasselbalch Equation

- Calculating the pH of a solution using pK: This equation can be used to determine the pH of different solutions in different chemical equations as well as in biological systems like enzymes and proteins.
- Calculating the Ionized and Unionized Concentrations of Chemicals: One of the most potent applications of the Henderson Hasselbalch equation is the ability to determine the concentrations of ionized and unionized chemicals. Generally, the amount of ionized and unionized species is detected using some spectroscopic technique, and thus, this equation is useful in the condition where spectroscopic studies are not feasible. The knowledge of the concentration of ionized and unionized chemicals is essential in fields like organic chemistry, analytical chemistry, and pharmaceuticals sciences.
- Calculating pK_a of a Molecule using pH: Determining the pK_a of a molecule is important as the pK_a is an essential characteristic of the chemistry of the structure of the molecule. Henderson Hasselbalch equation can be used to determine the pK_a when the ratio of ionized and unionized forms and the pH of the solution is known.

- Determination of Solubility: It has been observed that the Henderson Hasselbalch equation is important in determining the pH dependency of solubility. Based on the pH of a solution, the solubility of the solution can be determined, and there is also a close relationship between pH and solubility of various components in a solution.
- Calculating the Isoelectric Point of Protein: This equation can also be used in determining the isoelectric point of different proteins (pH at which proteins neither lose nor accept protons).

Limitations

- The most critical assumption of this equation is that the concentration of acid and its conjugate base will remain the same during the equilibrium.
- The significance of hydrolysis of water and its effect on the pH of the overall solution is neglected.
- Similarly, the hydrolysis of the base and dissociation of acid is also neglected.
- The assumption made in the equation might fail while dealing with strong acids or bases.

Titration Curve of Glycine

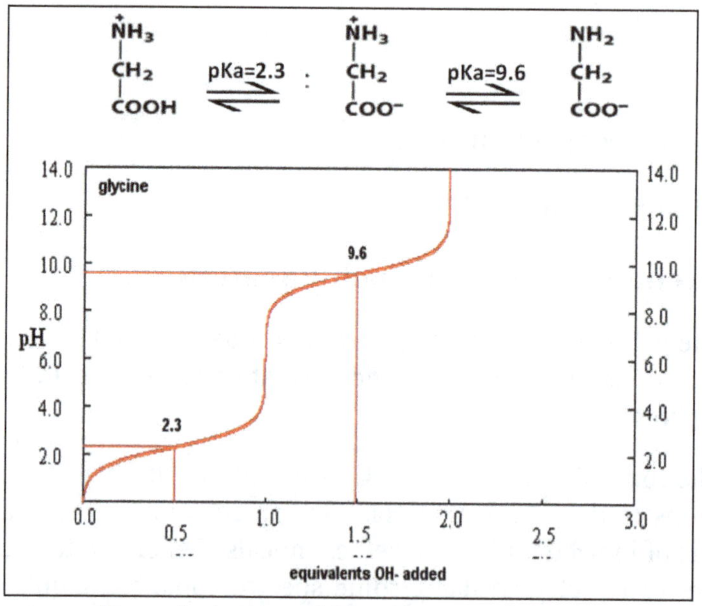

Figure: Ionization and Titration curve of glycine.

The amino acid glycine has no ionisable side groups. It has only two ionisable groups, a carboxyl group and amino group. When titrated with a strong base (NaOH), the titration curve so obtained is as shown in Figure. Two steps are involved corresponding to the deprotonation of the two groups with NaOH.

- At low pH, the amino acid is completely protonated and exists as $^{+}NH_3$-CH_2-COOH.

- As the titration continues, at first midpoint (pH = 2.34), there is equimolar concentration of $^{+}NH_3$-CH_2- COOH and $^{+}NH_3$-CH_2-COO.
- The mid-point of any titration is the point of inflection where pH = pKa of the protonated group. For glycine, the first point of inflection occurs at pH = 2.34 so pKa at this point equals 2.34. Thus, the carboxyl group of glycine has a pKa of 2.34.
- The second point of inflection is reached at pH = 5.97. At this point, deprotonation of carboxyl group is complete and that of amino group has just started. At this pH, glycine is present as a dipolar ion $^{+}NH_3$-CH_2-COO.
- As the titration continues, deprotonation of amino group takes place. At mid-point, the pH = 9.60 so pKa for amino group is 9.60. Also, at this point, there exists an equimolar concentration of proton donor and proton acceptor forms.
- As the titration proceeds further, around pH = 12, the deprotonated form NH_2 - CH_2 – COO^- is found in abundance.

Similarly titration curves for lysine and glutamic acid are shown below.

Titration Curve for Lysine

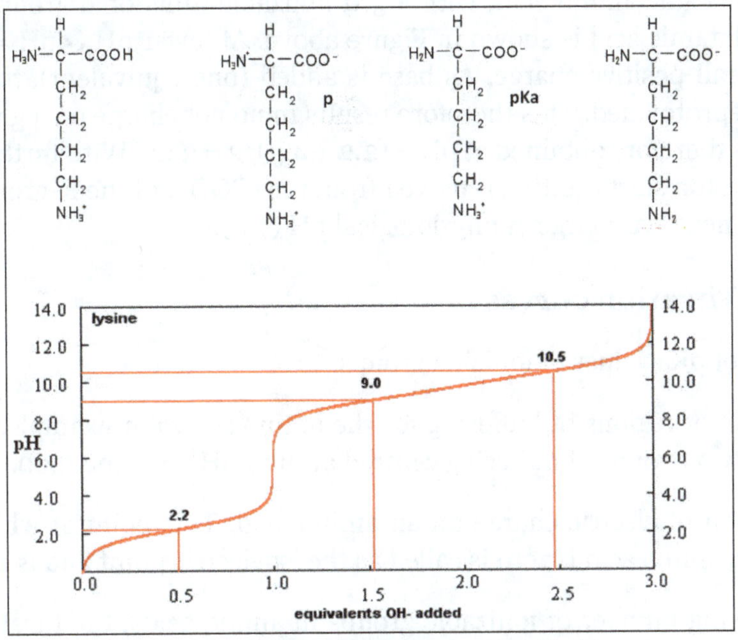

Figure: Ionization and Titration curve of lycine.

Lysine is a basic amino acid, with a 3rd ionizable functional group, the R group. The titration curve of lysine is shown in Figure above. As the concentration of the hydroxide ion increases, the hydrogen ion concentration decreases. At the first pKa, the dissociation of the alpha-carboxyl takes place. At the second pKa, the dissociation of the alpha-amino occurs while at the third pKa, the dissociation of the R-group amino takes place. At each pKa, the solution resists change in pH with addition of more and more hydroxide. The pI occurs where lysine has no net charge. Thus, pI= $(pK_2 + pK_3)/2$ = (9 + 10.5)/2. Thus pI is reached at pH 9.75.

Titration Curve for Glutamic Acid

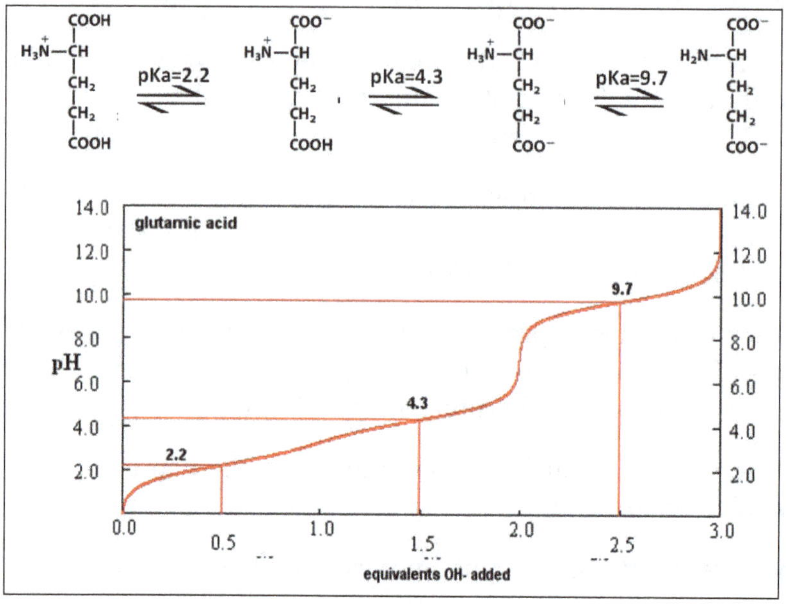

Figure: Ionization and Titration curve of glutamic acid.

Glutamic acid is an acidic amino acid, with a 3rd ionizable functional group, the R group. The titration curve of glutamic acid is shown in Figure above. At lower pH (<pH = 2.0), the glutamic acid carries an overall positive charge. As base is added (one equivalent), the group which has the lowest pKa is deprotonated. This therefore results in no net charge on a glutamic acid. The pI (isoelectric point) is therefore obtained at pH = (2.2 + 4.3)/2 = 3.25. With further addition of base, the second acidic proton starts getting removed from the -COOH of the R-group. This amino acid bears only a formal negative charge at physiological pH (7.0).

Significance of Titration curves

- Calculation of pKa values of ionizing groups.
- Identification of regions of buffering for the amino acid. For example, it can be said that glycine has two regions of buffering centred around pH 2.34 and 9.60, respectively.
- Determination of electric charge on an amino acid. This point at which the net electric charge on an amino acid is zero is called as the isoelectric point and is designated as pI.
- With increasing number of ionizable groups in amino acids, the titration curves become more and more complex due to increase in number of ionization steps.

Cofactor and Coenzyme

Coenzymes and cofactors are molecules or ions that are used by enzymes to help catalyse reactions. Coenzymes are typically organic molecules that contain functionalities not found in proteins, while cofactors are catalytically essential molecules or ions that are covalently bound to the enzyme.

Functional Groups in Proteins

The 20 common amino acids that comprise the building blocks of proteins are characterized by their side-chains, which can range from simple aliphatic groups such as that found in alanine to more complex, functionalized side groups such as that of histidine. All of the amino acids play important roles in determining the three-dimensional structure of proteins through hydrophobic and hydrophilic interactions, ionic interactions and formation of disulfide bonds. Only a smaller subset of these amino acid side-chains, however, have been implicated in direct roles in enzymatic catalysis. Amino acids that are implicated directly in enzymatic catalysis are those with functional groups on the side-chains. Protein functional groups include hydroxyl (serine, threonine, tyrosine), sulfhydryl (cysteine), ammonium (lysine and arginine), imidazolium (histidine), carboxylate (aspartate and glutamate), amide (asparagine and glutamine), and thioether (methionine) groups. These groups can function in a variety of ways in enzymatic catalysis.

Work in recent years has pointed to the presence of a new type of protein functional group: the amino acid radical. A number of enzymes have now been identified that contain a catalytically essential protein radical. The most notable and best characterized example at this point is the ribonucleotide reductase found in aerobic Escherichia coli, which contains a stable, catalytically essential tyrosyl radical. Other enzymes have been shown to require a stable tyrosyl radical for catalytic activity, including the photosynthetic oxygen-evolving complex of photosystem II as well as galactose oxidase. Other enzymes, including the ribonucleotide reductase from anaerobic E. coli and pyruvate formate-lyase contain a catalytically essential glycyl radical in their active forms. Because these protein-based radicals have distinct chemical properties and functionalities relative to the counterpart nonradical amino acid, they can be considered as protein functional groups distinct from the typical amino acid side-chains.

Catalytic Roles for Protein Functional Groups

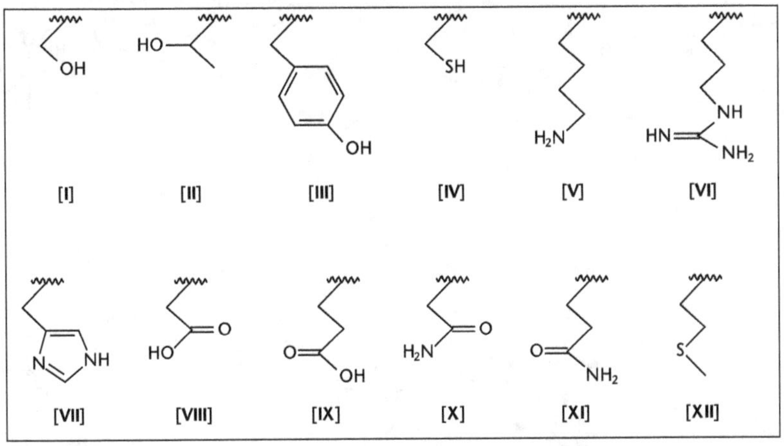

Catalysis of reactions is perhaps the most central process in biology, since the great majority of reactions required by living systems would occur far too slowly to be of any use to a living organism. From the simple hydrolysis of a peptide bond to the complex, multistep synthesis of purines and pyrimidines, it would be a slow world without enzymes! Uncatalysed reactions occur slowly because the substrates are relatively unreactive, and/or because one or more of the intermediates generated along the reaction pathway would be of high energy (i.e. unstable). The role of the catalyst, therefore, is to make the substrates more reactive, and/or to stabilize the reactive intermediates such that reaction will occur more rapidly.

Protein functional groups participate in catalysis by providing nucleophilic, electrophilic, acidic, or basic groups to aid in catalysis. A prevalent type of protein functional group is the nucleophile (the hydroxyl groups in [I], [II], and [III], and the sulfhydryl group in [IV], are all good nucleophiles), and nucleophilic catalysis is commonly observed in enzymes. A classic example of nucleophilic catalysis occurs in the serine proteases. Note in this mechanism that the role of the serine hydroxyl is to form, via nucleophilic attack, an acyl enzyme intermediate that is more readily hydrolysed than the substrate amide bond. Note also the role of aspartate and histidine as general acid–base catalysts in this mechanism: (1) they serve to deprotonate the serine, making it more nucleophilic; (2) they protonate the amine, making it a better leaving group; (3) they deprotonate water, making it a better nucleophile; and (4) they reprotonate the serine, making it a better leaving group. This mechanism is a perfect example of how protein functional groups can aid in catalysis of enzymatic reactions through both nucleophilic and general acid– base catalysis. The carboxylate, histidine, and ammonium functional groups can all participate in general acid–base catalysis, and the amine side-chains can participate in electrophilic catalysis via Schiff base formation.

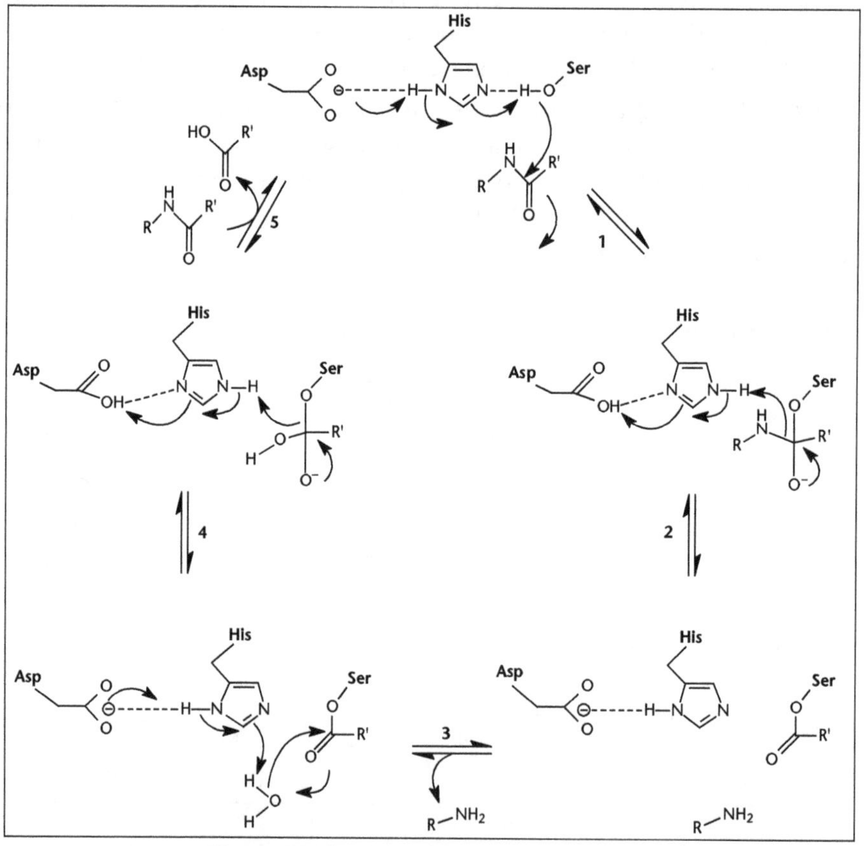

Figure: Mechanism of the serine proteases.

The role of general acid–base catalysis in each step of the mechanism are:

Step 1: Nucleophilic attack of the serine hydroxyl on the substrate.

Step 2: Collapse of tetrahedral intermediate to the acyl enzyme intermediate.

Step 3: Release of amine product.

Step 4: Deacylation resulting from nucleophilic attack of water to form a tetrahedral intermediate.

Step 5: Collapse of the tetrahedral intermediate and release of carboxyl product.

Amino acid radicals, the final type of protein functional group appear to function in the initiation of radical chemistry by long-range electron transfer and/or hydrogen atom abstraction. Examples include the tyrosyl radical present in the active form of E. coli ribonucleotide reductase, which has been proposed to generate a radical at the active site via long-range electron transfer. This active site radical, in turn, initiates ribonucleotide reduction by abstraction of the 3' hydrogen of the substrate to generate an intermediate substrate radical. Other stable protein radicals, for example the glycyl radical of pyruvate formate-lyase and the tyrosyl radical of the photosynthetic oxygen-evolving complex have been proposed to initiate catalysis in similar ways, via initial H atom abstraction from the substrate.

Limitations of Protein Functional Groups in Biocatalysis

The functional groups in proteins are well-suited to general acid–base catalysis, nucleophilic and electrophilic catalysis, and in a few instances radical initiation, but certainly do not account for all of the types of catalytic reactions of enzymes. For example, an important class of enzymatic reactions are redox reactions, and in general protein functional groups are not suited to redox chemistry. One notable exception is cysteine, which can be oxidized by one electron per cysteine residue to form a disulfide bond, and the disulfide bond can be re-reduced by two electrons to form two cysteine sulfhydryls. The only other exceptions are the few instances in which amino acid radicals have been identified as protein functional groups, in which case the amino acid radical is perfectly suited to doing oneelectron chemistry. Due to the limited number of different types of protein functional groups and the therefore limited chemistry in which they can be involved, it is not surprising that enzymes utilize the unique properties of a variety of nonprotein molecules and ions to assist in their catalytic chemistry.

Molecules or ions that are used by enzymes to help catalyse reactions are known as coenzymes or cofactors. Most coenzymes are organic molecules that contain functionalities not found in proteins, and therefore augment the functional groups and mechanisms available for enzyme catalysis. coenzymes are used in electrophilic catalysis, redox chemistry, rearrangements, group transfers, and other types of reactions. Some coenzymes, such as lipoic acid, contain functional groups analogous to those found in proteins, but offer other properties, such as facile translocation, that are not possible for protein functional groups. Coenzymes are typically derivatives of vitamins or bacterial growth factors; for example, thiamin pyrophosphate (TPP) is a derivative of vitamin B_1 (thiamin), and coenzyme B_{12} is a derivative of vitamin B_{12}. While coenzymes participate directly in enzymatic catalysis, they are not usually covalently bound to the enzyme with which they function. Coenzymes may, in fact, be only loosely associated with the enzyme during catalysis, or in other instances may be tightly, but noncovalently, bound to the enzyme.

The term cofactor is typically applied to catalytically essential molecules or ions that are covalently bound to the enzyme. Examples of enzyme cofactors include the unusual peptide-bound topa quinone in amine oxidase, in which a tyrosine residue has been oxygenated and oxidized to produce a catalytically essential and covalently bound cofactor, and the haem group found in many proteins and enzymes, including cytochrome c oxidase. In addition, the metal ions found in metalloenzymes, which are typically bound to the protein via coordination of amino acid sidechains, are cofactors in metalloenzyme catalysis. These metallocofactors can provide important characteristics to the enzyme, including sites for binding substrates via coordination, electrophilic centres to aid in catalysis, and multiple readily available oxidation states that facilitate involvement in redox and electron transfer chemistry.

Enzymatic Reactions that Require Coenzymes

Nicotinamide Coenzymes

The nicotinamide coenzymes include nicotinamide adenine dinucleotide (NAD^+) and nicotinamide adenine dinucleotide phosphate ($NADP^+$), which differ only in the presence or absence of a phosphate group on the 2'-C of the adenosyl moiety. Both NAD^+ and $NADP^+$ serve as reversible carriers of reducing equivalents in the cell, and are utilized in a wide variety of enzymatic reactions including, for example, alcohol dehydrogenase and dihydrofolate reductase. The redox-active moiety is the nicotinamide base, which can be reduced by two electrons to generate NADH or NADPH. In effect, the nicotinamide accepts a hydride ion equivalent from the substrate. Such a reaction requires a strongly electrophilic centre, something which is lacking among the functional groups of proteins, but present in the nicotinamide coenzymes due to the oxidized pyridinium ring. The reduced coenzymes, NADH and NADPH, in turn serve as reductants in other reactions, including the critical electron transfer processes that result in O_2 reduction and ATP synthesis.

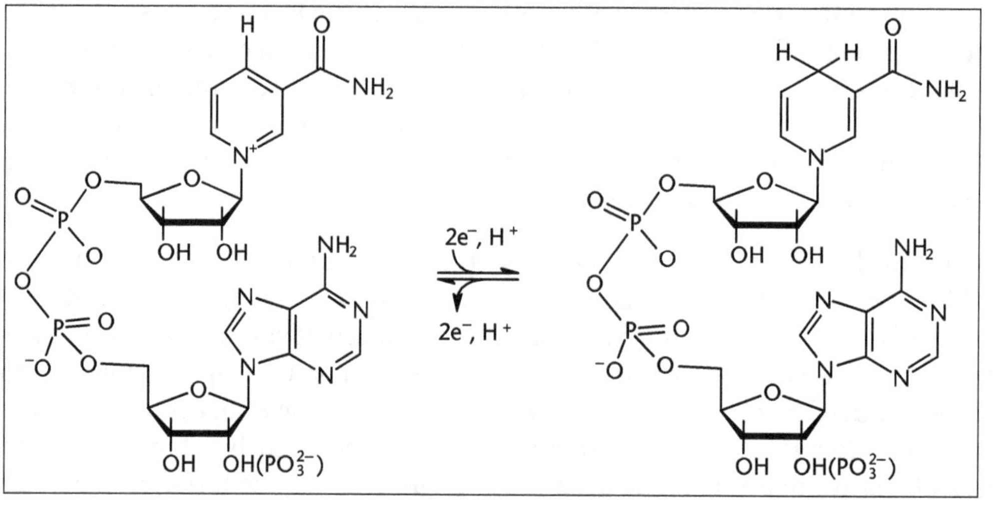

In many cases, such as those just discussed, the nicotinamide coenzymes are actually serving as cosubstrates, but they do serve as true coenzymes in some reactions, for example S-adenosylhomocysteinase. Here the coenzyme NAD^+ serves the same general function as in the reactions, i.e. it accepts a hydride ion equivalent from the substrate, but in this case the reduced coenzyme is an intermediate which is reoxidized to NAD^+ in generation of the product.

Flavin Coenzymes

The flavin coenzymes, flavin–adenine dinucleotide (FAD) and flavin mononucleotide (FMN, which lacks the second phosphate and the adenosyl moiety of), are derivatives of riboflavin (vitamin B_2).

The business end of the flavin coenzymes is the flavin moiety, and FMN and FAD serve analogous functions in enzymatic reactions. Like the nicotinamide coenzymes, the flavin coenzymes participate in redox reactions; the flavin coenzymes, however, are more versatile in that they can participate in either one- or two-electron redox chemistry. The flavin coenzymes can exist in an oxidized form (FMN and FAD), a one-electron reduced semiquinone form (FMN. and FAD.), and a two-electron reduced form ($FMNH_2$ and $FADH_2$). Because of this flexibility in oxidation states, the flavin coenzymes are able to participate in a wider variety of redox reactions, with a wider variety of redox partners, than the nicotinamide coenzymes. Enzymes that utilize flavin coenzymes (often called flavoproteins) include dehydrogenases, oxidases and oxygenases. The direct redox partners of flavoproteins can include twoelectron acceptors such as NAD^+ and $NADP^+$ or oneelectron acceptors such as the ferric haems, iron–sulfur clusters and quinones.

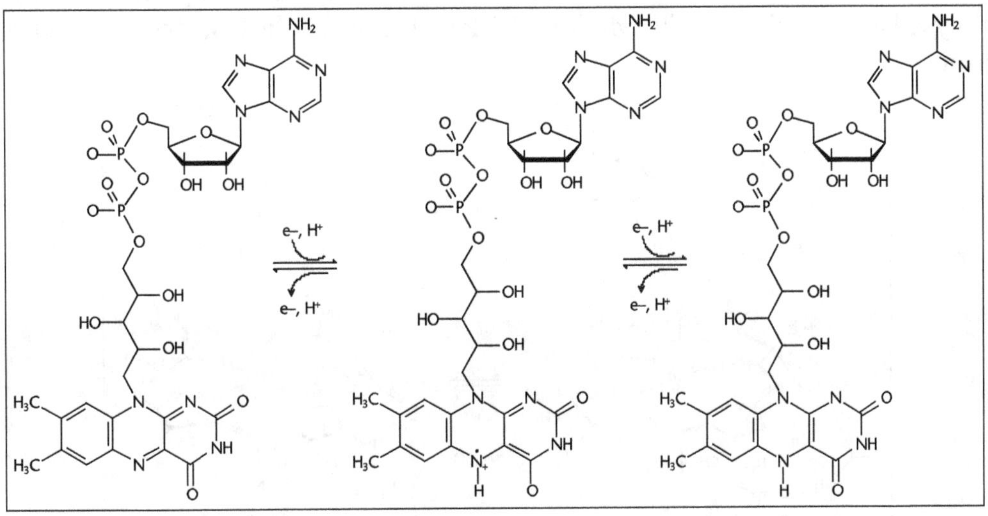

B_{12} Coenzymes

The B_{12} coenzymes, adenosylcobalamin [XVIII] and methylcobalamin, consist of cobalt(III) bound to a substituted corrin ring and an alkyl group (either adenosyl or methyl). These coenzymes are frequently involved in the catalysis of rearrangement reactions. By virtue of the direct bond between the cobalt and a carbon of the alkyl group, the B_{12} coenzymes are organometallic coenzymes, and are the only well-characterized organometallic species in biology. The B_{12} coenzymes function in rearrangements through reversible cleavage of the cobalt– carbon bond to generate a cobalt(II)–organic radical intermediate. The organic radical intermediate then abstracts a hydrogen atom from the substrate, which rearranges and reabstracts the hydrogen atom from the 5'- deoxyadenosine intermediate to regenerate the 5'-deoxyadenosyl radical intermediate. This radical intermediate can then re-form the bond to the cobalt(II) to regenerate the coenzyme. One way to think of the role of the B_{12} coenzymes is as reversible radical generators, coenzymes that can generate a radical on demand at the enzyme active site to aid in catalysis. In this sense, the B_{12} coenzymes are more elegant versions of the amino acid radicals.

While the amino acid radicals are always present in the active enzymes and thus are subject to potentially damaging side reactions, the B_{12} coenzyme radicals are 'stored' in a relatively innocuous form, the organometallic complex, and are generated only when and where needed through homolytic cobalt–carbon bond cleavage. Lysine 2, 3-aminomutase utilizes another coenzyme,

S-adenosylmethionine, to initiate radical chemistry via generation of an adenosyl radical intermediate. In this case, no cobalt is required for the reaction, but the enzyme does contain an iron–sulfur cluster cofactor. The mechanism of generation of the adenosyl radical in this system has yet to be elucidated.

Thiamin Pyrophosphate

Thiamin pyrophosphate (TPP), the pyrophosphorylated form of thiamin (vitamin B1), plays a central role in a number of enzymatic reactions that involve the cleavage of carbon–carbon bonds, including the critical a-ketoacid decarboxylations such as that catalysed by pyruvate decarboxylase. Decarboxylation reactions require an electron acceptor to dissipate the negative charge that builds up on the a-carbon, but α-ketoacids have no such intrinsic electron acceptor. In the case of most enzymecatalysed α-ketoacid decarboxylations, TPP serves as an electron acceptor to promote decarboxylation.

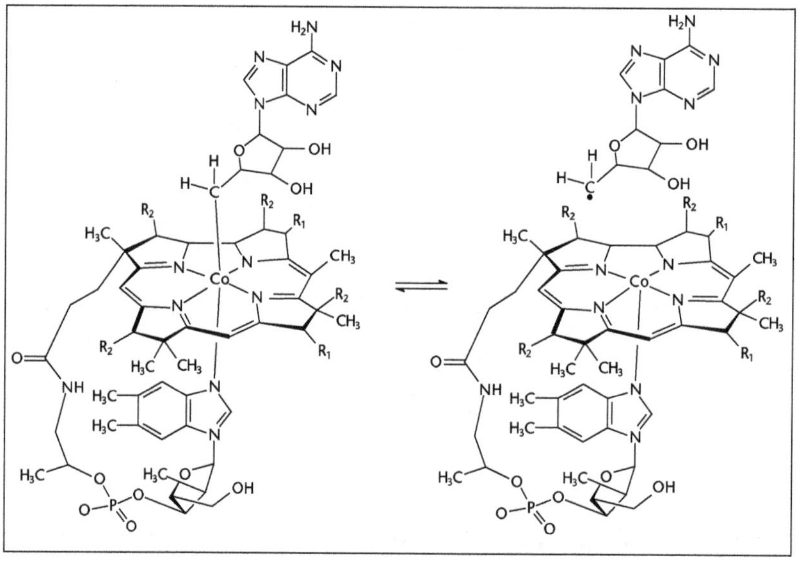

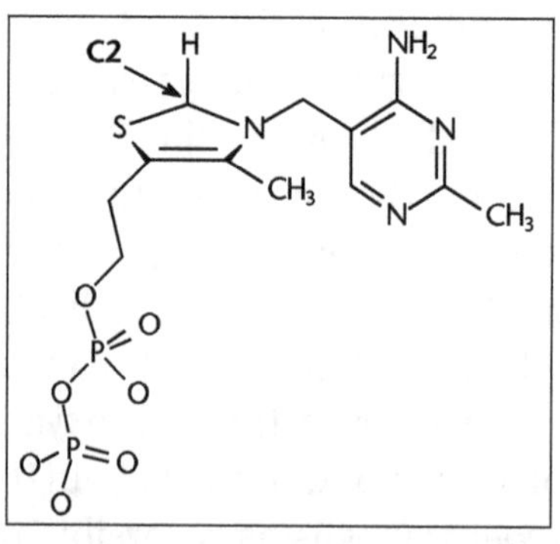

It was shown by Breslow that the thiazolium C_2 proton of TPP is very acidic, implicating the C_2 carbanion of the thiazolium ring in the catalytic mechanism. TPPdependent decarboxylation, for example in pyruvate decarboxylase, proceeds by attack of the thiazolium C_2 carbanion on the α carbon of the substrate to produce a tetrahedral intermediate. Decarboxylation then proceeds, with delocalization of the resulting negative charge on to the thiazolium ring. The decarboxylated product is then produced and the thiazolium carbanion regenerated.

Pyridoxal 5'-Phosphate

Like thiamin pyrophosphate, the role of pyridoxal 5'- phosphate (PLP) in enzyme catalysis is to serve as an electron sink, to dissipate negative charge that builds up during turnover. PLP serves this role in conjunction with a large number of enzymes that catalyse diverse reactions involving the α, β or γ carbons of α-amino acids. Also like TPP, PLP dissipates negative charge via a covalent intermediate, in this case an aldimine formed between the amine of the amino acid substrate and the aldehyde of PLP. In this way, PLP participates in the catalysis of decarboxylases, transaminases, aldolases, dehydratases and other enzymes that utilize a-amino acids as substrates.

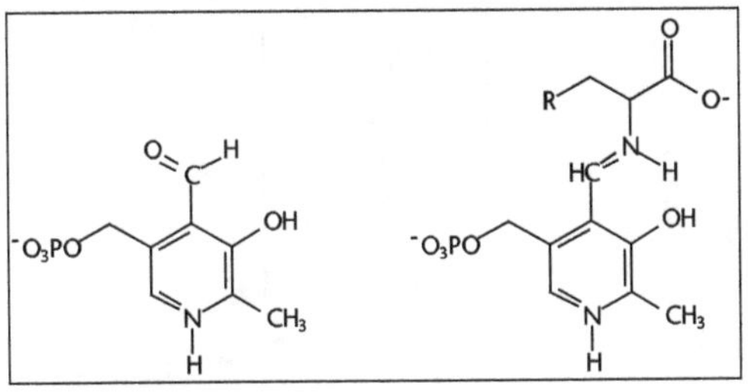

Phosphopantetheine Coenzymes

The phosphopantetheine coenzymes include the ubiquitous coenzyme A (CoA), in which 4'-phosphopantetheine is attached to an adenosyl moiety, and acyl carrier protein (ACP), in which 4'-phosphopantethiene is attached to the small protein through a serine hydroxyl group. The key feature of the phosphopantetheine coenzymes is a long, flexible chain terminating in a sulfhydryl group, and this sulfhydryl group is the key functional group of the coenzymes. The primary functions of the phosphopantetheine coenzymes are acyl group transfer and carboxylate activation, although other functions have been identified as well. These coenzymes serve as the intracellular currency of active acyl groups, and thus are central to many reactions, including fatty acid biosynthesis.

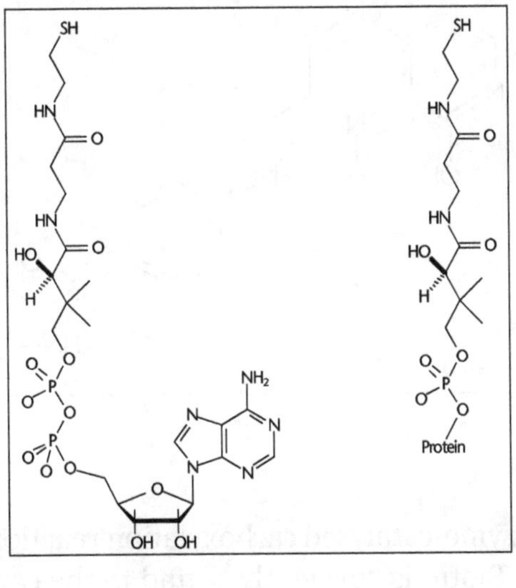

Lipoic Acid

Lipoic acid, like the pantetheinyl coenzymes, serves in the transfer of active acyl groups, with the key difference being that lipoic acid concomitantly acts as a mediator in the transfer of electrons.

The catalytically important functional group in lipoic acid is the disulfide, which can accept two electrons and two protons to be reduced to the dihydrolipoate state. It is in this reduced state that lipoic acid can accept activated acyl groups. In this way, lipoic acid couples the electron and acyl group transfers catalysed by α-ketoacid dehydrogenase multienzyme complexes. One of these, the pyruvate dehydrogenase complex, is a key enzyme in glucose metabolism.

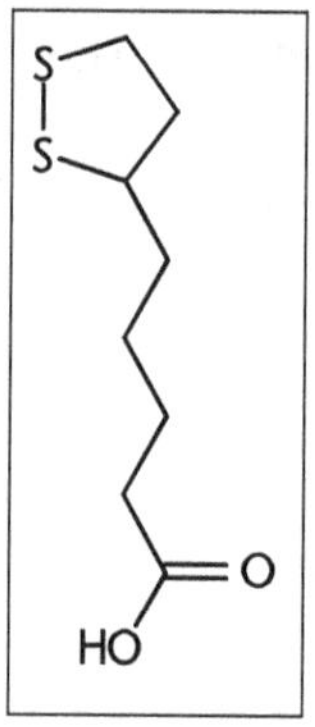

Folate Coenzymes

Tetrahydrofolate (FH_4) and related molecules serve as key coenzymes in the biosynthesis of purines and pyrimidines by supplying one-carbon units for the various synthetic steps. The one-carbon units supplied by N^{10}- formyl-FH_4, N^5, N^{10}-methenyl-FH_4, N^5, N^{10}-methyleneFH_4 and N^5 -methyl-FH_4 are essentially the cell's supply of formate and formaldehyde, kept in a relatively innocuous form as an FH_4 adduct until needed. In addition to the critical purine and pyrimidine biosynthetic functions performed by the tetrahydrofolates, N^5 -methyltetrahydrofolate is the methyl group donor in the biosynthesis of methionine.

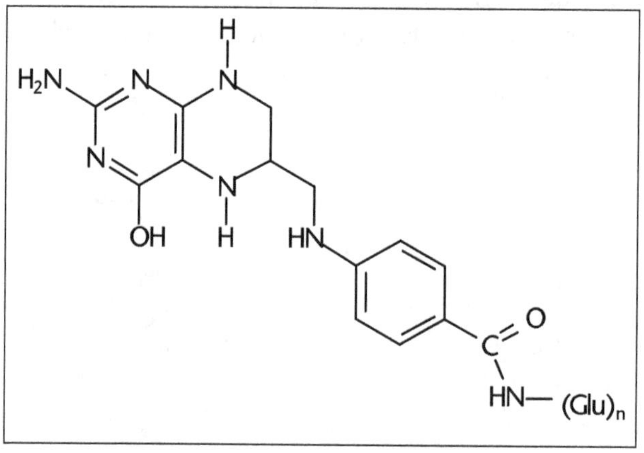

Biotin

Biotin plays a central role in enzyme-catalysed carboxylation reactions, most of which utilize bicarbonate and ATP as substrates. Biotin is covalently bound to the carboxylase enzymes with which it functions via an amide linkage between the biotin carboxyl and a lysyl ε-NH_2 group in the active site of the protein. The role of biotin in these carboxylase reactions is to act as mediator by accepting the activated carboxyl generated from bicarbonate and ATP and then to transfer this carboxyl to the appropriate substrate.

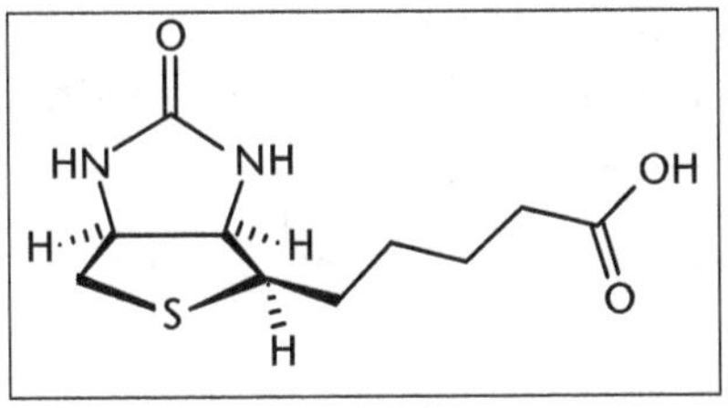

Organic Cofactors

A number of enzymes have been identified that contain covalently modified amino acids which function as coenzymes in catalysis. These include amine oxidase, which contains a topa quinone cofactor, and methylamine dehydrogenase, which contains a tryptophan tryptophylquinone, as well as other examples. These quinoproteins are typically redox catalysts such as oxidases and dehydrogenases.

Enzymatic Reactions that Require Metal Ion Cofactors

Thousands of enzymatic reactions, perhaps as many as one third of all those known, require metal ion cofactors in functions ranging from Lewis acid catalysis to redox catalysis to electron transfer. Metals add functionality to enzymes by providing strongly electrophilic centres, and in some cases multiple readily available oxidation states that can facilitate redox and electron transfer reactions. Most of the first row transition metals (the only exceptions being scandium, titanium, and perhaps chromium) as well as molybdenum, tungsten and magnesium are known to function as cofactors in enzymatic catalysis. Typically, these metal ion cofactors are bound to the enzyme via coordination of amino acid side-chains, although in some cases the metal is bound to a nonprotein ligand, e.g. a haem group, and the metal–ligand complex is bound to the enzyme through the metal or the ligand or both. In most cases, the metal is reasonably tightly bound to the enzyme, such that the metalloenzyme can be isolated with the metal ion intact, allowing the observation of enzymatic activity in the absence of added metal ion. Enzymes with more loosely bound metal ion cofactors, commonly called metalactivated enzymes, require the presence of the appropriate metal ion in the buffer in order to observe maximal catalytic activity.

Iron

Iron is the most abundant transition metal in the Earth's crust, and thus it is perhaps not surprising that it is ubiquitous in biological systems. Iron is a key player in some of the most central processes of biological systems, including oxygen transport and utilization, electron transfer, metabolism of nucleic acids and many other key biological molecules, degradation of biological pollutants, and many other reactions. A number of the enzymatic processes that require an iron cofactor are discussed below; these utilize both the accessibility of multiple redox states of iron (both Fe(II) and Fe(III) are readily accessible under physiological conditions) as well as the Lewis acidity of Fe(III) to aid in enzymatic catalysis. Iron cofactors can be bound directly to amino acid side-chains, for example in the intradiol catechol dioxygenases, where the iron is bound to two tyrosines, two histidines and one hydroxide to give a 5-coordinate environment. Iron is also commonly found bound to haem groups, for example in the ubiquitous cytochrome enzymes. In cytochrome P450, for example, iron is coordinated by the four pyrrole nitrogens of the protoporphorin IX in a square planar arrangement, and is also coordinated in one of its axial positions by a cysteine sulfhydryl and in the other by water, to give an overall distorted octahedral coordination environment.

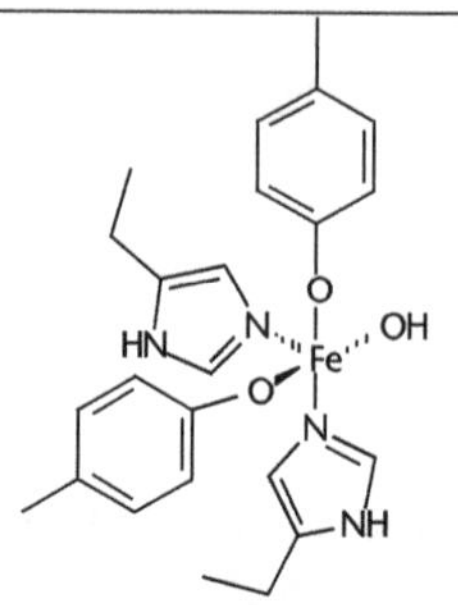

[XXVIII]

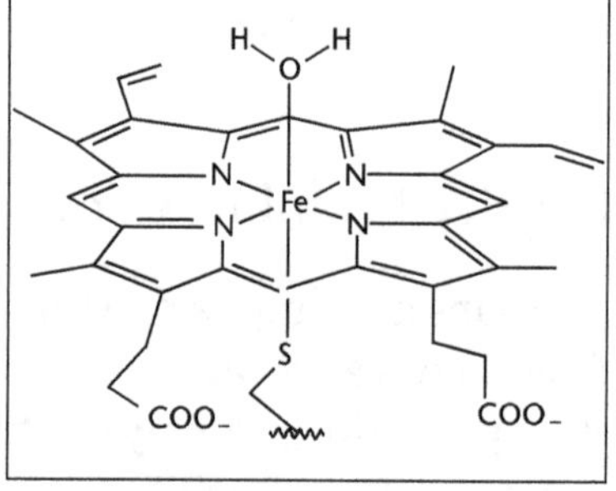

Some iron cofactors are found bound to a combination of amino acid side-chains and exogenous, nonsolvent, ligands. The most widespread examples of this are the iron–sulfur proteins, in which the iron atoms exist in clusters of two, three, four or more, coordinated by a combination of bridging inorganic sulfide and amino acid side-chains, usually cysteines.

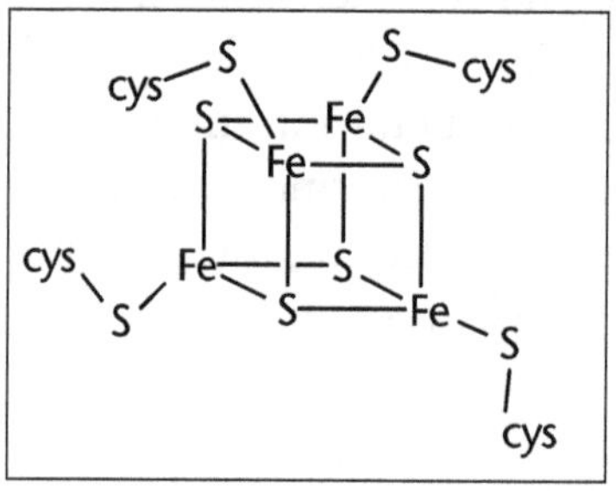

Metalloenzymes containing iron catalyse an enormous variety of reactions, most of which utilize the redox properties of iron and many of which involve reaction with oxygen and/or its reduced species, superoxide and peroxide. A common example of an iron-dependent enzymatic reaction is oxygenation, in which one or both atoms from molecular oxygen are incorporated into organic substrate. Oxygenases are central to many important metabolic processes, including the degradation of aromatic compounds, lipid metabolism, collagen formation, toxin degradation and alkane functionalization. Oxygenases also span the representative types of iron centres in proteins, with some containing mononuclear nonhaem iron centres, some binuclear iron centres, some haem iron centres, and some iron–sulfur clusters. In many of these enzymes, it is believed or has been clearly demonstrated that a primary role for the iron is to activate molecular oxygen for reaction with substrate. There is a strong thermodynamic driving force for the reaction of most organic molecules with dioxygen, yet such reactions do not occur spontaneously owing to the spin-forbidden nature of the reaction between the triplet dioxygen and singlet organic molecules. One way in which this spin restriction can be relaxed is via interaction of dioxygen with a metal centre such that the dioxygen is reduced to the superoxide or peroxide oxidation state; this type of 'oxygen activation' appears to be a primary role for iron in biological systems.

$$RH + O_2 + 2H^+ \xrightarrow[\text{monooxygenase}]{} ROH + H_2O$$

Iron also serves as a cofactor in oxidases, in which the oxidizing power of oxygen is utilized for a redox reaction, but oxygen is not incorporated into the product. One of the most interesting and functionally central oxidases is cytochrome c oxidase, which plays a central role in oxidative phosphorylation, the process by which the oxidizing power of oxygen is harnessed to synthesize ATP, the energy currency of the cell. Cytochrome c oxidase contains both copper and iron cofactors and is the site of reduction of oxygen to water. The ready accessibility of two oxidation states for both copper and iron is utilized by cytochrome oxidase in transferring electrons to oxygen.

Iron is also a central cofactor in the reactions catalysed by superoxide dismutase (SOD) and catalase. Both of these enzymes serve to protect the living system from the potential damaging effects of reactive reduced oxygen species. Iron has also been more recently shown to play a central role in the generation of catalytically essential amino acid radicals and in the initiation of radical mechanisms in certain enzymes. The best-characterized example of this type of iron metalloenzyme is ribonucleotide reductase, in which a binuclear iron(II) centre reacts with oxygen to produce a binuclear iron(III) centre and a tyrosyl radical, the latter being an absolute requirement for catalytic activity. Iron, in the form of an iron–sulfur cluster, is also centrally involved in the generation of the glycyl radical in ribonucleotide reductase from anaerobic E. coli and the glycyl radical of pyruvate formatelyase; again, both of these radicals are catalytically essential. Iron, again in the form of iron– sulfur clusters, is also involved in initiation of radical chemistry in lysine 2, 3-aminomutase.

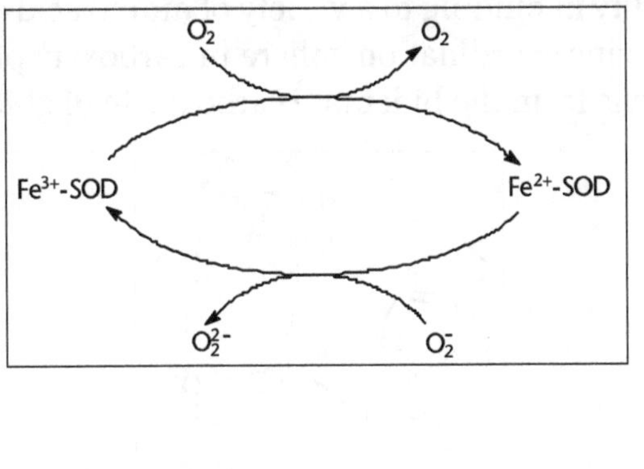

Copper

Copper, like iron, has two oxidation states (I and II) that are readily available under physiological conditions, and therefore can participate in catalysis of redox reactions in much the same way that iron does. One example is in the superoxide dismutases, which catalyse the disproportionation of superoxide to peroxide and oxygen. An alternative to the Fe-SOD is the Cu/Zn-SOD, which contains copper and zinc in a binuclear site. In this case it is copper that binds superoxide and serves the redox role in the mechanism, while the role of zinc appears to be merely structural.

There are other analogies in the biological chemistry of copper and iron, the most notable being the prevalence of both in oxygenase and oxidase chemistry. Copper is a key cofactor in dopamine β-monooxygenase and peptidylglycine α-amidating enzyme, both of which catalyse monooxygenations. Like iron in the iron oygenases, copper is believed to activate dioxygen, via redox chemistry, for reaction with organic substrate.

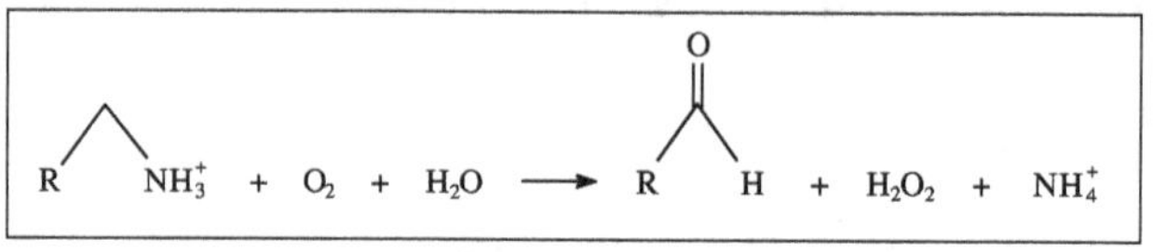

Copper is also a prevalent component of oxidases, including the cytochrome c oxidase mentioned above, as well as a variety of amine oxidases, and galactose oxidase. The amine oxidases also contain a covalent catalytically essential topaquinone cofactor. The similarities between iron and copper in biological systems go beyond their roles as enzyme cofactors to their roles in dioxygen transport and in electron transfer. What is clear from studies to date on metalloenzymes is that the unique chemical properties of iron and copper, including their strongly electrophilic character and the availability of multiple redox states, make them uniquely suited as cofactors to a variety of critical biochemical reactions.

Zinc

Among the transition metals playing key roles in biological systems, zinc is second only to iron in biological prevalence. Unlike iron and copper, zinc has only one oxidation state (+2) accessible under physiological conditions, and therefore does not participate in catalysis of redox chemistry or electron transfer reactions. Zinc is, however, quite flexible in its preference for ligands. This flexibility is due in part to zinc's affinity for both the soft (polarizable) protein ligands, such as cysteine thiols, and the harder (less polarizable) ligands, such as carboxylate, histidine imidazole and water. In addition, zinc can easily adopt coordination numbers of 4, 5, or 6. These properties give zinc tremendous flexibility in binding to a variety of amino acid side-chains and substrates in a variety of geometries. The zinc coordination sphere in carboxypeptidase contains two histidine imidazole ligands, two oxygens from the bidentate carboxylate of glutamate, and one water.

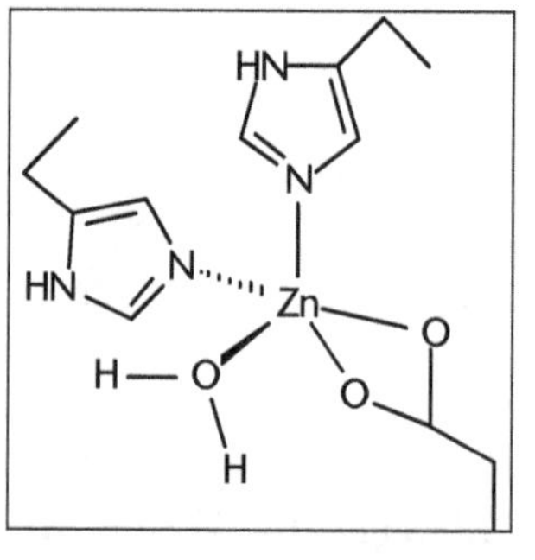

Zinc functions widely in Lewis acid catalysis and is also found to play important structural roles in enzymes and proteins. The types of enzymatic reactions in which zinc is found to play a role include peptidases and amidases, phosphatases, phospholipase, phosphotriesterase, deaminases and alcohol dehydrogenase, among others. One mechanism by which zinc is believed to facilitate some of these reactions is through binding and thereby lowering the pK_a of water, generating a localized high concentration of metal-bound hydroxide in the active site, which can act as a nucleophile in the hydrolytic reactions.

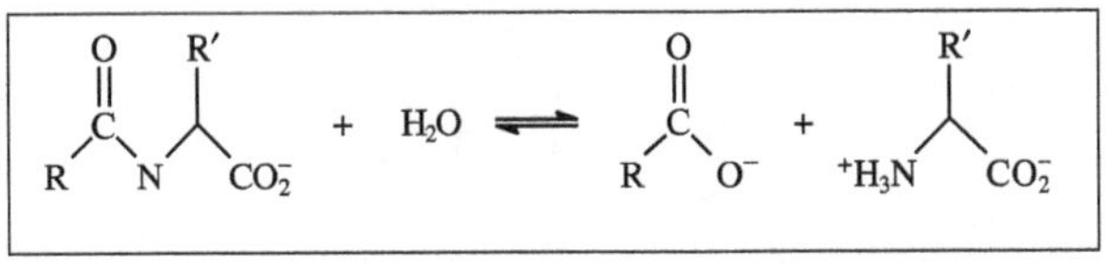

Other Metals

Although we have touched on the most common metal ion cofactors to enzymatic reactions, a number of others have been identified. Nickel has been found to be a required cofactor for urease and certain hydrogenases. Both vanadium and molybdenum have been identified as cofactors in nitrogenases and oxidases, and molybdenum has also been found to be a required cofactor for certain oxygen atom transfer reactions. A four-manganese cluster is the site of water binding and oxidation to O_2 in the photosynthetic oxygen-evolving complex. It is intriguing to note that this photosynthetic oxygen evolution requires a tyrosyl radical in addition to the manganese cluster, and this tyrosyl radical has been proposed to function in abstraction of hydrogen atoms from water. Manganese is also a cofactor in oxidase chemistry, while magnesium has been shown to be required in certain hydrolases and isomerases. Tungsten has been found to be a cofactor in certain dehydrogenases.

References

- Coenzymes-and-cofactors-228017458: researchgate.net, Retrieved 25, june 2020
- Enzyme-definition-and-classification: creative-enzymes.com, Retrieved 07, Jan 2020
- Role-of-isomerase-enzyme: infinitabiotech.com, Retrieved 13, April 2020
- Henderson-hasselbalch-equation: microbenotes.com, Retrieved 09, May 2020
- Translocase-introduction: creative-enzymes.com, Retrieved 19, Feb 2020
- Enzyme-technology-an-emerging-trend-in-biotechnology: longdom.org, Retrieved 11, July 2020

Chapter 2

Catalysis and Kinetics of Enzymes

The process of increase in the rate of a biological reaction by an enzyme is known as enzyme catalysis. A few of the significant topics studied in relation to enzyme catalysis and kinetics are reaction kinetics, Michaelis-Menten kinetics, multi-substrate systems, enzyme inhibition, etc. This chapter has been carefully written to provide an easy understanding of the varied facets of enzyme catalysis and kinetics.

Enzyme Catalysis

Catalysis is a phenomenon in which the rate of the reaction is altered with the help of a substance called a catalyst (the catalyst does not participate in the reaction; its concentration and composition remain unchanged). The substance used to change the rate of the reaction is called a catalyst. Enzymes are a class of catalysts that are responsible for facilitating and increasing the rate of many vital biochemical reactions in plants and animals. The catalysis in which enzymes act as a catalyst is called enzyme catalysis. Enzymes are complex compounds containing nitrogen. Animals and plants produce these compounds naturally in their bodies. Enzymes are proteins which have high molecular mass and form a heterogeneous mixture when dissolved in water. These proteins act very efficiently and are responsible for various reactions which occur in the body of living beings.

Characteristics of Enzyme Catalysis

- A single molecule of the enzyme catalyst can transform up to a million molecules of the reactant per second. Hence, enzyme catalysts are said to be highly efficient.
- These biochemical catalysts are unique to certain types of reactions, i.e. the same catalyst cannot be used in more than one reaction.
- The effectiveness of a catalyst is maximum at its optimum temperature. The activity of the biochemical catalysts declines at either side of the optimum temperature.
- Biochemical catalysis is dependent upon the pH of the solution. A catalyst works best at an optimum pH which ranges between 5-7 ph values.
- The activity of the enzymes usually increases in the presence of a coenzyme or an activator such as Na^+, Co^{2+}. The rate of the reaction increases due to the presence of a weak bond which exists between the enzyme and a metal ion.

Mechanism of Enzyme Catalyst

Enzymes consist of a number of cavities which are present on the outer surface. These cavities possess groups such as -COOH, -SH, etc. These centres are called as the active centre of the biochemical particle. The substrate which has the opposite charge of the enzyme fits into the cavities just as a key fits into a lock. Due to the existence of the active groups, the complex formed decomposes to give the products.

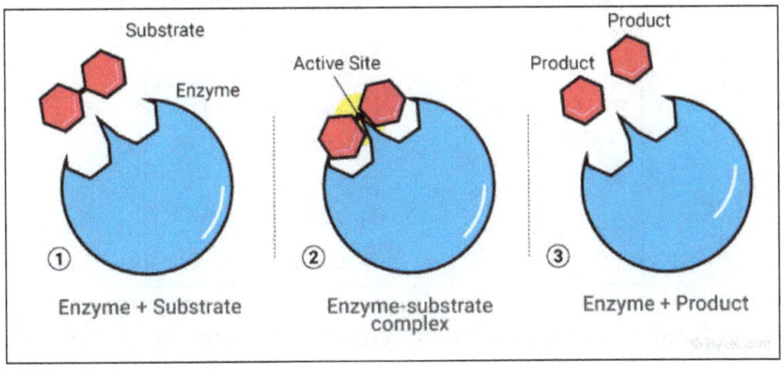

Hence this happens in two steps:

Step1: Combining of enzyme and the reactant.

$$E + R \rightarrow ER$$

Step 2: Disintegration of the complex molecule to give the product.

$$ER \rightarrow E + P$$

Types of Enzyme Catalysts

Although a huge number of reactions occur in living systems, these reactions fall into only half a dozen types. The reactions are:

- Oxidation and reduction: Enzymes that carry out these reactions are called oxidoreductases. For example, alcohol dehydrogenase converts primary alcohols to aldehydes.

  $$H_3CCH_2OH + NAD\ H_3CCHO + NADH + H^+$$

  In this reaction, ethanol is converted to acetaldehyde, and the cofactor, NAD, is converted to NADH. In other words, ethanol is oxidized, and NAD is reduced. (The charges don't balance, because NAD has some other charged groups.) Remember that in redox reactions, one substrate is oxidized and one is reduced.

- Group transfer reactions: These enzymes, called transferases, move functional groups from one molecule to another. For example, alanine aminotransferase shuffles the alpha-amino group between alanine and aspartate:

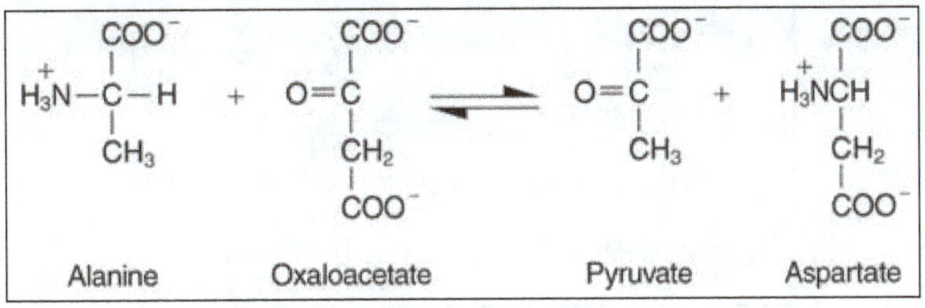

Other transferases move phosphate groups between ATP and other compounds, sugar residues to form disaccharides, and so on.

- Hydrolysis: These enzymes, termed hydrolases, break single bonds by adding the elements of water. For example, phosphatases break the oxygen-phosphorus bond of phosphate esters:

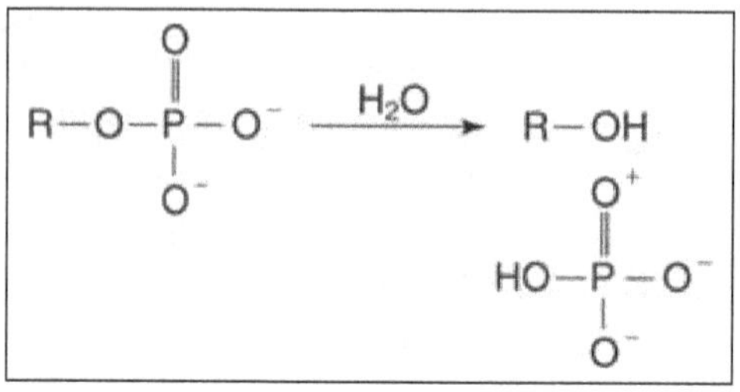

Other hydrolases function as digestive enzymes, for example, by breaking the peptide bonds in proteins.

- Formation or removal of a double bond with group transfer: The functional groups transferred by these lyase enzymes include amino groups, water, and ammonia. For example, decarboxylases remove CO_2 from alpha- or beta-keto acids:

Dehydratases remove water, as in fumarase (fumarate hydratase):

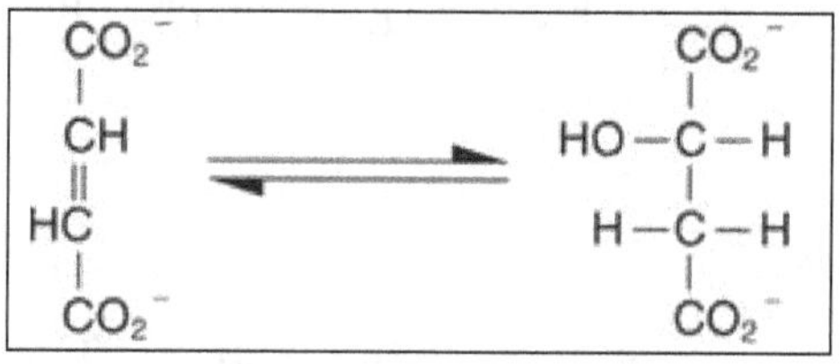

Deaminases remove ammonia, for example, in the removal of amino groups from amino acids:

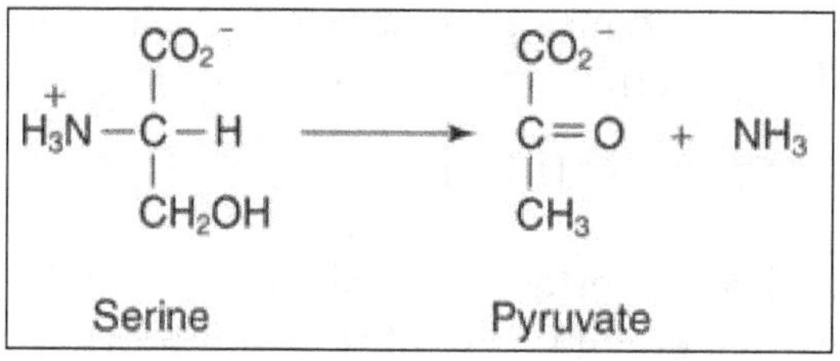

- Isomerization of functional groups: In many biochemical reactions, the position of a functional group is changed within a molecule, but the molecule itself contains the same number and kind of atoms that it did in the beginning. In other words, the substrate and product of the reaction are isomers. The isomerases (for example, triose phosphate isomerase, shown following), carry out these rearrangements.

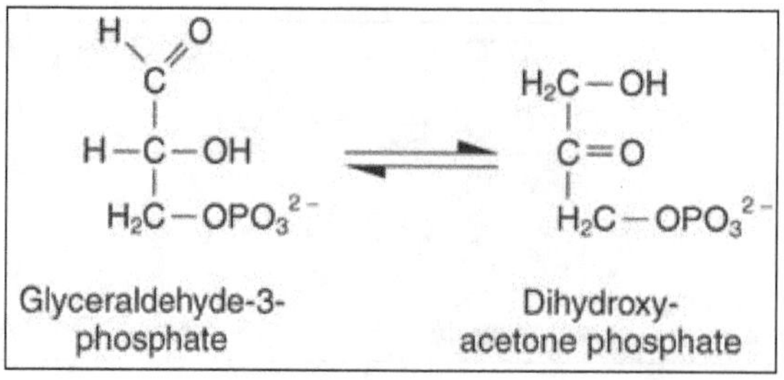

- Single bond formation by eliminating the elements of water. Hydrolases break bonds by adding the elements of water; ligases carry out the converse reaction, removing the elements of water from two functional groups to form a single bond. Synthetases are a subclass of ligases that use the hydrolysis of ATP to drive this formation. For example, aminoacyl-transfer RNA synthetases join amino acids to their respective transfer RNAs in preparation for protein synthesis; the action of glycyl-tRNA synthetase is illustrated in this figure:

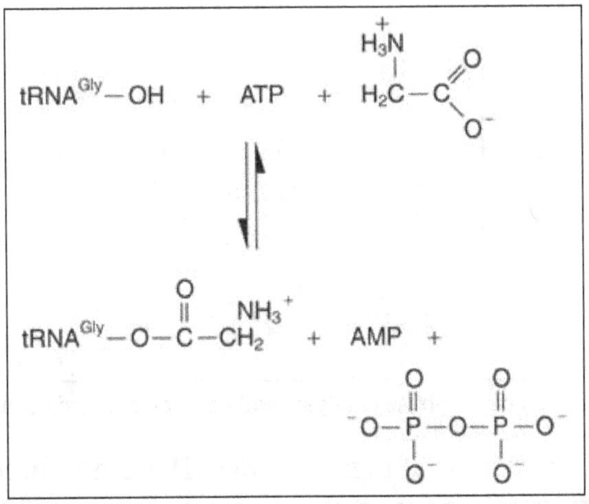

Acid-Base Catalysis

In acid-base catalysis, the chemical reaction is accelerated by the addition of an acid or a base, and the acid or base itself is not consumed in the reaction. Proton transfer is the commonest reaction that enzymes perform. Proton donors and acceptors, i.e. acids and base may donate and accept protons in order to stabilize developing charges in the transition state. This typically has the effect of activating nucleophile and electrophile groups, or stabilizing leaving groups. Many acid-base catalysis reactions involve histidine because it has a pK_a close to 7, allowing it to act as both an acid and a base.

Characteristics of Acid-Base Catalysis

The great majority of enzyme reactions–the chemical reactions on which life depends–take place in water, under "physiological" conditions, near pH 7. Yet this is where chemical reactions are normally at their slowest. Reactive compounds are readily hydrolyzed by adding strong acid or base in vitro. But this is not a practical proposition for an enzyme: proteins are denatured in the presence of strong acid or base.

The acid-base catalysis is illustrated by the pH–rate profiles shown in Figure: the three plots are representative of the behavior of most types of substrate undergoing most sorts of reaction in water. The lower curve (I) represents the reaction in vitro of a typical, unreactive compound: it shows only acid- and base-catalyzed reactions, and reaction is very slow at the minimum, near pH 7. For the most reactive compounds, an additional feature (curve II) is a pH-independent region, where the uncatalyzed reaction with water becomes faster than the acid- and base-catalyzed reactions near neutrality. Finally, curve III is the pH–rate profile for a typical enzyme-catalyzed reaction: it is much faster than the others (signified by the break in the ordinate), and it is also qualitatively quite different, since now the rate reaches a maximum near pH 7. Enzymes are 'designed' to operate near pH 7, and typically show pH optima in this region, with rates falling off at higher and lower pH values.

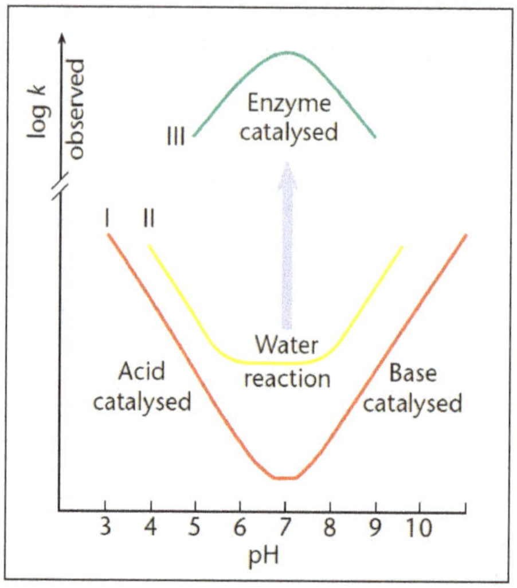

Figure: Specific acid–base catalysis and enzyme catalysis compared.

The acid- and base-catalyzed reactions at high and low pH are not directly relevant to catalysis by enzymes. They typically involve activation of the substrate by the addition or removal of a proton in a rapid pre-equilibrium, followed by the rate-determining reaction of the conjugate acid or base.

Mechanism of Acid-Base Catalysis

If the substrate is reactive (electrophilic) enough, activation by protonation is not necessary and the attack of water on the neutral molecule is rate determining near pH 7 (curve II of Figure). Formally this generates both a positive and a negative charge, and as the reaction proceeds both will be "delocalized" into the surrounding solvent via the network of hydrogen bonds. The nucleophile in a hydrolysis reaction is of course a water molecule. As the new C–O bond forms, positive charge develops on the nucleophilic oxygen, and the attached OH protons become more and more acidic, until they can be transferred to solvating water, acting formally as a general base (gb). The negative charge developing on the carbonyl oxygen may similarly be transferred via a hydrogen bond to another water molecule, acting this time as a general acid (ga). This mechanism is always available, though it does not always lead to reaction at an observable rate.

The direct reaction does have the entropic advantage of not involving a third molecule, and a carboxylate anion can–given the right conditions–displace better leaving groups than ethoxide. As long as the intermediate is more reactive than the starting material the result is catalysis of hydrolysis. This mechanism competes equally with general base catalysis for esters with a leaving group of pK_a of ~7, and becomes dominant for derivatives with better leaving groups.

If the leaving group is poor it can be made viable by protonation: complete protonation to form the conjugate acid if the group is sufficiently basic, but involving partial proton transfer in the case of a weakly basic group like OR or OH. This mechanism is involved, though not easily observed, in the breakdown of the tetrahedral addition intermediates involved in the acyl transfer reactions of esters and amides.

Brønsted Acid and Base Groups in Enzymes

The "catalytic machinery" of an enzyme consists of a small number of functional groups brought together in a well-defined three-dimensional arrangement by the tertiary structure of the protein to form the active site. The functional groups concerned are a subset of those available on the side-chains of the naturally occurring amino acids. Under physiological conditions near pH 7, only weak acids and bases, with pK_a within a unit or two of 7, can exist to a significant extent in the acidic or basic form. Thus the strongest acids available on amino acid side chains are the two carboxylic acids; but these will be present at pH 7 almost exclusively as the aspartate and glutamate anions–unless their pK_as are perturbed by their local environment. Similarly, the strongest bases are likely to be protonated unless similarly perturbed. Although such perturbations are not uncommon, the imidazole group of histidine, because it has a pK_a near 7, is at once the strongest acid and the strongest base normally available to enzymes under physiological conditions near pH 7.

A functional group involved in catalysis of a particular step of a particular reaction plays a very specific role, and so can be expected to be active in one particular ionic form: thus reactivity will depend on pH. When two (or more) such groups are involved it is common for one to be active as a general acid, in the protonated form, and the other as a general base or nucleophile, and thus in its basic form. The pH–rate dependence of the catalyzed reaction then reflects the fraction of the system with both groups in the active ionic form. At sufficiently high and low pH, a single species is present, as the free base or fully protonated form. But the fraction of the intermediate zwitterionic form increases to a maximum at a pH halfway between the two pK_as. If this is the reactive form, the pH–rate profile will also show a maximum at this pH. This is the simplest explanation for a pH optimum for an enzyme-catalyzed reaction.

Covalent Catalysis

Covalent Catalysis is one of the four strategies that an enzyme will use to catalyze a specific reaction, which involves the formation of a transient covalent bond between a substrate and a residues in the enzyme active site or with a cofactor. In covalent catalysis, an additional covalent intermediate is added to the reaction, and helps to reduce the energy of transition states in the later stages of the reaction. The most common covalent bonds are formed as a result of the attack by an enzyme nucleophilic group on an electrophilic moiety of the substrate that is bound at the active site.

Mechanism

In an enzymatic reaction, covalent catalysis occurs when the substrate become temporarily covalently attached to the enzyme during the catalytic reaction. In this reaction, the enzyme contains a reactive group, usually a nucleophilic residue or an electrophilic residue, which reacts with the substrate through a nucleophilic or electrophilic attack. The nucleophilic groups can be either RCOO-, RNH, ROH that are present on the side-chains of amino acid residues, or the nitrogen atoms of the imidazole ring of histidine residues. The electrophilic moieties of substrates may be acyl, phosphoryl, or glycosyl groups, so the covalent intermediates would be acyl-, phosphoryl-, and glycosyl-enzyme complexes. Enzyme molecules are poor in electrophilic groups, but electrophilic catalysis does occur with those enzymes that contain metals or prosthetic groups that act as electron sinks during catalysis. The charge loss in the reaction during transitional state will then accelerate the hydrolysis.

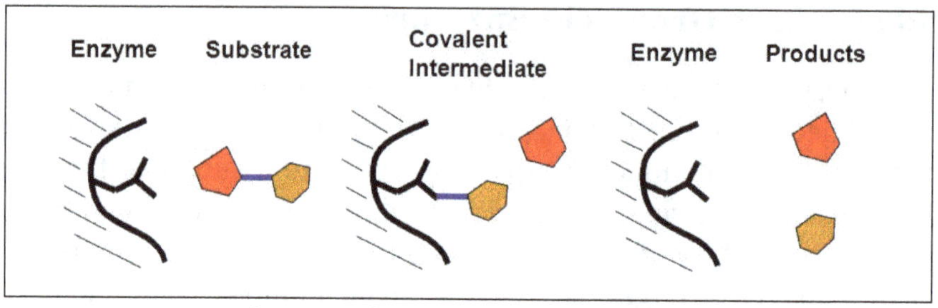

Figure: Mechanism of covalent catalysis.

Rather than lowering the activation energy for a reaction pathway, covalent catalysis provides an alternative pathway for the reaction (via to the covalent intermediate) and so is distinct from true catalysis. A true proposal of a covalent catalysis would require, for example, a partial covalent bond to the transition state by an enzyme group, and such effects do not contribute significantly to catalysis.

Covalent Intermediates Formed by Substrates and Cofactors

Some enzymes utilize non-amino acid cofactors such as pyridoxal phosphate (PLP) or thiamine pyrophosphate (TPP) to form covalent intermediates with reactant molecules. Such covalent intermediates function to reduce the energy of later transition states, similar to how covalent intermediates formed with active site amino acid residues allow stabilization, but the capabilities of cofactors allow enzymes to carryout reactions that amino acid side residues alone could not. Enzymes utilizing such cofactors include the PLP-dependent enzyme aspartate transaminase and the TPP-dependent enzyme pyruvate dehydrogenase.

Catalytic Triad of Enzymes

At a later stage in covalent catalysis, the covalent bond must be broken to regenerate the enzyme. This mechanism is utilized by the catalytic triad of enzymes such as proteases like chymotrypsin and trypsin, where an acyl-enzyme intermediate is formed. Chymotrypsin is a degradative protease of the digestive system. It catalyzes the cleavage of peptide bonds that are adjacent to large aromatic or nonpolar residues. It cleaves the peptide bond on the carboxyl terminus side of the protein. The chymotrypsin has three main catalytic residues termed as the catalytic triad. These are His57, Asp102 and Ser195. The nucleophile is the hydroxyl group on the serine. Upon deprotonation the serine residue becomes a powerful nucleophile due to its alkoxide that will attack the relatively unreactive carbon of the carbonyl in the protein.

Enzyme Kinetics

Enzyme kinetics deals with enzyme reactions which are time-dependent and explains the mechanisms of enzyme catalysis and its regulation.

Let's understand enzyme kinetics as a function for the concentration of the substrate available for the enzyme:

- Start the experiment with a series of tubes which contains substrate, [S].

- At time (t) zero, add some amount of the enzyme.
- Wait for few minutes.
- Then, measure the newly formed product concentration. We can also use spectrophotometer, If product absorbs light.
- At a time when the amount of substrate is greater than the amount of enzyme, then, the rate is the initial velocity of V_i.

If we plot V_i as a function of [S], following observations will be made:

- At low [S], the initial velocity, V_i, rises linearly with increasing [S].
- When [S] increases, V_i settle down (rectangular hyperbolais formed).
- The asymptote shows Vmax as the maximum velocity of the reaction.
- The substrate concentration which produces a V_i equal to one-half of Vmax is called the Michaelis-Menten constant (K_m).

K_m is (approximately) inversely related to the maximum reaction velocity, or strength of binding between the enzyme and its substrate. Lower the K_m value, higher is the affinity for its substrate.

Reaction Kinetics

Chemical reaction kinetics deals with the rates of chemical processes. Any chemical process may be broken down into a sequence of one or more single-step processes known either as elementary processes, elementary reactions, or elementary steps. Elementary reactions usually involve either a single reactive collision between two molecules, which we refer to as a a bimolecular step, or dissociation/isomerisation of a single reactant molecule, which we refer to as a unimolecular step. Very rarely, under conditions of extremely high pressure, a termolecular step may occur, which involves simultaneous collision of three reactant molecules. An important point to recognise is that many reactions that are written as a single reaction equation in actual fact consist of a series of elementary steps.

As a general rule, elementary processes involve a transition between two atomic or molecular states separated by a potential barrier. The potential barrier constitutes the activation energy of the process, and determines the rate at which it occurs. When the barrier is low, the thermal energy of the reactants will generally be high enough to surmount the barrier and move over to products, and the reaction will be fast. However, when the barrier is high, only a few reactants will have sufficient energy, and the reaction will be much slower. The presence of a potential barrier to reaction is also the source of the temperature dependence of reaction rates.

The huge variety of chemical species, types of reaction, and the accompanying potential energy surfaces involved means that the timescale over which chemical reactions occur covers many orders of magnitude, from very slow reactions, such as iron rusting, to extremely fast reactions, such as the electron transfer processes involved in many biological systems or the combustion reactions occurring in flames.

A study into the kinetics of a chemical reaction is usually carried out with one or both of two main goals in mind:

- Analysis of the sequence of elementary steps giving rise to the overall reaction. i.e. the reaction mechanism.
- Determination of the absolute rate of the reaction and/or its individual elementary steps.

Rate of Reaction

When we talk about the rate of a chemical reaction, what we mean is the rate at which reactants are used up, or equivalently the rate at which products are formed. The rate therefore has units of concentration per unit time, mol dm^{-3} s^{-1} (for gas phase reactions, alternative units of concentration are often used, usually units of pressure – Tor, mbar or Pa). To measure a reaction rate, we simply need to monitor the concentration of one of the reactants or products as a function of time. There is one slight complication to our definition of the reaction rate so far, which is to do with the stoichiometry of the reaction. The stoichiometry simply refers to the number of moles of each reactant and product appearing in the reaction equation. For example, the reaction equation for the well-known Haber process, used industrially to produce ammonia, is:

$$N_2 + 3H_2 \rightleftharpoons 2NH_3$$

N2 has a stochiometric coefficient of 1, H_2 has a coefficient of 3, and NH_3 has a coefficient of 2. We could determine the rate of this reaction in any one of three ways, by monitoring the changing concentration of N_2, H_2, or NH_3. Say we monitor N_2, and obtain a rate of $-\frac{d[N_2]}{dt}$ = x mol dm^{-3} s^{-1}. Since for every mole of N_2 that reacts, we lose three moles of H_2, if we had monitored H_2 instead of N2 we would have obtained a rate $-\frac{d[H_2]}{dt} = 3x$ mol dm^{-3} s^{-1}. Similarly, monitoring the concentration of NH_3 would yield a rate of 2x mol dm^{-3} s^{-1}. Clearly, the same reaction cannot have three different rates, so we appear to have a problem. The solution is actually very simple: the reaction rate is defined as the rate of change of the concentration of a reactant or product divided by its stochiometric coefficient. For the above reaction, the rate (usually given the symbol ν) is therefore,

$$v = -\frac{d[H_2]}{dt} = -3\frac{d[H_2]}{dt} = \frac{1}{2}\frac{d[NH_3]}{dt}$$

A negative sign appears when we define the rate using the concentration of one of the reactants. This is because the rate of change of a reactant is negative (since it is being used up in the reaction), but the reaction rate needs to be a positive quantity.

Rate Laws

The rate law is an expression relating the rate of a reaction to the concentrations of the chemical species present, which may include reactants, products, and catalysts. Many reactions follow a simple rate law, which takes the form,

$$v = k\,[A]^a\,[B]^b\,[C]^c \ldots$$

I.e. the rate is proportional to the concentrations of the reactants each raised to some power. The constant of proportionality, k, is called the rate constant. The power a particular concentration is raised to is the order of the reaction with respect to that reactant. Note that the orders do not have to be integers. The sum of the powers is called the overall order. Even reactions that involve multiple elementary steps often obey rate laws of this kind, though in these cases the orders will not necessarily reflect the stoichiometry of the reaction equation. For example,

$$H_2 + I_2 \rightarrow 2HI \qquad v = k\,[H_2][I_2]$$

$$3ClO^- \rightarrow ClO_3^- + 2Cl^- \qquad v = k\,[ClO^-]^2$$

Other reactions follow complex rate laws. These often have a much more complicated dependence on the chemical species present, and may also contain more than one rate constant. Complex rate laws always imply a multi-step reaction mechanism. An example of a reaction with a complex rate law is:

$$H_2 + Br_2 \rightarrow 2HBr \qquad v = \frac{[H_2][Br_2]^{1/2}}{1 + k'[HBr]/[Br_2]}$$

In the above example, the reaction has order 1 with respect to $[H_2]$, but it is impossible to define orders with respect to Br_2 and HBr since there is no direct proportionality between their concentrations and the reaction rate. Consequently, it is also impossible to define an overall order for this reaction.

To give you some idea of the complexity that may underlie an overall reaction equation, a slightly simplified version of the sequence of elementary steps involved in the above reaction is shown below.

$$Br_2 \rightarrow Br + Br$$

$$Br + H_2 \rightarrow H + HBr$$

$$H + Br_2 \rightarrow Br + HBr$$

$$Br + Br \rightarrow Br_2$$

As well as having rate laws for overall reactions, we can of course also write down individual rate laws for elementary steps. Elementary processes always follow simple rate laws, in which the order with respect to each reactant reflects the molecularity of the process (how many molecules are involved). For example,

Unimolecular decomposition $A \rightarrow B \qquad v = k\,[A]$

Bimolecular Reaction:

$$A + B \rightarrow P \qquad v = k\,[A][B]$$

$$A + A \rightarrow P \qquad v = k\,[A][A] = k\,[A]^2$$

Multi-step processes may follow simple or complex rate laws, and as the above examples have hopefully illustrated, the rate law generally does not follow from the overall reaction equation. This makes perfect sense, since the overall reaction equation for a multi-step process is simply the net result of all of the elementary reactions in the mechanism. The 'reaction' given in the overall reaction equation never actually takes place! However, even though the rate law for a multi-step reaction cannot immediately be written down from the reaction equation as it can in the case of an elementary reaction, the rate law is a direct result of the sequence of elementary steps that constitute the reaction mechanism. As such, it provides our best tool for determining an unknown mechanism. Conversely, if we do not know the reaction mechanism, we can carry out experiments to determine the orders with respect to each reactant and then try out various 'trial' reaction mechanisms to see which one fits best with the experimental data. At this point it should be emphasised again that for multi-step reactions, the rate law, rate constant, and order are determined by experiment, and the orders are not generally the same as the stoichiometric coefficients in the reaction equation.

A final important point about rate laws is that overall rate laws for a reaction may contain reactant, product and catalyst concentrations, but must not contain concentrations of reactive intermediates (these will of course appear in rate laws for individual elementary steps).

The Units of the Rate Constant

A point which often seems to cause endless confusion is the fact that the units of the rate constant depend on the form of the rate law in which it appears i.e. a rate constant appearing in a first order rate law will have different units from a rate constant appearing in a second order or third order rate law. This follows immediately from the fact that the reaction rate always has the same units of concentration per unit time, which must match the overall units of a rate law in which concentrations raised to varying powers may appear. The good news is that it is very straightforward to determine the units of a rate constant in any given rate law. Below are a few examples.

- Consider the rate law $v = k[H_2][I_2]$. If we substitute units into the equation, we obtain,

  $(\text{mol dm}^{-3}\ \text{s}^{-1}) = [k]\ (\text{mol dm}^{-3})\ (\text{mol dm}^{-3})$

  where the notation [k] means 'the units of k'. We can rearrange this expression to find the units of the rate constant, k.

  $$[k] = \frac{(\text{mol dm}^{-3}\text{s}^{-1})}{(\text{mol dm}^{-3})(\text{mol dm}^{-3})} = \text{mol}^{-1}\ \text{dm}^{-3}\text{s}^{-1}$$

- We can apply the same treatment to a first order rate law, for example:

  $v = k\ [CH_3N_2CH_3]\ (\text{mol dm}^{-3}\ \text{s}^{-1}) = [k]\ (\text{mol dm}^{-3})$

  $$[k] = \frac{(\text{mol dm}^{-3}\ \text{s}^{-1})}{(\text{mol dm}^{-3})} = \text{s}^{-1}$$

- As a final example, consider the rate law

  $v = k\ [CH_3CHO]^{3/2}.\ (\text{mol dm}^{-3}\ \text{s}^{-1}) = [k]\ (\text{mol dm}^{-3})^{3/2}$

$$[k] = \frac{\left(\text{mol dm}^{-3}\ \text{s}^{-1}\right)}{\left(\text{mol dm}^{-3}\right)^{3/2}} = \text{mol}^{-1/2}\text{dm}^{3/2}\text{s}^{-1}$$

An important point to note is that it is meaningless to try and compare two rate constants unless they have the same units.

Integrated Rate Law

A rate law is a differential equation that describes the rate of change of a reactant (or product) concentration with time. If we integrate the rate law then we obtain an expression for the concentration as a function of time, which is generally the type of data obtained in an experiment. In many simple cases, the rate law may be integrated analytically. Otherwise, numerical (computer-based) techniques may be used. Four of the simplest rate laws are given below in both their differential and integrated form.

Reaction	Order	Differential form	Integrated form
$A \rightarrow P$	zeroth	$\frac{d[A]}{dt} = -k$	$[A] = [A]_0 - kt$
$A \rightarrow P$	first	$\frac{d[A]}{dt} = -k[A]$	$\ln[A] = \ln[A]_0 - kt$
$A + A \rightarrow P$	second	$\frac{1}{2}\frac{d[A]}{dt} = -k[A]^2$	$\frac{1}{[A]} = \frac{1}{[A]_0} + 2kt$
$A + B \rightarrow P$	second	$\frac{d[A]}{dt} = -k[A][B]$	$kt = \frac{1}{[B]_0 - [A]_0} \ln \frac{[B]_0[A]}{[A]_0[B]}$

In the above $[A]_0$ and $[B]_0$ represent the initial concentrations of A and B i.e. their concentrations at the start of the reaction.

Half Lives

The half-life, $t_{1/2}$, of a substance is defined as the time it takes for the concentration of the substance to fall to half of its initial value. Note that it only makes sense to define a half-life for a substance not present in excess at the start of the reaction. We can obtain equations for the half-lives for reactions of various orders by substituting the values $t = t_{1/2}$ and $[A] = ½ [A]_0$. We obtain,

- Zeroth order reaction: $t_{1/2} = \frac{[A]_0}{2k}$
- First order reaction: $t_{1/2} = \frac{\ln 2}{k}$
- Second order reaction: $t_{1/2} = \frac{1}{k[A]_0}$

Determining the Rate Law from Experimental Data

A kinetics experiment consists of measuring the concentrations of one or more reactants or products at a number of different times during the reaction.

Isolation Method

The isolation method is a technique for simplifying the rate law in order to determine its dependence on the concentration of a single reactant. Once the rate law has been simplified, the differential or integral methods discussed here may be used to determine the reaction orders.

The dependence of the reaction rate on the chosen reactant concentration is isolated by having all other reactants present in a large excess, so that their concentration remains essentially constant throughout the course of the reaction. As an example, consider a reaction A + B → P, in which B is present at a concentration 1000 times greater than A. When all of species A has been used up, the concentration of B will only have changed by 1/1000, or 0.1%, and so 99.9% of the original B will still be present. It is therefore a good approximation to treat its concentration as constant throughout the reaction.

This greatly simplifies the rate law since the (constant) concentrations of all reactants present in large excess may be combined with the rate constant to yield a single effective rate constant. For example, the rate law for the reaction considered above will become:

$$v = k\,[A]^a\,[B]^b \approx k[A]^a[B]_0^b = k_{eff}[A]^a \text{ with } k_{eff} = k[B]_0^b$$

When the rate law contains contributions from a number of reactants, a series of experiments may be carried out in which each reactant is isolated in turn.

Differential Methods

When we have a rate law that depends only on the concentration of one species, either because there is only a single species reacting, or because we have used the isolation method to manipulate the rate law, then the rate law may be written:

$$v = k[A]^a$$

$$\log v = \log k + a \log[A]$$

A plot of logv against log[A] will then be a straight line with a slope equal to the reaction order, a, and an intercept equal to log k. There are two ways in which to obtain data to plot in this way.

- We can measure the concentration of the reactant [A] as a function of time and use this data to calculate the rate, $v = -d[A]/dt$, as a function of [A]. A plot of logv vs log[A] then yields the reaction order with respect to A.
- We can make a series of measurements of the initial rate v_0 of the reaction with different initial concentrations $[A]_0$. These may then be plotted as above to determine the order, a. This is a commonly used technique known as the initial rates method.

Integral Methods

If we have measured concentrations as a function of time, we may compare their time dependence with the appropriate integrated rate laws. Again, this is most straightforward if we have simplified the rate law so that it depends on only one reactant concentration. The differential rate law given in Equation $v = k[A]^a$ will give rise to different integrated rate laws depending on the value of a, The most commonly encountered ones are:

- Zeroth order integrated rate law: [A] = [A] – kt.

  A plot of [A] vs t will be linear, with a slope of –k.

- First order integrated rate law: $\ln[A] = \ln[A]_0 - kt$.

  A plot of ln [A] vs *t* will be linear with a slope of –k.

- Second order integrated rate law: $\frac{1}{[A]} = \frac{1}{[A]_0} + 2kt$.

  A plot of $\frac{1}{[A]}$ vs t will be linear with a slope of 2k.

If none of these plots result in a straight line, then more complicated integrated rate laws must be tried.

Half Lives

Another way of determining the reaction order is to investigate the behaviour of the half-life as the reaction proceeds. Specifically, we can measure a series of successive half-lives. t = 0 is used as the start time from which to measure the first half life, $t_{1/2}^{(1)}$. Then $t_{1/2}^{(1)}$ is used as the start time from which to measure the second half life, $t_{1/2}^{(2)}$, and so on.

- Zeroth order $t_{1/2} = \frac{[A]_0}{2k}$.

  Since at $t_{1/2}^{(1)}$, the new starting concentration is ½[A]$_0$, successive half-lives will decrease by a factor of two for a zeroth order reaction.

- First order $t_{1/2} = \frac{\ln 2}{k}$.

  There is no dependence of the half-life on concentration, so $t_{1/2}$ is constant for a first order reaction.

- Second order $t_{1/2} = \frac{1}{k[A]_0}$.

  The inverse dependence on concentration means that successive half-lives will double for a second order reaction.

Experimental Techniques

Experimental techniques have been developed to monitor reactions over timescales varying from hours or days all the way down to a few femtoseconds (1 fs = 10^{-15} s). While it is relatively simple to monitor the kinetics of a slow reaction (occurring over minutes to hours or longer), highly specialised techniques are required in order to study fast reactions, some of which will be considered here.

Whatever the details of the experimental arrangement, any kinetics experiment essentially consists of mixing the reactants and initiating reaction on a timescale that is negligible relative to that of the reaction, and then monitoring the concentration(s) of one or more reactants and/or products as a function of time. Because rate constants vary with temperature, it is also important to determine and control accurately the temperature at which the reaction occurs.

Most of the techniques we will look at are batch techniques, in which reaction is initiated at a single chosen time and concentrations are then followed as a function of time after initiation. We will also consider one or two examples of continuous techniques, in which reaction is continuously initiated and the time dependence of the reaction mixture composition is inferred from, for example, the concentrations in different regions of the reaction vessel.

Techniques for Mixing the Reactants and Initiating Reaction

For slow reactions, occurring over minutes to hours, reaction is usually initiated simply by mixing the reactants together by hand or with a magnetic stirrer or other mechanical device. For fast reactions, a wide range of techniques have been developed.

Flow Techniques

Flow techniques are typically used to study reactions occurring on timescales of seconds to milliseconds. In the simplest flow method, shown schematically on the left below, reactants are mixed at one end of a flow tube, and the composition of the reaction mixture is monitored at one or more positions further along the tube. If the flow velocity along the tube is known, then measurements at different positions provide information on concentrations at different times after initiation of reaction. In a variation on this method, shown on the right below, the detector may be in a fixed position, but a moveable injector may be used to inject one of the reactants into the flow tube at different positions relative to the detector in order to study the time dependence of the reaction mixture composition. Reactions of atomic or radical species may be studied using the discharge flow method, in which the reactive species is generated by a microwave discharge immediately prior to injection into the flow tube.

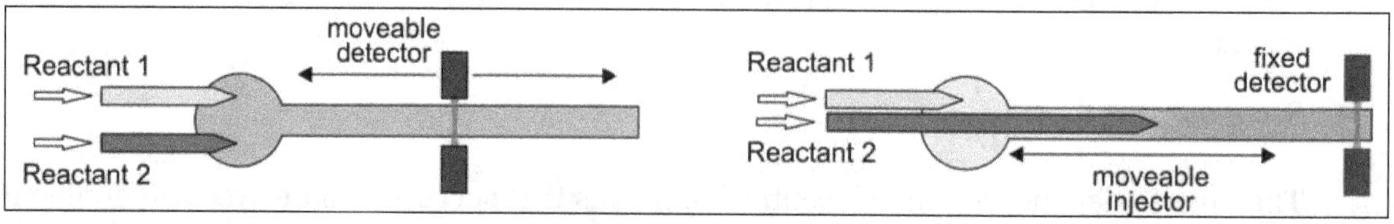

Continuous flow methods have the disadvantages that relatively large quantities of reactants are needed, and very high flow velocities are required in order to study fast reactions. These problems may be avoided by using a stopped flow technique. In this method, a fixed volume of reactants are rapidly flowed into a reaction chamber and mixed by the action of a syringe fitted with an end stop. The composition of the reaction mixture is then monitored spectroscopically as a function of time after mixing at a fixed position in the reaction chamber. Experimental systems may be designed to allow measurements to be made on very small sample volumes, making the stopped flow method popular for the study of biochemical kinetics e.g. enzyme action.

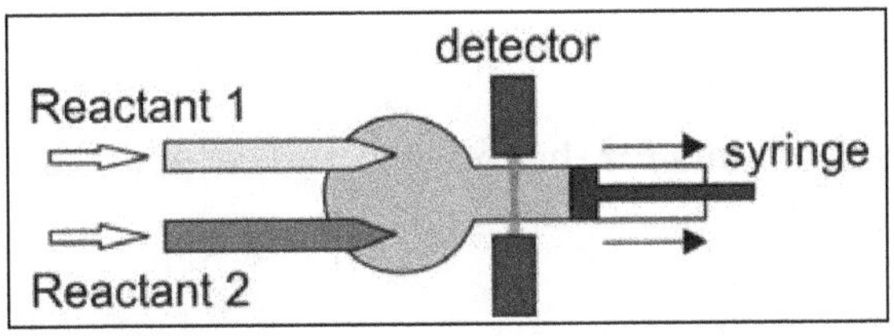

All flow techniques share the common problem that contributions from heterogeneous reactions at the walls of the flow tube can complicate the experiments. These can be minimised by coating the inner surface of the flow tube with an unreactive substance such as Teflon or halocarbon wax, and the relative contributions from the process under study and reactions involving the walls may be quantified by varying the diameter of the flow tube (and therefore the ratio of volume to surface area).

Flash Photolysis and Laser Pump Probe Techniques

In flash photolysis, reaction is initiated by a pulse of light (the 'flash') that dissociates a suitable precursor molecule in the reaction mixture to produce a reactive species, thereby initiating reaction. The concentration of the reactive species is then monitored as a function of time, usually spectroscopically using absorption spectroscopy or fluorescence techniques. The shortest timescale over which reactions may be studied using this technique is determined by the duration of the 'flash'. Originally, the flash was provided by a discharge lamp, with durations in the region of tens of microseconds to several milliseconds. However, in most modern experiments the flash is provided by a laser pulse, typically with a duration of a few nanoseconds (1 ns = 10^{-9} s). For studying extremely fast reactions, such as some of the electron transfer processes involved in photosynthesis, laser pulses as short as a few tens of femtoseconds (1 fs = 10^{-15} s) may be used.

Flash photolysis has the advantage that because reactants are produced from well-mixed precursors, there is no mixing time to reduce the time resolution of the technique. Also, because the reactants are generated and monitored in the centre of the reaction cell, there are no wall reactions to worry about as there are in flow methods.

Pulse radiolysis is a variation on flash photolysis in which a short pulse of high energy electrons (10^{-9} to 10^{-6} s in duration) is passed through the sample in order to initiate reaction.

For very fast processes, the 'pump-probe' technique is often used, in which pulsed lasers are employed both to initiate reaction (the 'pump') and to detect the products via a pulsed spectroscopic technique (the 'probe'). The time separation between the two pulses can be varied either electronically or with an optical delay line down to a resolution of around 10 femtoseconds (10^{-14} s).

Relaxation Methods

If we allow a chemical system to come to equilibrium and then perturb the equilibrium in some way, the rate of relaxation to a new equilibrium position provides information about the forward and reverse rate constants for the reaction. Since a system at chemical equilibrium is already well-mixed, relaxation methods overcome the mixing problems associated with many flow methods.

As an example, we will investigate the effect of a sudden increase in temperature on a system at equilibrium, an experiment known as a 'temperature jump'. Consider a simple equilibrium:

$$A \underset{k_{1r}}{\overset{k_{1f}}{\rightleftharpoons}} B$$

where k_{1f} and k_{1r} are the rate constants for the forward and reverse reactions at the initial temperature T_1. The rate of change of A is:

$$\frac{d[A]}{dt} = -k_{1f}[A] + k_{1r}[B]$$

At equilibrium, the concentration of A is constant, and so,

$$k_{1f}[A]_{eq,1} = k_{1r}[B]_{eq,1}$$

We now increase the temperature suddenly by a few degrees. This is often done by discharging a high voltage capacitor through the solution (~10–7 s), or by employing a UV or IR laser pulse or microwave discharge. After the temperature jump, the concentrations of A and B are initially at the values [A]eq,1 and [B]eq,1, but the system is not at the equilibrium composition for the higher temperature. The system relaxes back to the new equilibrium concentrations [A]eq,2 and [B]eq,2 at a rate determined by the new higher-temperature rate constants k2f and k2r. The new concentrations are given by:

$$k_{2f}[A]_{eq,2} = k_{2r}[B]_{eq,2}$$

If we define $x = [A] - [A]_{eq,2}$ as the deviation of the concentration from its new (higher temperature) equilibrium value (note that the deviation of [B] from its equilibrium value must therefore be –x), then during the relaxation the concentrations change as follows:

$$\begin{aligned} \frac{d[A]}{dt} &= -k_{2f}[A] + k_{2r}[B] \\ &= -k_{2f}\left([A]_{eq,2} + x\right) + k_{2r}\left([B]_{eq,2} - x\right) \\ &= -(k_{2f} + k_{2r})x \quad \left(\text{since } k_{2f}[A]_{eq,2} = k_{2r}[B]_{eq,2}\right) \end{aligned}$$

Since the rate of change of [A] is the same as the rate of change of x, we can integrate the rate law to give,

$$x = x_o \exp(-t/\tau) \quad \text{With, } \frac{1}{\tau} = k_{2f} + k_{2r}$$

We see that the rate at which the concentrations relax to their new equilibrium values is determined by the sum of the two new rate constants. The new equilibrium constant is given by the ratio of the two rate constants, $K = k_{2f}/k_{2r}$, so together a measurement of the rate of relaxation and the equilibrium constant allows the individual reaction rate constants for the forward and reverse reaction to be determined.

The details of the kinetic equations change for more complicated reactions, but the basic principle of the technique remains the same.

Shock Tubes

The shock tube method provides a way of producing highly reactive atomic or radical species through rapid dissociation of a molecular precursor, without the use of a discharge or laser pulse. The method is based on the fact that a very rapid increase in pressure (the shock) causes rapid heating of a gas mixture to a temperature of several thousand Kelvin. Since most dissociation reactions are endothermic, at high temperatures their equilibria are shifted towards products. A rapid increase in temperature therefore leads to rapid production of reactive species (the dissociation products) initiating the reaction of interest. A shock tube (shown schematically below) essentially consists of two chambers separated by a diaphragm. One chamber contains the appropriate mixture of reactants and precursor, the second an inert gas at high pressure. To initiate reaction, the diaphragm is punctured and a shock wave propagates through the reaction mixture. The temperature rise can be controlled by varying the pressure and composition of the inert gas. The composition of the reaction mixture after initiation is monitored in real time, usually spectroscopically.

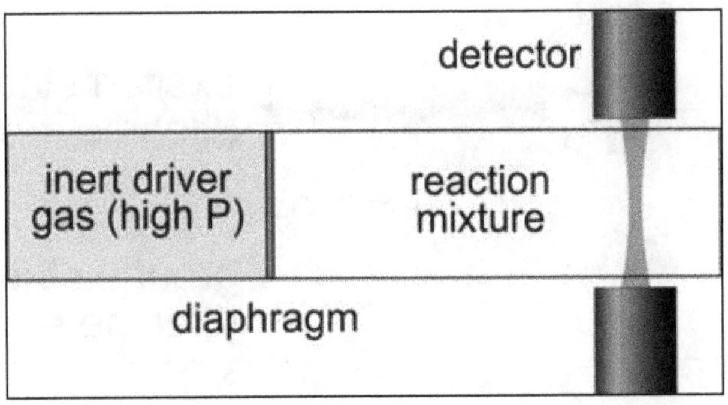

The shock tube approach is often used to study combustion reactions. Suitable precursors for such studies, together with the radical species obtained on dissociation using argon as the shock gas include:

$$HCN \rightarrow H + CN \qquad CH_4 \rightarrow CH_3 + H$$
$$SO_2 \rightarrow SO + O \qquad N_2O \rightarrow N_2 + O$$
$$CH_3 \rightarrow CH_2 + H \qquad C_2H_2 \rightarrow C_2H\_H$$
$$H_2S \rightarrow HS + H \qquad CF_3Cl \rightarrow CF_3 + Cl$$
$$NO \rightarrow N + O \qquad C_2H_4 \rightarrow C_2H_3 + H$$
$$NH_3 \rightarrow NH_2 + H \qquad C_2H_4 \rightarrow C_2H_2 + H_2$$

The method does have some major drawbacks, not least of which is the fact that the rapid heating is not selective for a particular molecules, and is likely to lead to at least partial dissociation of all of the species in the 'reactants' chamber. This leads to a complicated mixture of reactive species and often a large number of reactions occurring in addition to the reaction under study. Modelling the kinetics of such a system is often challenging, to say the least. Also, because each experiment is essentially a 'one off', no signal averaging is possible, and signal to noise levels are often low.

Compare this with laser pump-probe methods, in which hundreds or even thousands of traces may be averaged to obtain good signal to noise.

Lifetime Methods

In quantum mechanics, you learnt about the Heisenberg uncertainty principle, which relating the uncertainty in position and momentum, $\Delta x \Delta p \geq h/4\pi$. A similar uncertainty principle relates energy and time.

$$\Delta E \Delta t \geq h/4\pi \text{ or, since } E = h\nu,\ \Delta\nu\Delta t \geq 1/4\pi$$

The result of this relationship is that an atomic or molecular state has an uncertainty ΔE in its energy that is related to its lifetime Δt. The lifetime of most grounds states is effectively infinite, so that the uncertainty in their energy is negligible. However, excited states are short-lived, and their energy is therefore 'fuzzy'. Since photons corresponding to any energy within this uncertainty ΔE may be absorbed, this leads to spectral lines having a finite width known as the natural line width.

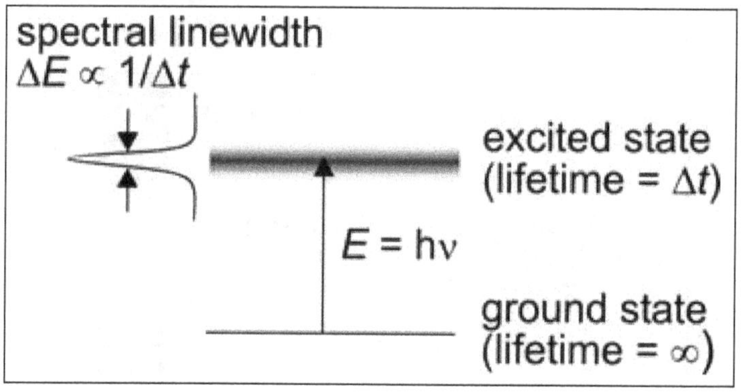

Kinetic processes involving excited states reduce their lifetime and cause further broadening. Many such processes have first order kinetics, for and in these cases the rate constant is simply equal to the reciprocal of the lifetime, $k = 1/\Delta t$. As a consequence, first order rate constants may be determined from measurements of spectral line widths, provided that other sources of line broadening are absent. Lifetime techniques covers a broad range of timescales, from around 10^{-15}s in photoelectron spectroscopy to around 1 s in NMR.

Techniques for Monitoring Concentrations as a Function of Time

For slow reactions, the composition of the reaction mixture may be analysed while the reaction is in progress either by withdrawing a small sample or by monitoring the bulk. This is known as a real time analysis. Another option is to use the quenching method, in which reaction is stopped a certain time after initiation so that the composition may be analysed at leisure. Quenching may be achieved in a number of ways. For example:

- Sudden cooling.
- Adding a large amount of solvent.
- Rapid neutralisation of an acid reagent.

- Removal of a catalyst.
- Addition of a quencher.

The key requirement is that the reaction must be slow enough (or the quenching method fast enough) for little reaction to occur during the quenching process itself.

Often, the real time and quenching techniques are combined by withdrawing and quenching small samples of the reaction mixture at a series of times during the reaction.

The composition of the reaction mixture may be followed in any one of a variety of different ways by tracking any chemical or physical change that occurs as the reaction proceeds. e.g.,

- For reactions in which at least one reactant or product is a gas, the reaction's progress may be followed by monitoring the pressure, or possibly the volume.
- For reactions involving ions, conductivity or pH measurements may often be employed.
- If the reaction is slow enough, the reaction mixture may be titrated.
- If one of the components is coloured then colourimetry may be appropriate.
- Absorption or emission spectroscopy is common (more on these later).
- For reactions involving chiral compounds, polarimetry (measurement of optical activity) may be useful.
- Other techniques include mass spectrometry, gas chromatography, NMR/ESR, and many more.

Fast reactions require a fast measurement technique, and as a consequence are usually monitored spectroscopically. A few commonly used techniques are outlined below.

Absorption Spectroscopy – Beer Lambert Law

Also known as spectrophotometry, absorption spectroscopy is widely used to track reactions in which the reactants and products have different absorption spectra. A monochromatic light source, often a laser beam, is passed through the reaction mixture, and the ratio of transmitted to incident light intensity, I/I_0, is measured as a function of time. The quantity $T = I/I_0$ is known as the transmittance, and may be related to the changing concentration of the absorbing species using the Beer Lambert law.

$$T = \frac{I}{I_0} = 10^{-\varepsilon cl} \text{ or } T = \frac{I}{I_0} = e^{-\alpha cl}$$

You may come across the Beer Lambert law in either of the forms above, or in log form,

$$\log(I/I_0) = -\varepsilon cl \text{ or } \ln(I/I_0) = -\alpha cl$$

In the above equations, c is the concentration of the absorbing species and l is the path length through the sample. ε and α are known as the molar absorption coefficient and molar exctinction coefficient, and are a measure of the strength of the spectral absorption. The quantity εcl is called the absorbance, A. Note that A = - logT. You may also see this quantity referred to as the optical density.

Resonance Fluorescence

Resonance fluorescence is a widely used technique for detecting atomic species such as H, N, O, Br, Cl or F. The light source is a discharge lamp filled with a mixture of helium and a molecular precursor for the atom of interest. A microwave discharge inside the lamp dissociates the precursor and produces a mixture of ground state and excited state atoms. The lamp then emits radiation at characteristic frequencies as the excited state atoms emit photons to relax down to the ground state. This radiation may be used to excite atoms of the same species present in a reaction mixture, and monitoring the intensity of radiation emitted from these atoms as they relax back to the ground state provides a measure of their concentration in the reaction mixture. To ensure that the detected light originates from atoms in the reaction mixture and not the lamp, the detector – usually a photomultiplier tube – is placed at right angles to the direction in which radiation exits the lamp.

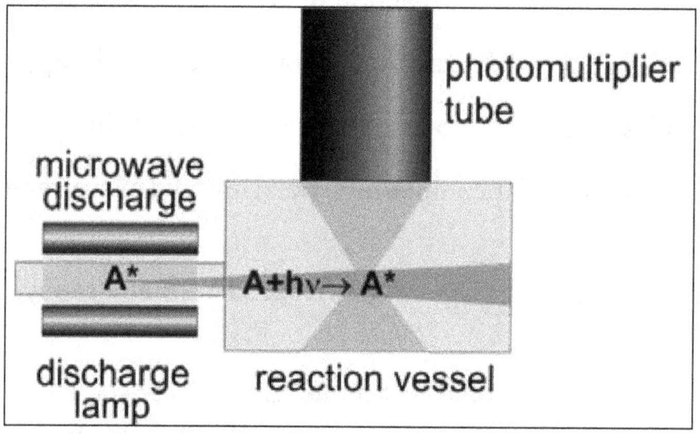

Laser-Induced Fluorescence

In laser-induced fluorescence a laser is used to excite a chosen species in a reaction mixture to an electronically excited state. The excited states then emit photons to return to the ground state, and the intensity of this fluorescent emission is measured. Because the number of excited states produced by the laser pulse is proportional to the number of ground state molecules present in the reaction mixture, the fluorescence intensity provides a measure of the concentration of the chosen species.

Temperature Control and Measurement

For any reaction with non-zero activation energy, the rate constant is dependent on temperature. The temperature dependence is often modelled by the Arrhenius equation,

$$k = A \exp(-E_a/RT)$$

where E_a is the activation energy for the reaction, and A is a constant known as the pre-exponential factor.

This temperature dependence means that in order to measure an accurate value for k, the temperature of the reaction mixture must be maintained at a constant, known value. If activation energies are to be measured as part of the kinetic study, rate constants must be measured at a series of temperatures. The temperature is most commonly monitored using a thermocouple, due to its wide range of operation and potential for automation; however, standard thermometers are also commonly used.

There are numerous ways in which the temperature of a reaction mixture may be controlled. For example, reactions in the liquid phase may be carried out in a temperature-controlled thermostat, while reactions in the gas phase are usually carried out inside a stainless steel vacuum chamber, in which thermal equilibrium at the temperature of the chamber is maintained through collisions of the gas molecules with the chamber walls. High temperatures up to 1300 K may be obtained using conventional heaters. Low temperatures may be achieved by flowing cooled liquid through the walls of the reaction vessel, and very low temperatures may be reached by using cryogenic liquids such as liquid nitrogen (~77 K) or liquid helium (~4 K). Extremely low temperatures (down to a few Kelvin), such as those relevant to reactions in interstellar gas clouds, may be obtained by preparing the reactant gases in a supersonic expansion.

Complex Reactions

In kinetics, a 'complex reaction' simply means a reaction whose mechanism comprises more than one elementary step. Disagreement of a predicted rate law with the experimental data is enough to rule out the corresponding proposed mechanism, while agreement inspires some confidence that the proposed mechanism is the correct one. It should be noted though that agreement between the predicted and measured kinetics is not always enough to assign a mechanism. The proposed mechanism must be able to account for all other properties of the reaction, which may include quantities such as the product distribution, product stereochemistry, kinetic isotope effects, temperature dependence, and so on.

The types of complex mechanisms that we will cover are: consecutive (or sequential) reactions; competing reactions; pre-equilibria; unimolecular reactions; third order reactions; enzyme reactions; chain reactions; and explosions.

Consecutive Reactions

The simplest complex reaction consists of two consecutive, irreversible elementary steps e.g.,

$$A \xrightarrow{k_1} B \xrightarrow{k_2} C$$

An example of such a process is radioactive decay. This is one of the few kinetic schemes in which it is fairly straightforward to solve the rate equations analytically, so we will look at this example in some detail. We can see immediately that the following initial conditions hold.

at $t = 0$, $[A] = [A]_0$

$[B] = 0$

$[C] = 0$

with at all times $[A] + [B] + [C] = [A]_0$.

Using this information, we can set up the rate equations for the process and solve them to determine the concentrations of [A], [B], and [C] as a function of time. The rate equations for the concentrations of A, B, and C are:

$$\frac{d[A]}{dt} = -k_1[A]$$

$$\frac{d[B]}{dt} = k_1[A] - k_2[B]$$

$$\frac{d[C]}{dt} = -k_2[B]$$

Integrating $\frac{d[A]}{dt} = -k_1[A]$ gives:

$$[A] = [A]_0 \exp(-k_1 t)$$

Substituting this into $\frac{d[B]}{dt} = k_1[A] - k_2[B]$ gives $\frac{d[B]}{dt} + k_2[B] = k_1[A]_0 \exp(-k_1 t)$, a differential equation with the solution.

$$[B] = \frac{k_1}{k_2 - k_1}\{\exp(-k_1 t) - \exp(-k_2 t)\}[A]_0$$

We will consider two special cases for a pair of sequential reactions:

Case 1: $k_1 >> k_2$

In this case, all of the A initially present is rapidly converted into B, which is then slowly used up to form C. Since k_2 becomes negligible in comparison with k_1, the equation for [C] becomes:

$$[C] = \{1 - \exp(-k_2 t)\} [A]_0$$

i.e. the rate of production of C (and therefore the overall rate of the two-step reaction) becomes independent of k_1 (apart from at the very beginning of the reaction). In other words, the second step is the rate determining step.

Case 2: $k_2 >> k_1$

In this case, B is consumed as soon as it is produced, and since k_1 becomes negligible in comparison with k_2, the equation for [C] simplifies to,

$$[C] = \{1 - \exp(-k_1 t)\} [A]_0$$

i.e. the overall rate now depends only on k_1, and the first step is rate determining.

The way in which the concentrations of A, B and C vary with time for each of the two cases considered above is shown in the figures below.

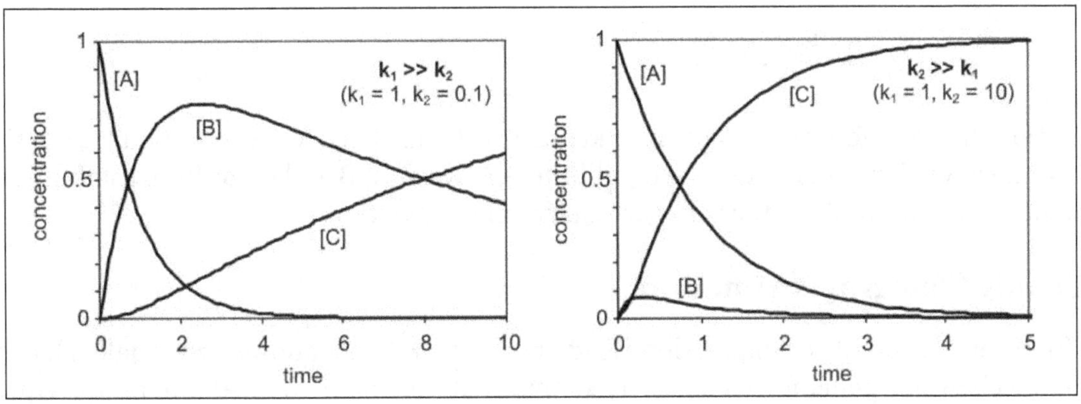

Pre-Equilibria

A situation that is only slightly more complicated than the sequential reaction scheme described above is,

$$A + B \underset{-k_1}{\overset{k_1}{\rightleftharpoons}} C \overset{k_1}{\rightarrow} D$$

The rate equations for this reaction are:

$$\frac{d[A]}{dt} = \frac{d[B]}{dt} = -k_1[A][B] + k_{-1}[C]$$

$$\frac{d[C]}{dt} = k_1[A][B] - k_2[C]$$

$$\frac{d[D]}{dt} = k_2[C]$$

These cannot be solved analytically, and in general would have to be integrated numerically to obtain an accurate solution. However, the situation simplifies considerably if $k_{-1} >> k_2$. In this case, an equilibrium is reached between the reactants A and B and the intermediate C, and the equilibrium is only perturbed very slightly by C 'leaking away' very slowly to form the product D.

If we assume that we can neglect this perturbation of the equilibrium, then once equilibrium is reached, the rates of the forward and reverse reactions must be equal. i.e.,

$$k_1[A][B] = k_{-1}[C]$$

Rearranging this equation, we find,

$$\frac{k_1}{k_{-1}} = \frac{[C]}{[A][B]} = K$$

The equilibrium constant K is therefore given by the ratio of the rate constants k_1 and k_{-1} for the forward and reverse reactions. The rate of the overall reaction is simply the rate of formation of the product D, so,

$$v = \frac{d[D]}{dt} = k_2[C] = k_2K[A][B]$$

The reaction therefore follows second order kinetics, with an effective rate constant $k_{eff} = k_2K$. Note that this rate law will not hold until the equilibrium between A, B and C has been established, and so is unlikely to be accurate in the very early stages of the reaction.

The Steady State Approximation

Apart from the two simple examples described above, the rate equations for virtually all complex reaction mechanisms generally comprise a complicated system of coupled differential equations that cannot be solved analytically. In state-of-the-art kinetic modelling studies, fairly sophisticated software is generally used to obtain numerical solutions to the rate equations in order to determine the time-varying concentrations of all species involved in a reaction sequence. However, very good approximate solutions may often be obtained by making simple assumptions about the nature of reactive intermediates.

Almost by definition, a reactive intermediate R will be used up virtually as soon as it is formed, and therefore its concentration will remain very low and essentially constant throughout the course of the reaction. This is true at all times apart from at the very start of the reaction, when [R] must necessarily build up from zero to some small non-zero value, and at the very end of the reaction in the case of a reaction that goes to completion, when [R] must return to zero. During the period of time when [R] is essentially constant, because d[R]/dt is so much less than the rates of change of the reactant and product concentrations, it is a good approximation to set d[R]/dt = 0. This is known as the steady state approximation.

Steady state approximation: if a reactive intermediate R is present at low and constant concentration throughout (most of) the course of the reaction, then we can set d[R]/dt = 0 in the rate equations.

As we shall see, applying the steady state approximation has the effect of converting a mathematically intractable set of coupled differential equations into a system of simultaneous algebraic equations, one for each species involved in the reaction. The algebraic equations may be solved to find the concentrations of the reactive intermediates, and these may then be substituted back into the more general equations to obtain an expression for the overall rate law.

Take the case where $k_2 >> k_{-1}$, so that C is now a reactive intermediate and there is no stable equilibrium between A, B and C. The reaction equation is:

$$A = B \underset{k_{-1}}{\overset{k_1}{\rightleftharpoons}} C \overset{k_2}{\rightarrow} D$$

We can apply the steady state approximation (SSA) to C, to obtain,

$$\frac{d[C]}{dt} = 0 = k_1[A][B] - k_{-1}[C] - k_2[C]$$

This may be solved to give [C] in terms of the reactant concentrations [A] and [B].

$$[C] = \frac{k_1}{k_{-1} + k_2}[A][B]$$

The overall rate is the rate of formation of the product, D, giving,

$$v = \frac{d[D]}{dt} = k_2[C] = \frac{k_1 k_2}{k_{-1} + k_2}[A][B]$$

In the limiting case where k–1 is much smaller than k2, we can neglect k–1 in the denominator and the rate becomes simply k1[A][B] i.e. the rate of the overall reaction is the same as the rate of the first elementary step. This is not all that surprising. If k_2 is much larger than k1 and k–1 then as soon as the $A + B \rightarrow C$ step has occurred, C is immediately converted into products, and there is virtually no chance for the reverse $C \rightarrow A + B$ reaction to occur. The initial elementary step is rate determining, and therefore dominates the kinetics.

In general, the steps required in order to use the SSA to obtain an overall rate law for a complex reaction are:

- Write down a steady state equation for each reactive intermediate.
- Solve the set of equations to obtain expressions for the concentrations of each intermediate in terms of the reactant and product concentrations. A couple of hints:
    - If the equation contains only one reactive intermediate, it may simply be rearranged to give the concentration of that intermediate in terms of reactant and product concentrations. This can often be substituted into other equations to obtain the corresponding expressions for other reactive intermediates.
    - If the equations depend on more than one reactive intermediate, and share terms, look for sums or differences of the equations that will simplify matters. Often a SSA problem that initially appears extremely complicated becomes trivial when you simply add together two of the steady state equations.
- Write down an expression for the overall rate (usually the rate of change of one of the products). This will generally involve the concentrations of one or more reactive intermediates.
- Substitute your expressions from step 2 into your overall rate equation to obtain an overall rate equation that depends only on reactant and product concentrations. Concentrations of reactive intermediates must not appear in the final rate law.

'Unimolecular' Reactions – the Lindemann-Hinshelwood Mechanism

A number of gas phase reactions follow first order kinetics and apparently only involve one chemical species. Examples include the structural isomerisation of cyclopropane to propene, and the decomposition of azomethane ($CH_2N_2CH_3 \rightarrow C_2H_6 + N_2$, with experimentally determined rate law $v = k[CH_3N_2CH_3]$) The mechanism by which these molecules acquire enough energy to react remained a puzzle for some time, particularly since the rate law seemed to rule out a bimolecular step. The puzzle was solved by Lindemann in 1922, when he proposed the following mechanism for 'thermal' unimolecular reactions.

$$A + M \underset{k_{-1}}{\overset{k_1}{\rightleftharpoons}} A^* + M$$

$$A^* \overset{k_2}{\rightarrow} P$$

The reactant, A, acquires enough energy to react by colliding with another molecule, M (note that in many cases M will actually be another A molecule). The excited reactant A* then undergoes unimolecular reaction to form the products, P. To determine the overall rate law arising from this mechanism, we can apply the SSA to the excited state A*.

$$\frac{d[A^*]}{dt} = 0 = k_1[A][M] - k_{-1}[A^*][M] - k_2[A^*]$$

Rearranging this expression yields the concentration [A*].

$$[A^*] = \frac{k_1[A][M]}{k_{-1}[M] + k_2}$$

The overall rate of reaction is then,

$$v = \frac{d[P]}{dt} = k_2[A^*] = \frac{k_1 k_2 [A][M]}{k_{-1}[M] + k_2}$$

High Pressure

At high pressure there are many collisions, and collisional de-excitation of A* is therefore much more likely than unimolecular reaction of A* to form products. i.e. $k^{-1}[A^*][M] >> k_2[A^*]$. In this limit, we can neglect the k_2 term in the denominator of Equation $v = \frac{d[P]}{dt} = k_2[A^*] = \frac{k_1 k_2 [A][M]}{k_{-1}[M] + k_2}$, and the rate law becomes:

$$v = \frac{k_1 k_2}{k_{-1}}[A]$$

Which is first order, with a first order rate constant $k_{uni} = k_1 k_2 / k_{-1}$. This mechanism therefore explains the observed first order kinetics at reasonable pressures, when the unimolecular step is rate determining.

Low Pressure

At low pressures there are few collisions, and A* will generally undergo unimolecular reaction before it undergoes collisional de-excitation. i.e. k2 >> k–1[A*][M]. In this case, we can neglect the k–1[M] in the denominator of Equation $v = \frac{d[P]}{dt} = k_2[A^*] = \frac{k_1 k_2 [A][M]}{k_{-1}[M] + k_2}$, and the rate law is now:

$$v = k_1[A][M]$$

We see that at low pressures the kinetics are second order. This is because formation of the excited species A*, a bimolecular process, is now the rate determining step.

Often, the rate law for such reactions is written,

$$v = k[A] \text{ With, } k = \frac{k_1 k_2 [M]}{k_{-1}[M] + k_2}$$

k is the first order rate constant that would be observed experimentally in the high pressure case. If experimental measurements of the rate constant as a function of pressure (equivalent to [M]) are available, the Lindemann-Hinshelwood mechanism may be tested. Taking the reciprocal of our expression for k gives,

$$\frac{1}{k} = \frac{k_{-1}}{k_1 k_2} + \frac{1}{k_1[M]}$$

A plot of 1/k against 1/[M] should therefore be linear, with an intercept of $k_{-1}/(k_1 k_2)$ and a slope of $1/k_1$. An example of such a plot is shown below. Usually there is a reasonable fit between theory and experiment at low pressure, but a pronounced deviation at high pressure, with experimental values of k being larger than the values predicted by the Lindemann-Hinshelwood mechanism, as shown below.

It turns out that while the general idea of a collisional activation process is correct, the true mechanism of 'unimolecular' reactions is slightly more involved. The principal failing of the Lindemann Hinshel wood mechanism is that it assumes that any excited reactant A* will undergo unimolecular reaction to produce products. In practice, however, excitation is generally required in a degree of freedom that is coupled to the reaction coordinate in some way e.g. vibrational excitation in a bond that breaks during the reaction. More sophisticated theories of unimolecular reactions have been developed which take this and other factors into account, and provide much better agreement with experiment.

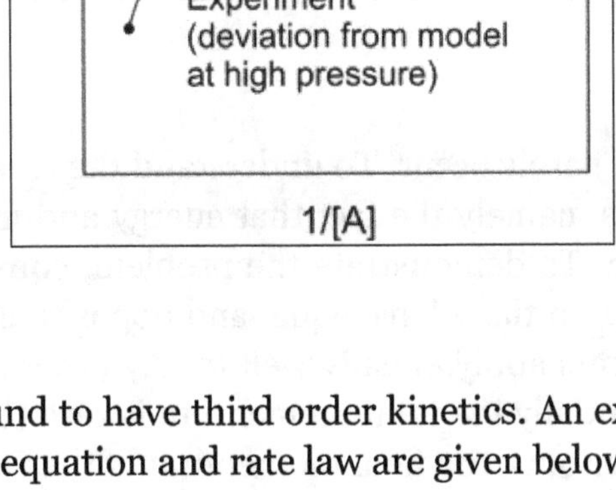

Third Order Reactions

A number of reactions are found to have third order kinetics. An example is the oxidation of NO, for which the overall reaction equation and rate law are given below.

$$2NO + O_2 \rightarrow 2NO_2$$

$$\frac{d[NO_2]}{dt} = k[NO]^2[O_2]$$

One possibility for the mechanism of this reaction would be a three-body collision (i.e. a true termolecular reaction). However, such collisions are exceedingly rare, and certainly too unlikely to explain the observed rate at which this reaction proceeds. An added complication is that the rate of this reaction is found to decrease with increasing temperature, an indication of a complex mechanism. An alternative mechanism that leads to the same rate law is a two-step process involving a pre-equilibrium.

$$NO + NO \underset{k_2}{\overset{k_1}{\rightleftharpoons}} (NO)_2$$

$$(NO)_2 + O_2 \xrightarrow{k_2} 2NO_2$$

The overall rate is,

$$v = \frac{1}{2}\frac{d[NO_2]}{dt} = k_2[(NO)_2][O_2]$$

However, from the pre-equilibrium, we have,

$$K = \frac{[(NO)_2]}{[NO]^2}$$

so, $[(NO)_2] = K[NO]^2$

And the overall rate is,

$$v = k_2K[NO]^2[O_2]$$

i.e. third order, as required.

Very common situations in which third order kinetics are observed are reactions in which two reactants combine to form a single product. Such reactions require a so-called 'third body' to take away some of the excess energy from the reaction product. An example is the formation of ozone,

$$O + O_2 \rightarrow O_3$$

As written, this reaction would barely occur. To understand the reason for this we need to turn to some basic classical mechanics, namely the fact that energy and momentum must both be conserved in the reactive collision. To demonstrate the problem, consider the somewhat contrived situation in which the O and O_2 initially have equal and opposite momenta, and collide head-on to react (the following arguments apply equally well to any other situation. Since total momentum must be conserved, and initially the total momentum is zero (because the momenta of the O and O_2 exactly cancel each other out), the final momentum of the O_3 product must be zero i.e. the product must be stationary. Now consider conservation of energy. Since we are forming a bond, the reaction is exothermic, so by conservation of energy, the total kinetic energy possessed by the O_3 product must be the sum of the reaction exothermicity and the kinetic energies of the reactants. We have already determined that conservation of momentum requires the O_3 to be stationary, so all of this kinetic energy must go into vibrational motion of the O_3. Highly vibrationally excited molecules are extremely unstable, and the O_3 will very quickly fall apart back into reactants. The only way for the O_3 to survive is for it to transfer some of its vibrational energy to another molecule M (known as a third body) in a collision. The energy may end up as internal excitation (rotation or vibration) of M, or simply as kinetic energy as the two molecules fly away from each other after the collision. The actual mechanism is therefore,

$$O + O_2 \rightleftharpoons O_3^*$$

$$O_3^* + M \rightarrow O_3 + M$$

The overall reaction is usually written $O + O_2 + M \rightarrow O_3 + M$.

A third body is only required for reactions in which a single product is formed from two or more reactants, since this is the only time that conservation of momentum forces a large amount of energy into the product. If two products are formed, they can both carry away almost arbitrary amounts of energy as kinetic energy while still conserving the total momentum.

Enzyme Reactions – the Michaelis-Menten Mechanism

An enzyme is a protein that catalyses a chemical reaction by lowering the activation energy. Enzymes generally work by having an active site that is carefully designed by nature to bind a particular reactant molecule (known as the substrate). An example of a substrate bound at the active site of an enzyme is shown on the left. The activation energy of the reaction for the enzyme-bound substrate is lower than for the free substrate molecule, often due to the fact that the interactions involved in binding shift the substrate geometry closer to that of the transition state for the reaction. Once reaction has occurred, the product molecules are released from the enzyme.

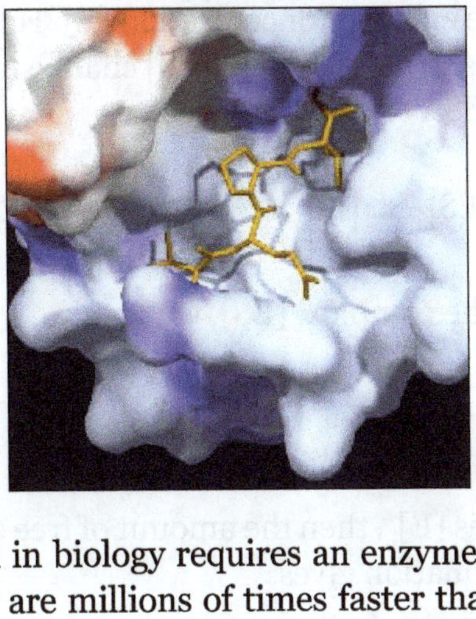

Virtually every chemical reaction in biology requires an enzyme in order to occur at a significant rate (enzyme-catalysed reactions are millions of times faster than the corresponding uncatalysed reactions), and each enzyme is specific to a particular reaction. Many drugs work by binding to a carefully targeted enzyme in place of the normal substrate molecule, thereby inhibiting enzyme activity and slowing the reaction rate. Enzyme kinetics is an extremely important and complex field, but the basic kinetics of a simple enzyme catalysis process may be modelled quite simply as follows.

In an enzyme-catalysed reaction, a substrate S is converted to products P in a reaction that is catalysed by an enzyme E. For many such reactions, the rate is found experimentally to follow the Michaelis-Menten equation:

$$v = \frac{v_{max}[S]}{K_M + [S]}$$

The constant K_M is called the Michaelis constant, and v_{max} is the maximum rate, which is found to be linearly proportional to the total enzyme concentration.

$$v_{max} = k_{cat}[E]_o$$

The constant of proportionality k_{cat} is known as the turnover number, and represents the maximum number of molecules of substrate that each enzyme molecule can convert into products (or 'turn over') per second. We shall see later that this occurs when the substrate is present in large excess.

Any kinetic model for enzyme catalysis must explain the fact that the rate depends on the enzyme concentration [E], even though there is no net change in its concentration over the course of the reaction. The simplest trial mechanism involves formation of a bound enzyme-substrate complex ES, followed by conversion of the complex into the products plus free enzyme (which may then go on to catalyse further reaction).

$$E + S \underset{-k_1}{\overset{k_1}{\rightleftharpoons}} ES \overset{k_2}{\rightarrow} P + E$$

This mechanism is a somewhat special case of applying the steady state approximation. Usually a good rule of thumb is that in order for the SSA to be valid, the concentration of the reactive intermediate (in this case ES) must be much less than the concentration of the reactants. In this case [ES] is not much less than the free enzyme concentration [E]. However, because [E] is regenerated in the second step of the mechanism, both [E] and [ES] change much more slowly than [S] and [P] so the SSA is valid. Applying the SSA to [ES], we have,

$$\frac{d[ES]}{dt} = 0 = k_1[E][S] - k_{-1}[ES] - k_2[ES]$$

Solving for the concentration of the complex, [ES], we obtain,

$$[ES] = \frac{k_1[E][S]}{k_{-1} + k_2}$$

If the total enzyme concentration is $[E]_o$, then the amount of free enzyme must be $[E] = [E]_o - [ES]$. Substituting this into the above equation gives,

$$[ES] = \frac{k_1([E]_o - [ES])[S]}{k_{-1} + k_2},$$

which rearranges to give,

$$[ES]=\frac{k_1[E]_0[S]}{k_{-1}+k_2+k_1[S]}$$

The overall rate of reaction is then found from the rate of formation of product, P.

$$v=\frac{d[P]}{dt}=k_2[ES]=\frac{k_2k_1[E]_0[S]}{k_{-1}+k_2+k_1[S]}=\frac{k_2[S][E]_0}{K_M+[S]}=k[E]_0$$

where,

$$k=\frac{k_2[S]}{K_M+[S]}$$

and

$$K_M=\frac{k_2+k_{-1}}{k_1}$$

is called the Michaelis constant.

We have chosen these particular combinations of rate constants in order to get our rate equation into the same form as the Michaelis-Menten equation. The two equations now agree if $k_2 = k_{cat}$.

We can see from the above treatment that the rate of enzyme-catalysed reaction (enzymolysis) depends linearly on the enzyme concentration, but in a more complicated way on the substrate concentration. The dependence on [S] simplifies under certain conditions.

- When $[S] >> K_M$, then $v = k_2[E]_0 = k_{cat}[E]_0$ and the overall rate is independent of the substrate concentration. This is because there is so much substrate present that its concentration remains essentially constant as the reaction proceeds. Under these conditions, the rate of reaction is a maximum ($v = v_{max}$) and the enzyme is saturated with substrate.
- When $[S] << K_M$, the reaction rate is $v = (k_2/K_M)\,[E]_0[S]$, and the rate is first order in both $[E]_0$ and [S].

To determine k_2 and K_M from experimental rate data, we invert the expression for k.

$$k=\frac{k_2[S]}{K_M+[S]}$$

becomes,

$$\frac{1}{k}=\frac{K_M}{k_2[S]}+\frac{1}{k_2}$$

A plot of 1/k against 1/[S] has a slope of K_M/k_2 and an intercept of $1/k_2$. Therefore, K_M = (slope)/(intercept) and k_2 = 1/(intercept). This is called a Lineweaver-Burke plot. Usually, the initial rates method is used to measure k, since this method is not susceptible to complications that may arise from secondary reactions of the products.

Chain Reactions

Chain reactions are complex reactions that involve chain carriers, reactive intermediates which react to produce further reactive intermediates. The elementary steps in a chain reaction may be classified into initiation, propagation, inhibition, and termination steps. As an example, consider the infamous reaction by which chlorofluorocarbons (CFCs) destroy ozone:

$$C_nF_mCl + h\nu \rightarrow \cdot C_nF_m + \cdot Cl \quad \text{Initiation}$$

$$\cdot Cl + O_3 \rightarrow \cdot ClO + O_2 \quad \text{Propagation}$$

$$\cdot ClO + O \rightarrow \cdot Cl + O_2 \quad \text{Propagation}$$

$$\cdot Cl + CH_4 \rightarrow \cdot CH_3 + HCl \quad \text{Termination}$$

Initiation Step

Reaction is initiated either thermally or photochemically. The first reactive intermediates/chain carriers (in this case a ·Cl radical) are produced.

Propagation Step

Reaction of a radical leads to formation of another radical – another reactive intermediate. In the first propagation step above, ·Cl reacts to form ·ClO; in the second step ·ClO reacts to form ·Cl.

Termination Step

Chain carriers are deactivated. Often this occurs through radical-radical recombination, reaction with walls, or reaction with another molecule to create an inactive product.

Some chain reactions involve inhibition steps, in which product molecules are destroyed. Inhibition steps are also sometimes referred to as retardation or de-propagation steps.

The chain length in a chain reaction is defined as the number of propagation steps per initiation step, or alternatively as the rate of propagation divided by the rate of initiation. Chain lengths can be very long – in the above example a single ·Cl radical can destroy around 106 molecules of ozone.

$$\text{chain length, n} = \frac{\nu(\text{propagation})}{\nu(\text{initiation})} = \frac{\nu(\text{propagation})}{\nu(\text{termination})}$$

The above reaction is an example of a cyclic chain reaction. The ·Cl essentially acts as a catalyst and is continuously regenerated until it is removed by a termination step. It is also possible to have non-cyclic chain reactions, involving many reactive species and elementary steps. Non-cyclic chain reactions can have extremely complicated kinetic mechanisms.

Chain reactions in which each propagation step produces only one reactive intermediate are called linear chain reactions.

Linear Chain Reactions

The hydrogen-bromine reaction has become the 'benchmark' system for illustrating the kinetics of linear chain reactions, and we will use this reaction as our main example. We will also compare the kinetics of the hydrogen-bromine reaction with the analagous reactions of chlorine and iodine.

The Hydrogen–Bromine Reaction

The kinetics of the reaction between H_2 and Br_2 were determined experimentally by Bodenstein around 100 years ago. The overall reaction has the equation:

$$H_2 + Br_2 \rightarrow 2HBr$$

Bodenstein determined the following rate law for the reaction:

$$\frac{d[HBr]}{dt} = k[H_2][Br_2]^{1/2}$$

The measured order of ½ with respect to Br_2 indicated that the reaction proceeded via a complex reaction mechanism rather than a simple bimolecular collision. Further investigation showed that this rate law in fact only holds for the early stages of the reaction, and that the true rate law takes the form:

$$\frac{d[HBr]}{dt} = \frac{k[H_2][Br_2]^{1/2}}{1 + k'[HBr]/[Br_2]}$$

Any proposed mechanism for the reaction must agree with both of these observations. The reaction can be initiated by either thermally-induced or photon-induced dissociation of Br_2.

$$Br_2 + M \rightarrow Br + Br + M \quad \text{or} \quad Br_2 + h\nu \rightarrow Br + Br$$

We will concentrate on the thermal mechanism for the purposes of deriving a rate law for the overall reaction, but the steps following the initiation step are the same for both cases. The currently accepted mechanism is:

$Br_2 + M \xrightarrow{k_1} Br\cdot + Br\cdot + M$ Initiation

$Br\cdot + H_2 \underset{k_{-2}}{\overset{k_2}{\rightleftharpoons}} H\cdot + HBr$ Propagation / Inhibition

$H\cdot + Br_2 \xrightarrow{k_3} Br\cdot + HBr$ Propagation

$Br\cdot + Br\cdot + M \xrightarrow{k_4} Br_2 + M$ Termination

The reaction chain contains two radical chain carriers, H· and Br·. In the second step, because the H-H bond is stronger than the H-Br bond, once an appreciable amount of HBr has built up the reverse (inhibition) reaction becomes possible. In order to arrive at an overall rate law for the reaction, we apply the steady state approximation to the two chain carriers.

$$\frac{d[H]}{dt}=0=k_2[Br][H_2]-k_{-2}[H][Br]-k_3[H][Br_2]$$

$$\frac{d[Br]}{dt}=0=2k_1[Br_2][M]-k_2[Br][H_2]+k_{-2}[H][HBr]+k_3[H][Br_2]-2k_4[Br]^2[M]$$

We can solve these two equations to obtain expressions for the concentrations of H and Br in terms of the reactant and product concentrations and the various rate constants. The two equations each depend on both carrier concentrations, and also share terms. We can simplify the equations by adding them together to give:

$$0=2k_1[Br_2][M]-2k_4[Br]^2[M]$$

which can be rearranged to give the steady state concentration of Br atoms.

$$[Br]=\left(\frac{k_1[Br_2]}{k_4}\right)^{1/2}$$

Note that $0=2k_1[Br_2][M]-2k_4[Br]^2[M]$ implies that the rate of initiation is the same as the rate of termination, as expected under steady state conditions (this is a good check that we have made no mistakes up to this point). This result also leads to a considerable simplification in Equation $\frac{d[Br]}{dt}=0=2k_1[Br_2][M]-k_2[Br][H_2]+k_{-2}[H][HBr]+k_3[H][Br_2]-2k_4[Br]^2[M]$, which now becomes:

$$0=-k_2[Br][H_2]+k_{-2}[H][HBr]+k_3[H][Br_2]$$

and may be rearranged to give an expression for the steady state H atom concentration.

$$[H]=\frac{k_2[Br][H_2]}{k_2[HBr]+k_3[Br_2]}=\frac{k_2[H_2]}{k_2[HBr]+k_3[Br_2]}\left(\frac{k_1[Br_2]}{k_4}\right)^{1/2}$$

We are now ready to determine the overall reaction rate.

$$\frac{d[HBr]}{dt}=k_2[Br][H_2]-k_{-2}[H][HBr]+k_3[H][Br_2]$$

Substituting in our expressions for [H] and [Br] gives:

$$\frac{d[HBr]}{dt}=\frac{2k_2(k_1/k_4)^{1/2}[Br_2]^{1/2}[H_2]}{1+(k_{-2}/k_3)[HBr]/[Br_2]}$$

We see that this agrees with the measured rate law, Equation $\frac{d[HBr]}{dt} = \frac{k[H_2][Br_2]^{1/2}}{1 + k'[HBr]/[Br_2]}$. In the early stages of the reaction, the concentration of the HBr product is much lower than that of the reactant Br_2, and the second term in the denominator becomes negligible. The rate law then reduces to:

$$\frac{d[HBr]}{dt} = 2k_2 (k_1/k_4)^{1/2} [Br_2]^{1/2} [H_2]$$

again reproducing the experimental observations. The proposed mechanism therefore fits well with the experimental measurements.

The Hydrogen–Chlorine Reaction

The mechanism for the hydrogen-chlorine reaction is essentially identical to that for the hydrogenbromine reaction:

$Cl_2 + M \xrightarrow{k_1} Cl\cdot + Cl\cdot$ Initiation

$Cl\cdot + H_2 \xrightarrow{k_2} H\cdot + HCl$ Propagation

$H\cdot + Br_2 \xrightarrow{k_3} Cl\cdot + HCl$ Propagation

$Cl\cdot + Cl\cdot + M \xrightarrow{k_4} Cl_2 + M$ Termination

The reaction may also be initiated photochemically. Unlike the H_2 + Br_2 reaction, both propagation steps are very efficient, and the inhibition step is very slow (and has therefore been omitted from the above mechanism), so the overall reaction rate is much faster. Chain lengths up to 10^6 are possible, and coupled with the exothermicity of the reaction; this can lead to a thermal explosion.

This reaction provides a good example of a case where the steady state approximation breaks down. Since both propagation steps are very efficient, when the radical concentrations [H] and [Cl] are much lower than the reactant concentrations [H_2] and [Cl_2], reactive collision of a Cl atom with H_2 (propagation) is much more likely than a terminating collision with another Cl atom. This means that the reaction may be very advanced before the steady state condition – that the rates of initiation and termination are equal – is reached. In practice, the situation is usually simplified somewhat due to the extreme sensitivity of the reaction to inhibition by contaminants such as O_2. Oxygen reacts with H and Cl radicals to form inert radicals, providing an alternative termination pathway and increasing the overall rate of termination. If O_2 is present at a concentration of around 1% or greater, the following termination steps dominate to the point where the steady-state approximation becomes valid.

$$H\cdot + O_2 + M \xrightarrow{k_5} \cdot HO_2 + M$$

$$Cl\cdot + O_2 + M \xrightarrow{k_6} \cdot ClO_2 + M$$

If we replace the termination step 4 of our above mechanism with the termination steps above, we can apply the SSA to obtain expressions for the steady state [H] and [Cl] concentrations and for the overall rate.

$$\frac{d[HCl]}{dt} = \frac{2k_a[H_2][Cl_2]}{[O_2]([H_2] \quad k_b[Cl_2])}$$

with,

$$k_a = \frac{k_1 k_2}{k_5}$$

and

$$k_b = \frac{k_3 k_6}{k_2 k_5}$$

The Hydrogen-Iodine Reaction

We might expect the hydrogen-iodine to have a similar mechanism to the bromine and chlorine analogues. However, the second step in the mechanism ($I + H_2 \rightarrow H + HI$) occurs much too slowly at normal temperatures for this mechanism to be viable. Various kinetic mechanisms operate at different temperatures, for example:

$I_2 + M \rightleftharpoons I + I + M$ Pre-equilibrium

$I + I + H_2 \rightarrow 2HI$ Termolecular reaction

Unlike the complicated rate laws followed by the chlorine and bromine reactions, the hydrogeniodine reaction follows a simple 'bimolecular' rate law.

Comparison of the Hydrogen-Halogen Reactions

The key difference between the reactions of Cl_2, Br_2 and I_2 with hydrogen lies in the exothermicity of the atomic halogen reactions with H_2 (step 2 in the chain reaction sequences).

Cl $\Delta H = 4.4$ kJ mol^{-1}

Br $\Delta H = 69.6$ kJ mol^{-1}

I $\Delta H = 137.7$ kJ mol^{-1}

The iodine reaction is so slow that it does not occur at accessible temperatures, ruling out the chain mechanism observed for the other two halogens. The chlorine reaction is fast, and the bromine reaction slow. While the reverse (inhibition) step $H + HX \rightarrow X + H2$ is thermodynamically favourable for both Cl and Br, it is unimportant for Cl due to the very low steady state concentration of H atoms. In contrast, the inhibition step has a considerable effect on the overall kinetics of the bromine reaction.

Explosions and Branched Chain Reactions

An explosion occurs when a reaction rate accelerates out of control. As the reaction speeds up, gaseous products are formed in larger and larger amounts, and more and more heat is generated. The rapid liberation of heat causes the gases to expand, generating extremely high pressures, and it is this sudden formation of a huge volume of expanded gas that constitutes the explosion. The pressure wave travels at very high speeds, often much faster than the speed of sound, and the 'bang' associated with an explosion is the result of a supersonic shock wave.

There are two different mechanisms that may lead to an explosion. These are related to the fact that the overall reaction rate depends on both the magnitude of the rate constant and the amounts of reactants present in the reaction mixture.

If the heat generated in a reaction due to the reaction exothermicity cannot be dissipated sufficiently rapidly, the temperature of the reaction mixture increases. This increases the rate constant, and therefore the reaction rate, producing more heat and accelerating the reaction rate still further, and so on until an explosion results. Such explosions are known as thermal explosions, and in principle may occur whenever the rate of heat production by a reaction mixture exceeds the rate of heat loss to the surroundings (often the walls of the reaction vessel).

The second category of explosions arise from chain branching within a chain reaction, and are known as chain branching explosions (or sometimes, somewhat misleadingly, isothermal explosions). In this case, one or more steps in the reaction mechanism produce two or more chain carriers from one chain carrier, increasing the number of chain carriers, and therefore the overall reaction rate.

In practice, both mechanisms often occur simultaneously, since any acceleration in the rate of an exothermic reaction will eventually lead to an increase in temperature. However, chain branching is not a requirement for an explosion. As an example, detonation of TNT (2, 4, 6-trinitrotoluene) is simply the result of an extremely fast chemical decomposition that generates huge quantities of gas. The reaction $2H_2(g) + O_2(g) \rightarrow 2H_2O(g)$ provides an example of a reaction in which both mechanisms are important.

The Hydrogen–Oxygen Reaction: $2H_2(g) + O_2(g) \rightarrow 2H_2O(g)$

Though this may at first sight appear to be a very simple reaction, the mechanism is in fact extremely complex, and is still not fully understood. It is known to be a branched chain reaction, involving carriers such as H·, ·O· and ·OH. A simplified version of the thermal mechanism is given below:

$$H_2 + \text{wall} \xrightarrow{k_1} H\cdot + H\cdot + \text{wall} \quad \text{Initiation}$$

$$H\cdot + O_2 \xrightarrow{k_2} \cdot OH + \cdot O\cdot \quad \text{Branching}$$

$$\cdot O\cdot + H_2 \xrightarrow{k_3} \cdot OH + H\cdot \quad \text{Branching}$$

$$\cdot OH + H_2 \xrightarrow{k_4} H\cdot + H_2O \quad \text{Propagation}$$

$$H\cdot + O_2 + M \xrightarrow{k_5} \cdot HO_2 + M \quad \text{Termination}$$

$$H\cdot + \text{wall} \xrightarrow{k_6} \text{H-wall} \qquad \text{Termination}$$

$$\cdot O\cdot + \text{wall} \xrightarrow{k_7} \text{O-wall} \qquad \text{Termination}$$

$$\cdot OH + \text{wall} \xrightarrow{k_8} \text{OH-wall} \qquad \text{Termination}$$

Each of reactions 2 and 3 produce two radical chain carriers for each chain carrier consumed. The two steps combine to give an overall branching coefficient of 3, corresponding to the hypothetical reaction $H\cdot + O_2 + H_2 \rightarrow H\cdot + \cdot OH + \frac{1}{2}OH$. Even so, the mixture is not explosive under all conditions, mainly because steps 1, 2 and 3 are endothermic ($\Delta H_1 = 427$ kJ mol^{-1}, $\Delta H_2 = 71$ kJ mol^{-1}, $\Delta H_3 = 17$ kJ mol^{-1}) and therefore slow at low temperatures. The efficiency of the branching steps increases with increasing temperature, and as a result, the reaction displays a complex dependence on temperature and pressure, as shown below.

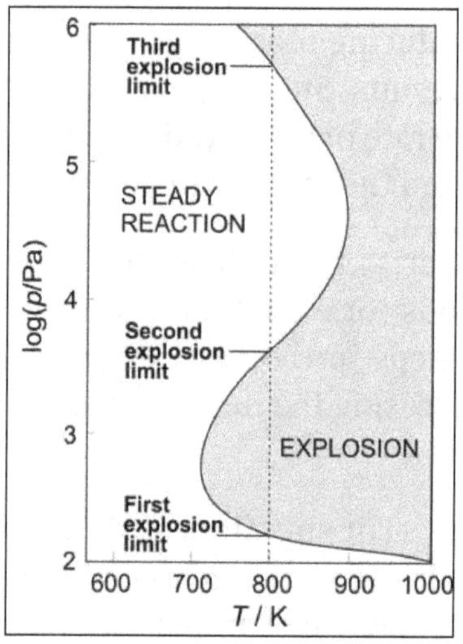

At very low pressures the mean free path in the gas is large, and chain carriers can reach the walls and combine. Collisions with walls are more likely than collisions with other gas phase molecules, so that overall $(v_5 + v_6 + v_7 + v_8) = (v_1 + v_2 + v_3)$ i.e. termination balances initiation, and steady reaction occurs.

At higher pressures, the chain carriers react before reaching the walls and the gas phase branching ratios become too high for wall termination to control (this is largely due to the fact that the rate of gas phase reaction is proportional to p_2, while the rate of wall reactions increases in proportion to p). Now we have $(v_1 + v_2 + v_3) > (v_5 + v_6 + v_7 + v_8)$, and the mixture becomes explosive. This is known as the first explosion limit.

At higher pressures, three body collisions start to become important. Termination step 5 can now match steps 1 to 3 in efficiency, and we again have $(v_5 + v_6 + v_7 + v_8) = (v_1 + v_2 + v_3)$ and steady reaction. This is the second explosion limit.

If the pressure is increased still further, the reaction rate increases so much and such a large amount of heat is generated that a thermal explosion results. This is the third explosion limit.

The factors that affect the explosion limits are fairly straightforward to understand in terms of their effects on the relative rates of initiation, propagation/branching, and termination steps. For example:

- Temperature: Increasing the temperature increases the efficiency of both endothermic reaction steps and steps for which there is activation barrier (steps 1-4 in the above mechanism). The termination steps are less sensitive to temperature, and may even be slowed down since they tend to be exothermic. As a result, the first explosion limit is lowered as the rates of steps 2 and 3 outpaces those of steps 6-8 more readily. The second explosion limit is increased because a higher pressure is needed for the termolecular step 5 to become important. The third limit is decreased, since at higher temperature more heat is produced, and also the heat that is produced is harder to lose from the system.
- Surface/Volume ratio: The shape and size of the reaction vessel can have a considerable effect on the explosion limits. Increasing the surface to volume ratio favours processes that involve collisions with the vessel walls over gas phase processes, which in this case means the initiation step 1 and termination steps 6, 7 and 8. The high efficiency of the branching steps means that 1 is unimportant in determining the overall rate, and the increased efficiency of the termination steps increases the pressure at which the first explosion limit is reached. The second limit has no dependence on the vessel walls and is unchanged. The third limit increases because it becomes easier to lose heat from the system due to the greater number of collisions with the walls.
- Overall pressure: Adding an inert gas to the mixture decreases the mean free path of the gas molecules, and disfavours collisions with the walls. This lowers the first explosion limit since the termination steps 6, 7, and 8 become less efficient. The second limit is also decreased because the inert gas can act as the third body M, increasing the rate of step 5, a termination process. The third limit is lowered due to the reduced heat transfer from the gas to the vessel walls.

Temperature Dependence of Reaction Rates

The Arrhenius Equation and Activation Energies

It is found experimentally that the rate constants for many chemical reactions follow the Arrhenius equation:

$$k = A\exp(-E_a/RT)$$

or equivalently,

$$\ln k = \ln A - \frac{E_a}{RT}$$

Where A is the pre-exponential factor and E_a is the activation energy. These parameters may be determined from experimental rate data by plotting lnk against 1/T. This is known as an Arrhenius plot, and has an intercept of lnA and a slope of $-E_a/R$. For most reactions, the Arrhenius equation works fairly well over at least a limited temperature range. However, there are often deviations.

These are generally due to the temperature dependence of the pre-exponential factor, but may also be due to more exotic effects, such as the influence of quantum mechanical tunnelling mechanisms on the reaction rate at low temperatures.

For an elementary reaction, both E_a and A have definite physical meanings; in particular, the activation energy may be interpreted as the energy difference between the reactants and the transition state involved in the collision and associated chemical rearrangement.

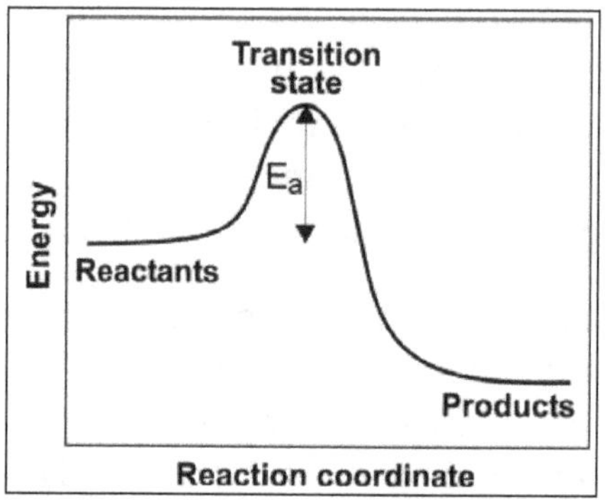

When the Arrhenius equation is applied to the overall kinetics of a multi-step reaction, E_a simply becomes an experimental parameter describing the temperature dependence of the overall reaction rate. E_a may vary with temperature, and may take positive or negative values. In this context, we may define the activation energy as:

$$E_a = RT^2 \frac{d\ln k}{dT}$$

This is a more general definition of the activation energy than the Arrhenius equation, and the two definitions become equivalent in the case when E_a is independent of temperature (all you need to do to prove this is to integrate the above equation, treating E_a as a constant). With the above definition, we can determine E_a at a given temperature from the slope (at the temperature of interest) of a plot of lnk against T, even if the Arrhenius plot is not a straight line.

There are a few observations that follow from Equation $E_a = RT^2 \frac{d\ln k}{dT}$.

- The higher the activation energy, the stronger the temperature dependence of the rate constant.
- A reaction with no temperature dependence has an activation energy of zero (this is common in ion-molecule reactions and radical-radical recombinations).
- A negative activation energy implies that the rate decreases as the temperature increases, and always indicates a complex reaction mechanism. An example of a reaction with a negative activation energy was the oxidation of NO to form NO_2, which has the mechanism.

$$NO + NO \underset{k_{-1}}{\overset{k_1}{\rightleftharpoons}} (NO)_2$$

$$(NO)_2 + O_2 \xrightarrow{k_2} 2NO_2$$

At higher temperatures, the intermediate complex $(NO)_2$ becomes more unstable and has a shorter lifetime. There is therefore less time for the O_2 to react with it to form the NO_2 products, and the reaction rate therefore decreases. Another way of thinking about this is that formation of the complex is exothermic, and increasing the temperature will therefore shift the pre-equilibrium to the left (by Le Chatelier's principle), again reducing the overall rate of reaction.

Overall Activation Energies for Complex Reactions

When dealing with complex reactions, the Arrhenius equation can often be used to estimate the overall activation energy from knowledge of the activation energies of individual steps. For example, in the above reaction, the overall rate law is:

$$v = \frac{k_1 k_2}{k_{-1}}[NO]^2[O_2] = k[NO]^2[O_2]$$

where k is the observed third order rate constant. The temperature dependence of k is:

$$k = \frac{k_1 k_2}{k_{-1}} = \frac{A_1 \exp\left(\frac{-E_a^{(1)}}{RT}\right) A_2 \exp\left(\frac{-E_a^{(2)}}{RT}\right)}{A_{-1} \exp\left(\frac{-E_a^{(1)}}{RT}\right)} = \frac{A_1 A_2}{A_{-1}} \exp\left(\frac{-E_a^{(1)} - E_a^{(2)} + E_a^{(-1)}}{RT}\right)$$

We can therefore identify that for the overall reaction,

$$A = \frac{A_1 A_2}{A_{-1}} \text{ and } E_a = E_a^{(1)} + E_a^{(2)} - E_a^{(-1)}$$

Catalysis

As well as quantifying the temperature dependence of a rate constant, the Arrhenius equation also provides a 'mathematical' explanation for the effect of a catalyst. A catalyst works by reducing the activation energy for a reaction. From the appearance of $-E_a$ in the exponent of the Arrhenius equation, it is clear that this will have the effect of increasing the rate constant.

Simple Collision Theory

As the name suggests, simple collision theory represents one of the most basic attempts to develop a theory capable of predicting the rate constant for an elementary bimolecular reaction of the form $A + B \rightarrow P$. We begin by considering the factors we might expect a reaction rate to depend upon. Obviously, the rate of reaction must depend upon the rate of collisions between the reactants. However, not every collision leads to reaction. Some colliding pairs do not have enough energy to overcome the activation barrier, and any theory of reaction rates must take this energy requirement into account. Also, it is highly likely that reaction will not even take place on every collision for which the energy requirement is met, since the reactants may need to collide in a particular orientation (e.g. SN2 reactions) or some of the energy may need to be present in a particular form (e.g. vibration in a bond coupled to the reaction coordinate). In summary, there are three aspects to a successful reactive collision, and we might expect an expression for the rate of a bimolecular reaction to take the following form.

v = (encounter rate) (energy requirement) (steric requirement).

We will now consider each of these factors in more detail.

Encounter Rate

The rate of collisions between molecules A and B present at number densities n_A and n_B is:

$$Z_{AB} = \sigma_C \left(\frac{8kT}{\pi\mu}\right)^{1/2} n_A n_B = \sigma_C \left(\frac{8kT}{\pi ì}\right)^{1/2} N_A{}^2 [A][B]$$

Energy Requirement

For a Maxwell-Boltzmann distribution of molecular speeds, the fraction of collisions for which the energy is high enough to overcome the activation barrier is $\exp(-E_a/RT)$.

Steric Requirement

Experimentally, measured rates are often found to be up to an order of magnitude smaller than those calculated from simple collision theory, suggesting that features such as the relative orientation of the colliding species is important in determining the reaction rate. We account for the disagreement between experiment and theory by introducing a steric factor, P, into our expression for the reaction rate. Alternatively, we can replace the collision cross section, σ_C, with a reaction cross section σ_R, where $\sigma_R = P\sigma_C$. Usually, P is considerably less than unity, but values greater than one are also possible. An example is the 'harpoon reaction' between Rb and Cl_2. The reaction mechanism involves an electron transfer at large separations to form $Rb^+ + Cl_2-$, after which the electrostatic attraction between the two ions guarantees reaction. P is large because the reaction cross section is determined by the electron transfer distance, which is much larger than the collision diameter.

Combining these three terms, the simple collision theory expression for the reaction rate is:

$$v = P \quad {}_C \left(\frac{8kT}{\pi\mu}\right)^{1/2} \exp\left(-\frac{E}{RT}\right) n_A n_I$$

and we can identify the second order rate constant as:

$$v = P\, \sigma_C \left(\frac{8kT}{\pi\mu}\right)^{1/2} \exp\left(\frac{-E_a}{RT}\right)$$

Simple collision theory provides a good first attempt at rationalising the Arrhenius temperature dependence seen for many reaction rate constants. However, at a quantitative level the predictions of the theory are far from accurate. There are a number of ways in which the model breaks down:

- It does not account for the fact that, unless the collision is head on, not all of the kinetic energy of the two reactants is available for reaction. Conservation of angular momentum means that only the kinetic energy corresponding to the velocity component along the relative velocity vector of the reactants actually contributes to the collision energy.

- The energy stored in internal degrees of freedom in the reactants (vibrations, rotations etc) has been ignored. For reactions involving large molecules, this often leads to a large discrepancy between simple collision theory and experiment, though this is partly corrected for by the inclusion of the steric factor, P.

Michaelis-Menten Kinetics

The Michaelis–Menten equation ($v = \frac{v_{max}[A]}{k_m + [A]}$), as presented by Michaelis and Menten and further developed by Briggs and Haldane, is fundamentally important to enzyme kinetics. The equation is characterized by two constants: the Michaelis–Menten constant (K_m) and the indirectly obtained catalytic constant, kcat. Although derived from a simple, single-substrate, irreversible reaction, the Michaelis–Menten equation also remains valid for more complex reactions.

The simple conversion of substrate (A) into product (P) catalyzed by the enzyme (E) is described below. As outlined by the induced-fit hypothesis, the first step is substrate binding and the second step is the catalytic step.

$$E + A \underset{k_{-1}}{\overset{k_1}{\rightleftarrows}} EA \xrightarrow{k_2} E + P$$

Following Eq. $v = \frac{\delta[P]}{\delta t} = -\frac{\delta[A]}{\delta t} = k_1[A]$, we can define the formation of the product in terms of the dissociation rate (k_2) of the enzyme–substrate complex, commonly denoted as kcat, and the concentration of the enzyme–substrate complex ([EA]).

$$v = k_{cat}[EA]$$

It is assumed that the dissociation rate (k_{cat} in Eq. $v = k_{cat}[EA]$ or k_2 in Eq. $E + A \underset{k_{-1}}{\overset{k_1}{\rightleftarrows}} EA \xrightarrow{k_2} E + P$) of the enzyme–substrate complex (EA) is slow compared to association (k_1) and redissociation (k_{-1}) reactions and that the reverse reaction (P → A) is negligible. Figure shows how the consumption of substrate, the production of product, and the concentration of the free enzyme and the enzyme–substrate complex change over the course of the reaction. During a very brief initial period, the enzyme–substrate complex is formed and reaches a concentration at which its consumption is matched by its formation. The [EA] then remains almost constant for a considerable time; this period is known as the steady state, and it is this steady-state condition that the Michaelis– Menten equation describes. Eventually, the reaction enters a third phase characterized by substrate depletion in which the [EA] gradually falls.

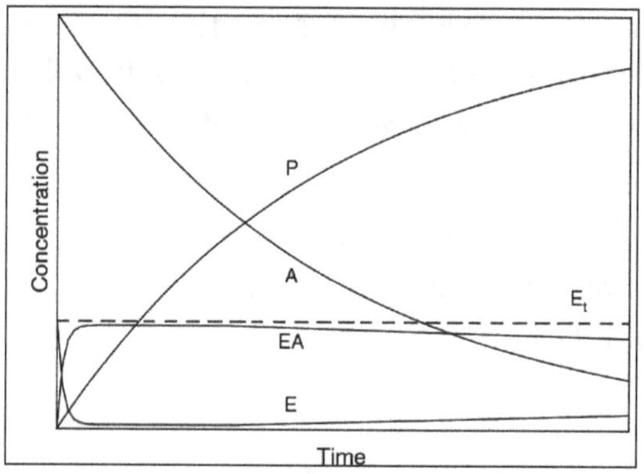

Figure shows change in substrate (A), product (P), free enzyme (E), enzyme–substrate complex (EA), and total enzyme (Et) concentration over time for the simple reaction described in Eq. $E + A \underset{k_{-1}}{\overset{k_1}{\rightleftarrows}} EA \overset{k_2}{\rightarrow} E + P$.

After a very brief initial period, the concentration of the enzyme–substrate complex reaches a steady state in which consumption and formation of the enzyme–substrate complex are balanced. As substrate is consumed, the concentration of the enzyme–substrate complex falls slowly and the concentration of the free enzyme rises. The amounts of enzyme and enzyme–substrate are greatly exaggerated for clarity.

At steady state, the enzyme–substrate concentration is stable, i.e.,

$$\frac{\delta[EA]}{\delta t} = 0$$

And therefore the formation of the ES complex (association reaction) and the breakdown of the ES complex (the sum of the re-dissociation and dissociation reactions) are equal.

$$k_1[E][A] = k_{-1}[EA] + k_{cat}[EA]$$

Rearrangement of Eq. $k_1[E][A] = k_{-1}[EA] + k_{cat}[EA]$ yields:

$$\frac{k_1[E][A]}{k_{-1} + k_{cat}} = [EA]$$

The three rate constants can now be combined as one term. This new constant, K_m, is known as the Michaelis–Menten constant,

$$k_m = \frac{k_{-1} + k_{cat}}{k_1}$$

And Eq. $\frac{k_1[E][A]}{k_{-1} + k_{cat}} = [EA]$ can be rewritten as:

$$\frac{[E][A]}{K_m} = [EA]$$

The concentration of enzyme in Eq. $\frac{[E][A]}{K_m} = [EA]$ refers to the unbound enzyme. The amount of free enzyme (E) and enzyme that is bound to the substrate (EA) varies over the course of a reaction, but the total amount of enzyme (Et) is constant such that,

$$E = E_t - EA$$

Substituting into Eq. $\frac{[E][A]}{K_m} = [EA]$ yields:

$$\frac{([E_t]-[EA])[A]}{K_m} = [EA]$$

which can be rearranged to yield:

$$\frac{[E_t][A]}{K_m + [A]} = [EA]$$

Substituting into Eq. $v = k_{cat}[EA]$ gives:

$$v = \frac{k_{cat}[E_t][A]}{K_m + [A]}$$

The maximum possible reaction rate (v_{max}) would be achieved when all the available enzyme is bound to the substrate and involved in catalysis, i.e.,

$$[EA] = [E_t]$$

Substituting Eq. $[EA] = [E_t]$ into Eq. $v = k_{cat}[EA]$ under conditions of saturating substrate concentration yields,

$$v_{max} = k_{cat}[Et]$$

and substituting Eq. $v_{max} = k_{cat}[Et]$ into Eq. $v = \frac{k_{cat}[E_t][A]}{K_m + [A]}$ yields what is widely recognized as the Michaelis–Menten equation.

$$v = \frac{v_{max}[A]}{k_m + [A]}$$

Key Parameters of the Michaelis–Menten Equation

K_m (mol.l^{-1})

Assuming a stable pH, temperature, and redox state, the K_m for a given enzyme is constant, and this parameter provides an indication of the binding strength of that enzyme to its substrate. Michaelis–Menten kinetics assumes that kcat is very low when compared to k1 and k–1. Therefore, following Eq. $k_m = \frac{k_{-1} + k_{cat}}{k_1}$, a high K_m indicates that the re-dissociation rate (k−1) is markedly greater than the association rate and that the enzyme binds the substrate weakly. Conversely, a low K_m indicates a higher affinity for the substrate (E1 in figure below). However, as Eq. $k_m = \frac{k_{-1} + k_{cat}}{k_1}$ shows, a large K_m could also be the result of very large k_{cat}. Therefore, care should be taken when using K_m as a proxy for the dissociation equilibrium constant of the enzyme–substrate complex.

k_{cat} (s^{-1})

The k_{cat}, also thought of as the turnover number of the enzyme, is a measure of the maximum catalytic production of the product under saturating substrate conditions per unit time per unit enzyme. The larger the value of k_{cat}, the more rapidly catalytic events occur. Values for k_{cat} differ markedly, e.g., 2.5 s^{-1} for rubisco (EC 4.1.1.39) with CO_2 as a substrate to c. 1,150 s^{-1} for fumarase (EC 4.2.1.2) with fumarate as a substrate.

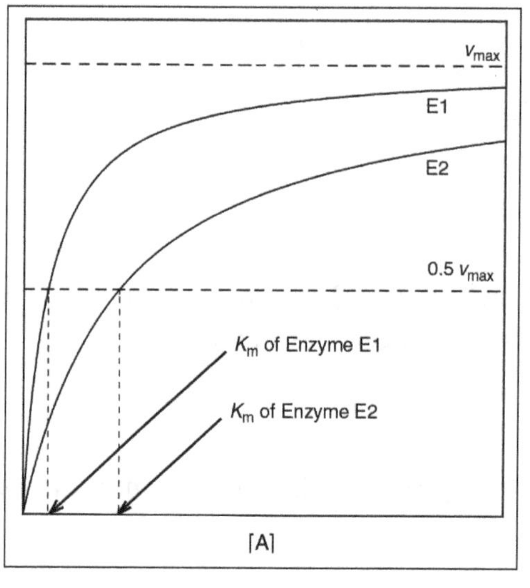

This figure shows change in velocity (v) with the concentration of substrate A ([A]) for the reaction shown by Eq. $E + A \underset{k_{-1}}{\overset{k_1}{\rightleftarrows}} EA \overset{k_2}{\rightarrow} E + P$ catalyzed by two enzymes E1 and E2.

The substrate concentration at the point at which the reaction has half its maximum velocity (0.5 v_{max}) is equal to the K_m. Enzyme E2 has a K_m four times greater than enzyme E1 but the same v_{max}.

Enzyme Efficiency (s^{-1} $(mol.l^{-1})^{-1}$)

The ratio of k_{cat}/K_m is defined as the catalytic efficiency and can be taken as a measure of substrate specificity. When the k_{cat} is markedly greater than k_{-1}, the catalytic process is extremely fast and the efficiency of the enzyme depends on its ability to bind the substrate. Based on the laws of diffusion, the upper limit for such rates, as determined by the frequency of collisions between the substrate and the enzyme, is between 10^8 and 10^9. Some enzymes actually have efficiencies that approach this range, indicating that they have near-perfect efficiency, e.g., fumarase, 2.3×10^8 $s^{-1}(mol.l^{-1})^{-1}$.

v_{max}

The v_{max} is the maximum velocity that an enzyme could achieve. The measurement is theoretical because at given time, it would require all enzyme molecules to be tightly bound to their substrates. As shown in Figure, v_{max} is approached at high substrate concentration but never reached.

Graphical Determination of Michaelis–Menten Parameters

Since the Michaelis–Menten parameters provide useful information for the network modeler, we need to consider the methods used to estimate K_m, k_{cat}, and v_{max}. There are a number of practical approaches to measuring reaction rates. Briefly, we need some way to follow the consumption of substrate or formation of product over time. We could simply mix enzyme with substrate and follow the formation of product in a progress reaction or conduct several experiments at multiple substrate concentrations and measure initial velocity at each substrate concentration. However, the graphical evaluation of nonlinear plots to obtain Michaelis–Menten parameters relies on accurate curve fitting. The problems associated with evaluating enzyme kinetics using a nonlinear plot can be avoided by using one of the three common linearization methods to obtain estimates for K_m and v_{max}. However, these methods are not without problems either. Errors in the determination of v at low substrate concentration are greatly magnified in Lineweaver–Burke and Eadie–Hofstee plots and to a lesser extent in Hanes plots. Despite this disadvantage, and in contrast to nonlinear plots, changes in enzyme kinetics, for example, due to the action of an inhibitor, are readily apparent on linear plots. Clearly, selection of a linear or nonlinear plot should be based on an understanding of the sources of error in the experiment and consistent with the goal of that experiment.

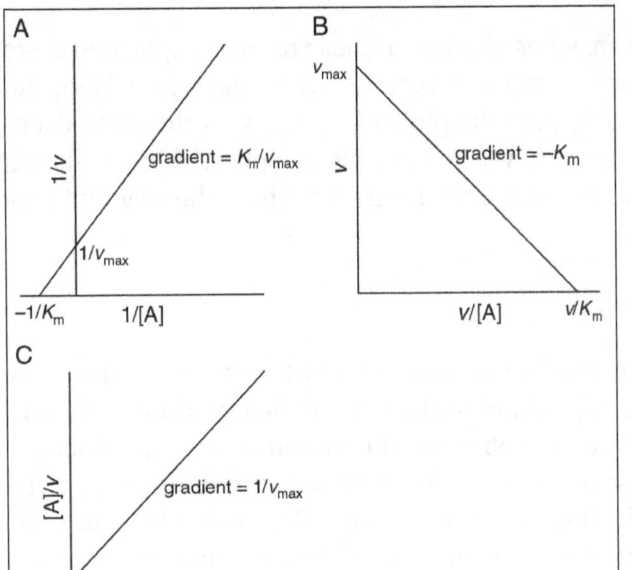

This figure shows linear representations of the Michaelis–Menten equation (Eq. $v = \frac{v_{max}[A]}{k_m + [A]}$). Lineweaver–Burk (A), Eadie–Hofstee (B), and Hanes (C) plots. The intercepts with the x- and y-axis and the gradient can be used to determine K_m and v_{max}.

Multisubstrate Systems

The Michaelis –Menten model of enzyme kinetics was derived for single substrate reactions. Enzymatic reactions requiring multiple substrates and yielding multiple products are more common and yielding multiple products are more common than single-substrate reaction. In these types of reactions, the all the substrates involved are bound to the enzyme before catalysis of the reaction takes place to release the products. Sequential reactions can be either ordered or random. In contrast to the Michealis-Menton kinetics where a binary Enzyme-Substrate complex is generated in the mechanism ([ES], in bisubstrate enzyme reactions, a ternary complex of the enzyme and two substrates is generated:

$$A + B \overset{E}{\rightleftharpoons} P + Q$$

Bisubstrate reactions account for ~60% of the known enzymatic reactions. Multi-substrate reactions follow complex rate equations that describe how the substrates bind and in what sequence. The analysis of these reactions is much simpler if the concentration of substrate A is kept constant and substrate B varied. Under these conditions, the enzyme behaves just like a single-substrate enzyme and a plot of v by [S] gives apparent K_M and V_{max} constants for substrate B. If a set of these measurements is performed at different fixed concentrations of A, these data can be used to work out what the mechanism of the reaction is. For an enzyme that takes two substrates A and B and turns them into two products P and Q, there are two types of mechanism: ternary complex and ping–pong.

How do you resolve the enzymes kinetics of these more complicated systems? The answer is fairly straightforward. You keep one of the substrates (B, for example) fixed, and vary the other substrate (A) and obtain a series of hyperbolic plots of v_0 vs A at different fixed B concentrations. This would give a series of linear 1/v vs. 1/A double-reciprocal plots (Lineweaver-Burk plots) as well. The pattern of Lineweaver-Burk plots depends on how the reactants and products interact with the enzyme.

Sequential Mechanism

In this mechanism, both substrates must bind to the enzyme before any products are made and released. The substrates might bind to the enzyme in a random fashion (A first then B or vice-versa) or in an ordered fashion. An abbreviated notation scheme developed by W.W. Cleland is shown in Figure below for the sequential random and sequential ordered mechanisms. For both mechanisms, Lineweaver-Burk plots at varying A and different fixed values of B give a series of intersecting lines. Derivative curves can be solved to obtain appropriate kinetic constants. In ordered sequential reactions, all the substrates are first bound to the enzyme in a defined order or sequence. The products, too, are released after catalysis in a defined order or sequence.

An example is the lactate dehydrogenase enzyme, which is a protein that catalyzes glucose metabolism. In this ordered mechanism, the coenzyme, NADH, always binds first, with pyruvate binding afterward. During the reaction, the pyruvate is reduced to lactate while NADH is oxidized to NAD^+ by the enzyme. Lactate is then released first, followed by the release of NAD^+.

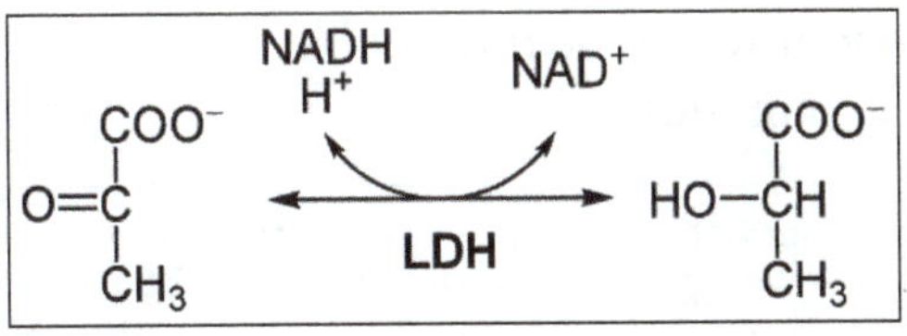

Figure: Reaction of the lactate dehydrogenase: pyruvate (left) is oxidized to lactate (right) by expense of NADH.

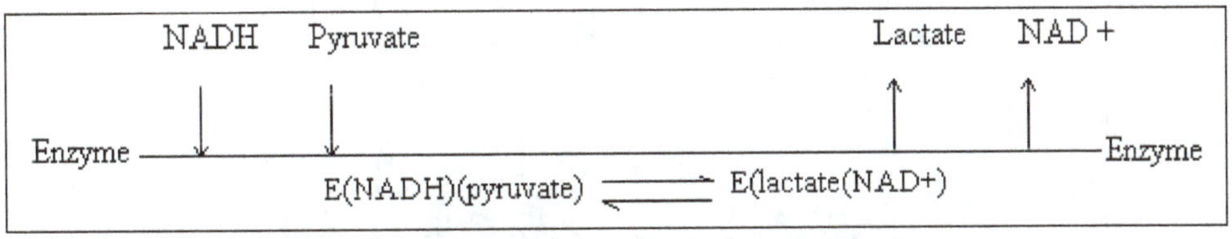

Figure: Ordered Sequential Mechanism for the lactate dehydrogenase enzyme.

This is a characteristic of a ternary complex, which consists of three molecules that are bound together. Before catalysis, the substrates and coenzyme are bound to the enzyme. After catalysis, the complex consists of the enzyme and products, NAD^+ and lactate.

In random sequential reactions, the substrates and products are bound and then released in no preferred order, or "random" order. An example is the creatine kinase enzyme, which catalyzes creatine and ATP (the two substrates) to form phosphocreatine and ADP (teh Products) in Figure below. In this case, either substrates may bind first or either products may be released first. A ternary complex is still observed for random sequential reactions. Before catalysis, the complex is generated that includes the enzyme, ATP, and creatine. After catalysis, the complex consists of the enzyme, ADP, and phosphocreatine.

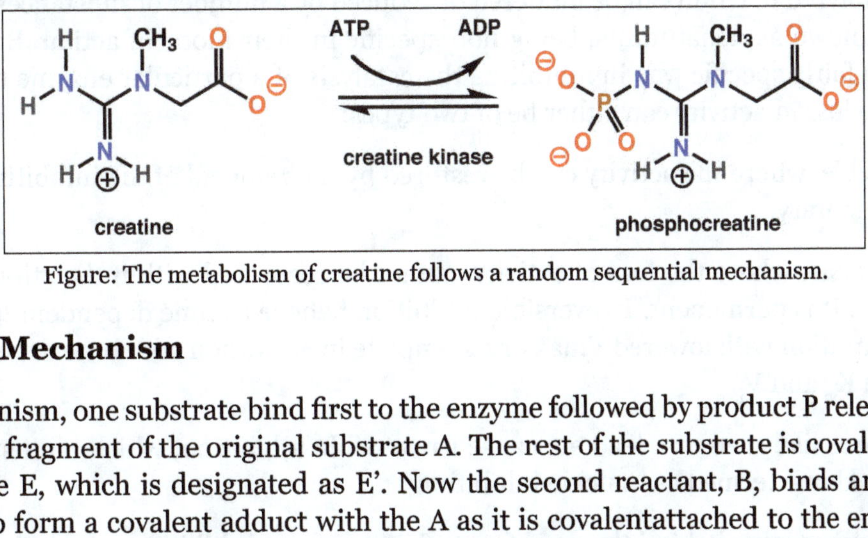

Figure: The metabolism of creatine follows a random sequential mechanism.

Ping-Pong Mechanism

In this mechanism, one substrate bind first to the enzyme followed by product P release. Typically, product P is a fragment of the original substrate A. The rest of the substrate is covalently attached to the enzyme E, which is designated as E'. Now the second reactant, B, binds and reacts with the enzyme to form a covalent adduct with the A as it is covalentattached to the enzyme to form product Q. This is now released and the enzyme is restored to its initial form, E. This represents a ping-pong mechanism. An abbreviated notation scheme is shown below for the ping-pong mechanisms. For this mechanism, Lineweaver-Burk plots at varying A and different fixed values of B give a series of parallel lines. An example of this type of reaction might be low molecular weight protein tyrosine phosphatase against the small substrate p-initrophenylphosphate (A) which binds to the enzyme covalently with the expulsion of the product P, the p-nitrophenol leaving group. Water (B) then comes in and covalently attacks the enzyme, forming an adduct with the covalently bound phosphate releasing it as inorganic phosphate. In this particular example, however, you cannot vary the water concentration and it would be impossible to generate the parallel Lineweaver-Burk plots characteristic of ping-pong kinetics.

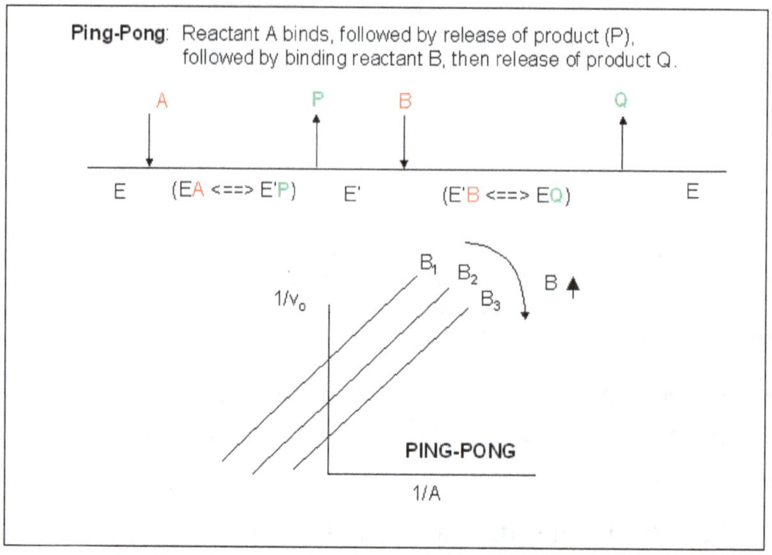

Figure: The Ping-Pong mechanism and associated Lineweaver-Burk.

Enzyme Inhibition

An enzyme catalyzed reaction can be hindered or reduced by a number of substances. Some others like urea are known as denaturants, being non-specific in their mode of action. But if any compound act in a fairly specific way in inhibiting the catalysis of a particular enzyme they are called inhibitors. The loss in activity can either be of two types:

- Reversible, where the activity can be restored by the removal of the inhibiting compound. It's temporary.
- Irreversible, where the loss of activity cannot be recovered within the stipulated time of interest. It is permanent. Irreversible inhibition behave as time dependent loss of enzyme concentration with lowered Vmax or incomplete in activation with time dependent change in both K_m and V_{max}.

Heavy metal ions like mercury, lead etc cause irreversible inhibition, which bind strongly to the amino acid backbone termed as "suicide inhibition".

There are several reasons behind the need for studying enzyme inhibition mechanisms. They are:

- Exploring potential mechanisms in multi-substrate reactions.
- Studying the relative binding affinity of competitive inhibitors to the enzyme active site, in the absence of 3-D structure information.
- For understanding various control mechanisms-how the balance of protease enzymes and their inhibitors in tissues achieve homeostasis.

- For various commercial applications like pesticide, insecticide, weed-killers, pharmaceutical compounds like drugs etc.

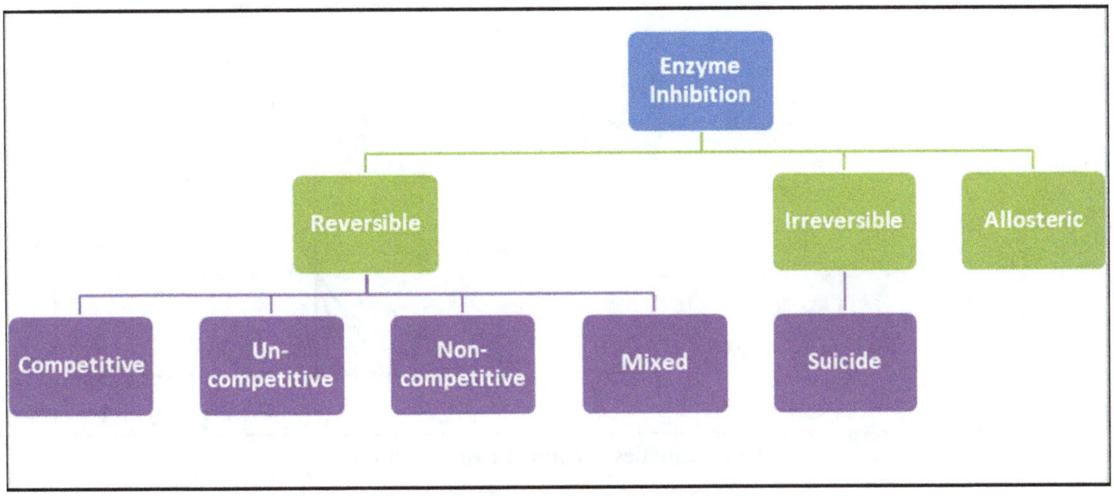

Figure: Types of enzyme inhibition.

Reversible Inhibition

In this type of inhibition, the hindrance is temporary and thus noncovalent interactions like hydrogen bonds, ionic bonds or hydrophobic bonds form between inhibitors and the enzyme. Even though these are weak bonds, multiple such bonds cause strong and specific binding. Despite absence of any chemical reactions, the inhibitors can easily remove or exchange by dilution or dialysis. After removing the inhibitor, enzyme can be fully restored in reversible inhibition. Equilibrium is established between free inhibitor and enzyme-inhibitor [EI] complex.

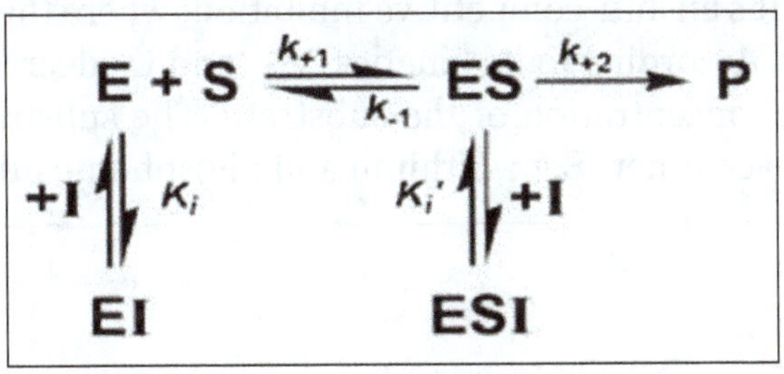

Figure: Mechanism of reversible inhibition.

Reversible inhibitions are of different types. The classification is based according to the effect of varying the concentration of the enzyme's substrate on the inhibitor.

Competitive Inhibition: In this type of reversible inhibition, both the substrate and its inhibitor cannot bind to the enzyme at the same time to the allosteric/active site. This normally occurs due to the structural similarity of substrate and the inhibitor, which results with affinity for the active site. The inhibition can be recovered by the presence of high concentration of substrate, outcoming the competing inhibitor. V_{max} of the reaction is unchanged, while k_d, the dissociation constant is apparently increased. Competitive inhibitors can also be used to find the enzyme active site.

E.g.: N-(phosphonacetyl)-L-asparate also known as PALA is a competitive inhibitor for asparatetranscarbamoylose.

E.g.: Malonate is a competitive inhibitor of enzyme succinate dehydrogenase, and competes with succinate.

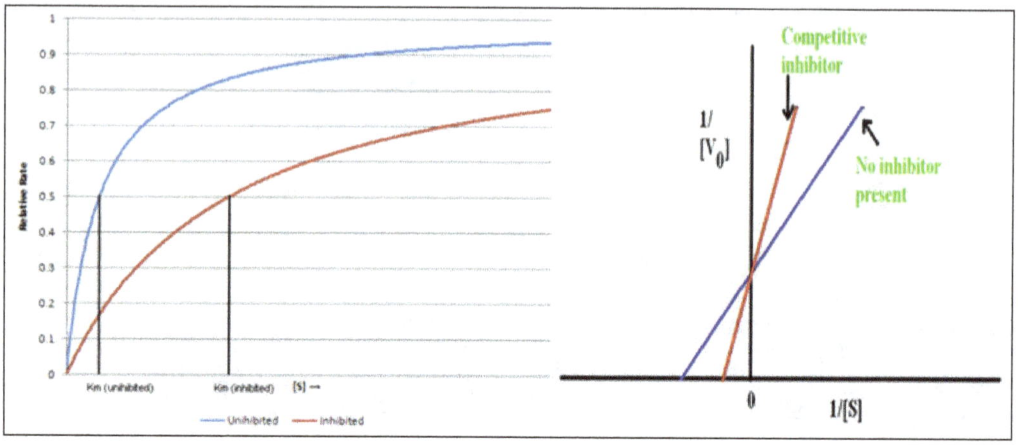

Figure: Kinetics of competitive inhibition.

Table: Clinical use of competitive inhibition.

Drugs	Target Enzyme	Therapeutic use
STATINS, Atorvastatin Simvastatin	HMG CoA reductase	Involved in the reduction of plasma cholesterol level- Anti-hyperlipidemic agents.
Allopurinol	Xanthine oxidase	Used in case of prevention of gout attacks.
Methotrexate	Dihydrofolate reductase	Used as canecr treatment drug.
Captopril, Enalapril	Angiotensin converting enzyme	Treatment of high blood pressure.
Discoumarol	Vit K-epoxide reductase	Used as an anti-coagulant.

Uncompetitive Inhibition: It's an anti-competitive inhibition; where the inhibitor binds only to the substrate-enzyme complex. According to its kinetics, V_{max} and K_m decrease. This type of inhibition works best in case of high concentration of the substrate. The substrate and the uncompetitive inhibitor do not resemble each other. E.g.:-Lithium and phosphoinositide cycle.

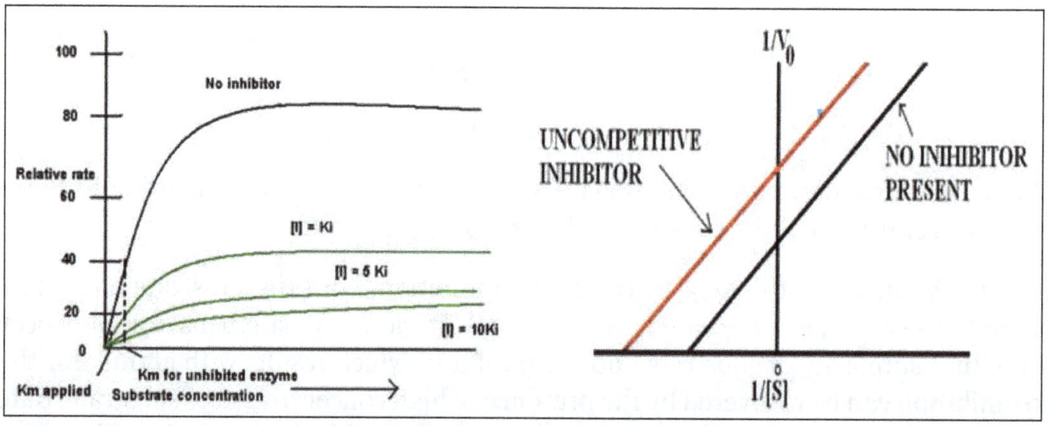

Figure: Kinetics of un-competitive inhibition.

Non-competitive Inhibition: A non-competitive inhibitor is one which reacts with enzyme-substrate or [ES] complex. It does not affect the binding of the substrate, but slows down the reaction rate for formation of the enzyme-product [EP] complex. The only factor on which the extent of hindrance or inhibition depends is the inhibitor concentration. There will be a decrease in V_{max} but k_m will remain the same.

- E.g.: Alanine non-competitively inhibits the enzyme pyruvate kinase.

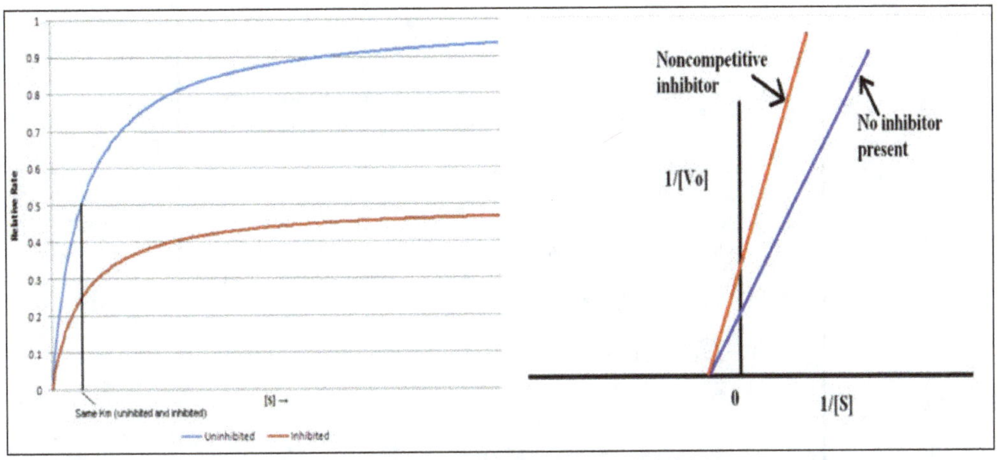

Figure: Kinetics of non-competitive inhibition.

Table: Examples of common non-competitive inhibitors.

Inhibitor	Enzyme inhibited
Heavy metals – Ag^{2+}, Hg^{2+}, Pb^{2+}	Heavy metals bind with cysteinyl SH group of enzyme.
Pepstatin	Pepsin.
Soybean trypsin inhibitor	Trypsin.
Ethanol/narcotic drugs	Acid phosphatase.

Mixed Inhibition: In this type of inhibitor, inhibitor is capable of binding to both free enzyme as well as enzyme- substrate complex. In this case, V_{max} and k_{max} varies. Mixed inhibitor binds to the allosteric site. This type of inhibition cannot overcome by increasing substrate concentration S, but can be reduced. The inhibitor binding to the allosteric site changes the structural confirmation to reduce the affinity of the substrate. E.g.: Mixed inhibition is observed on case of oxidoreductase activity of xanthine.

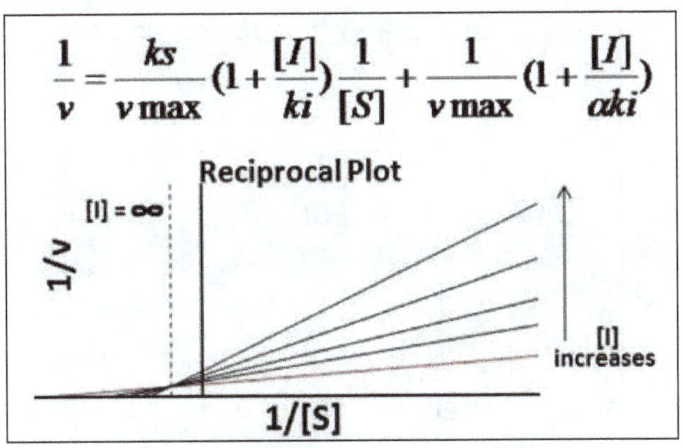

Figure: Kinetics of mixed inhibition.

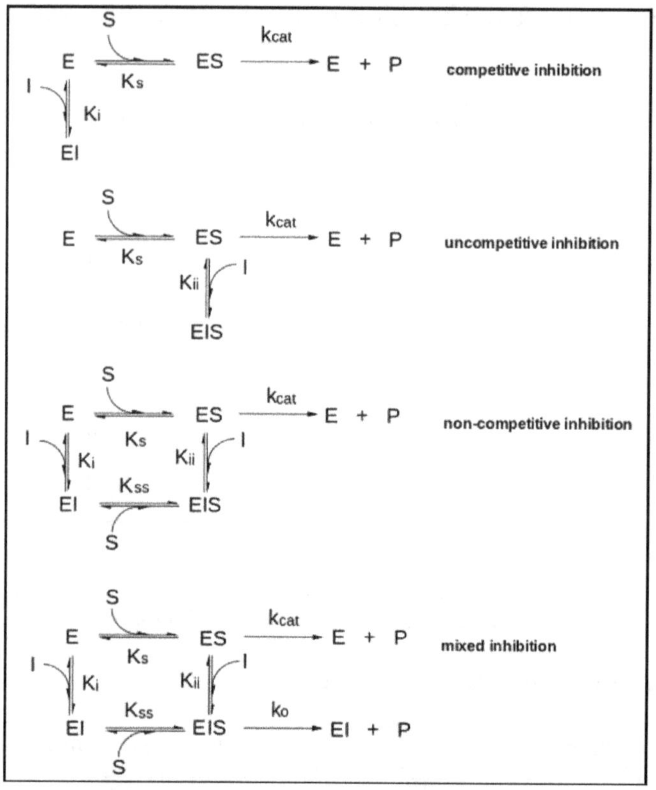

Figure: Different types of reversible inhibitions.

Irreversible Inhibition: In this type of inhibition, the hindrance is of permanent nature by modifying enzyme covalently. These types of inhibitors often contain electrophilic functional groups like fluorophosphates, aldehydes, haloalkanes, alkenes, nitrogen mustards, phenyl sulfonates, Michael acceptors etc, which react with aminoacid sidechains having nucleophilic residues.

These inhibitors are very specific in the mechanism of inactivation for a particular class of enzyme-They do irreversible inhibition by specially altering the active site. They display inhibition which is time-dependent. Their potency cannot be characterized by IC50 value. These inhibitors increase K_m and decrease V_{max}.

E.g.: Diisopropylflurophosphate (DFP) is an irreversible protease inhibitor.

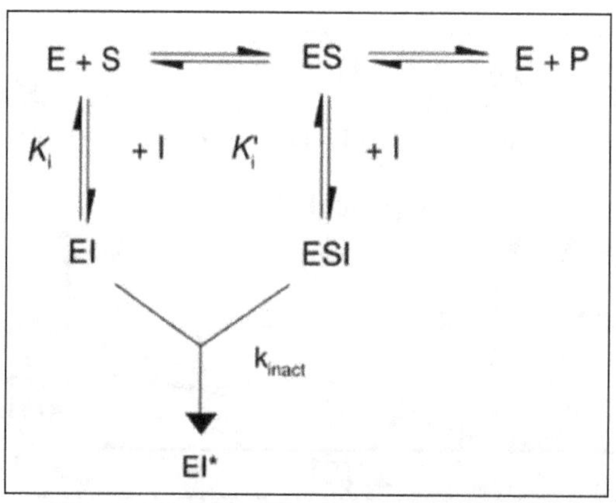

Figure: Kinetics of irreversible inhibition.

Table: Therapeutic uses of irreversible inhibitors.

Inhibitors	Enzyme inhibited	Therapeutic uses
Disulfiram	Aldehyde dehydrogenase	Treats alcoholism
Cyanide ions	Cytochrome oxidase	Inhibition of the respiratory Chain
Fluoride ions	Enolase	Inhibition of Glycolysis
Melathion	Acetyl choline esterase	Used as an organophosphorus insecticide
Di-isopropyl fluorophos-phate	Serine proteases, Acetyl choline esterase	Used as a nerve gas
British Anti Lewisite (BAL)	Reaction with the thiol (-SH) group of the enzyme	Used as an antidote in case of poisoning due to heavy metal

Suicide inhibition: This is another type of irreversible inhibition. In this case, the target enzyme converts the inhibitor compound into a reactive form in its active site. They are also known as mechanism based inhibitors or transition state analogs.

E.g.: DFMO [α-difluromethyl ornithine], an analogue of ornithine inhibits ornithine decarboxylase.

E.g.: Allopurinol is a suicide inhibitor of xanthine oxidase Aspirin inhibits cyclooxygenase.

Table: Therapeutic uses of suicide inhibitors.

Drugs	Product	Target Enzyme	Therapeutic use
Allopurinol	Alloxanthin	Xanthine oxidase	Treatment of gout
5-Fluorouracil	Fluoro-deoxy uridylate	Thymidylate synthase	Cancer treatment
Aspirin	Acetylates the serine residue present in the cyclooxygenase active site	Cyclooxygenase	Used as a non-steroidal anti-inflammatory drug
Difluro methyl ornithine (DFMO)	Forms an irreversible covalent complex with the co-enzyme	Ornithine decarboxylase	Used for treating Sleeping sickness (trypanosomiasis)

Table: Differences between reversible & irreversible inhibitions.

Reversible	Irreversible
Binds via non-covalent interactions	Binds via covalent interactions
Do not perform any chemical changes	Inhibitor binds to the substrate and prevent catalytic activity of enzymes
As there is no bonding between the inhibitor and substrate, reversible inhibition can be reversed,	Irreversibility due to strong covalent bonding

Allosteric Inhibition: Allosteric inhibition is a type of enzyme regulation, in which allosteric inhibitor binds to a site other than the active site of the enzyme. This additional site to which effector binds is called allosteric site. When these effectors bind to the protein, results with conformational change and cause enhancement in activity is known as allosteric activators. When they decrease the activity of the protein, they are known as allosteric inhibitors. Allosteric enzymes are K or V types.

References

- The-chemistry-behind-enzyme-catalysis: byjus.com, Retrieved 03, April 2020
- Six-types-of-enzyme-catalysts, biology: cliffsnotes.com, Retrieved 30, June 2020
- Acid-base-catalysis, resource: creative-enzymes.com, Retrieved 09, March 2020
- A-Enzyme-Kinetics, A-Physical-Chemistry-for-Life-Scientists: libretexts.org Retrieved 23, August 2020

Chapter 3

Enzyme: Source, Production and Purification

The choice of a source of enzymes is a crucial step in the process of production of enzymes. Large scale production of enzymes is achieved by fermentation. It is the metabolic process of breakdown of an organic substrate by the micro-organisms through the action of enzymes. This chapter covers in detail topics such as enzyme source, enzyme production, microbial enzymes and enzyme purification.

Enzyme Source

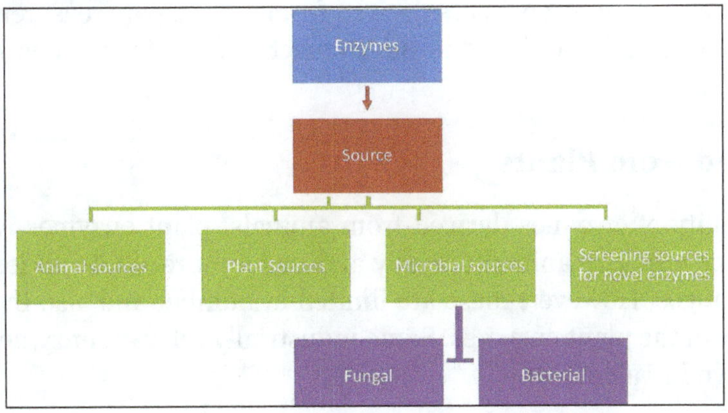

Enzymes in their biologically active form can be isolated from any living organism. A very wide range of sources are currently being used for production of enzymes commercially. There are three major sources of enzymes: animals, plants or microbial organisms (as outlined in Figure). Microbial sources contribute about 80% of the overall enzyme production. Over half of the microbial enzymes that are being used industrially are sourced from fungi and yeast, while one-third of the enzyme production is derived from bacteria. The remaining (~12%) is obtained from animal and plant sources. Non-microbial sources account for a larger proportion enzymes being used in chemical analysis and clinical diagnosis.

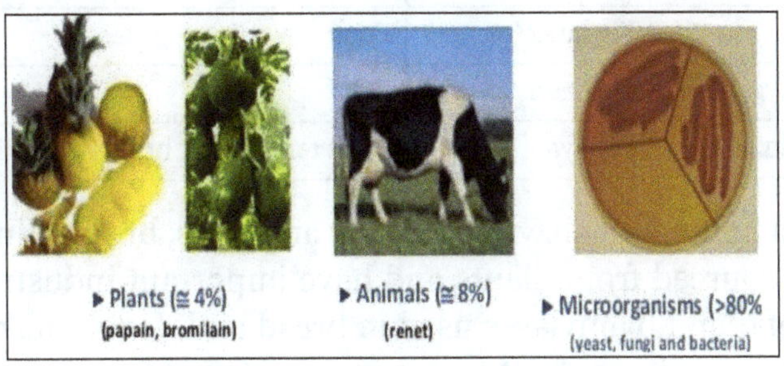

Figure: Major sources of enzymes.

Enzymes Sourced from Animals

Enzymes were initially obtained strictly from animals. The processes included slaughtering of pigs or cows, followed by extraction of enzymes from their pancreas. However, such enzymes suffered drawbacks of being relatively unstable at lower pH (acidic environment of the stomach) or at temperatures different from temperature of the source animals. Some industrially relevant enzymes which are sourced from animals are shown in Table below.

Table: Industrial enzymes isolated from animals.

Enzyme	Source	Areas of Industrial application
Catalase	Liver	Food
Lipase	Pancreas	Food
Chymotrypsn	Pancreas	Medicine
Trypsin	Pancreas	Leather
Rennet	Abomasum	Cheese making (Food)

While lipase, trypsin and chymotrypsin are derived from animal pancreas, rennet is another complex of enzymes that is usually produced in the stomachs of ruminant mammals and are used in the cheese production.

Enzymes Sourced from Plants

Owing to the instability of enzymes derived from animals, plant enzymes were discovered and used next. These enzymes were animal-friendly and much more stable under low pH conditions and temperature changes. However, these are limited in number and also there is not much of a broader applicability of the plant enzymes. Some industrially relevant enzymes which are sourced from plants are shown in Table below.

Table: Industrial enzymes isolated from plants.

Enzyme	Source	Areas of industrial application
Actinidin	Kiwi fruit	Food
α-Amylase	Malted barley	Brewing, bakery, bread making
β-Amylase	Malted barley	Brewing, bakery, bread making
Bromelain	Pineapple latex	Brewing
β-Glucanase	Malted barley	Brewing
Ficin	Fig latex	Food
Lipoxygenase	Soybean	Food
Papain	Papaya	Meat tenderizer, brewing, baking

Table above shows that a number of enzymes namely amylases, bromelain, glucanase, ficin, lipoxygenase and papain are sourced from plants and have important industrial applications. Malted barley acts as a source of α and β amylases used in bread and dough making, while papain is obtained from the green unripe papaya fruit and is in meat tenderizer, protein processing, brewing and baking. Bromelain is another important enzyme extracted from the pineapple fruit. It has applications in pharmaceutical, baking and protein processing industries. Ficin is purified from the latex of the fig tree, Ficus glabatra or Ficus carica. All three of them, namely papain, bromelain and ficin are protein hydrolyzing enzymes and exhibits broad specificity protein hydrolysis over a wide pH range.

Drawbacks in Animal and Plant based Enzymes

- Stability issues (outlined above).
- These enzymes have limited areas of application, primarily in food industry.
- Additionally, it is said that plant and animal based enzymes contain potentially harmful materials from plant and animal tissues (for, e.g. phenolic compounds from plants; endogenous enzyme inhibitors and proteases) as compared to microbes, which could interfere in enzymatic reactions.

Attempts are being made to overcome some of these difficulties by the use of animal and plant cell culture.

Enzymes Sourced from Microbes

Microbial enzymes provide an important source for isolation of numerous enzymes that are applicable to various domains of enzymatic catalysis. They can be sourced from fungus, bacteria, yeast etc. through their fermentation.

What makes microbial enzymes different and much more in demand as compared to animal and plant derived enzymes? It is their stability in different environments over plant and animal based enzymes complemented by the ability to tailor these microbial enzymes to our benefit which makes these enzymes so powerful. Microbes are preferred to plants and animals as sources of enzymes because of the following factors:

- Relatively cheaper to produce- cost effective source.
- Volume of enzyme to be produced can be controlled.
- Raw material required for media for fermenting microbes is easily affordable and easily arranged.
- Lesser amount of toxic components (antibiotics/ steroids) present as compared to animal and plan tissues.
- Stability over a broad pH and temperature range.

Types of Microbial Sources

Fungal Sources

These are enzymes which are obtained by fungal fermentation. Some of the industrially important enzymes which are sourced from fungus are given in table.

Table: Industrial enzymes isolated from fungus.

Enzyme	Source	Areas of industrial application
α-Amylase	Aspergillus	Baking; cereal and starch; beverages;
Tannas	Aspergillus	Beverages
Glucoamylase	Aspergillus	Starch; beverage making; confectionery; bakery; dietary foods
Catalase	Aspergillus	Food; cheese making; beverages
Cellulase	Trichoderma	Waste remediation, beverages; dietary foods
Dextranase	Penicillium	Food
Glucose oxidase	Aspergillus	Food; beverages
Lactase	Aspergillus Kluyveromyces	Dairy; dietary foods Dairy; dietary foods
Lipase	Rhizopus Candida	Food; hydrolysis of fats and oils Food; cheese; fats and oils
Rennet	Mucor miehei	Cheese
Pectinase	Aspergillus	Fruit juice extraction; oil extraction; textile processing; wastewater treatment; animal feed
Protease	Aspergillus	Baking, detergents; meat tenderization; leather tanning; skin dehairing; pharmaceutcals
Raffinase	Mortierella Saccharomyces	Food; beverages Food; sugar and honey
Invertase	Saccharomyces	Confectionery

It is evident that fungus has been source to a plethora of industrial hydrolytic enzymes ranging from amylases, glucoamylases, tannase cellulase, dextranase, glucose oxidase, lactase, lipase, protease, rennet, invertase, pectinase etc. Aspergillus sp. appears to be an important fungal source of enymes. For example, lactase, protease, pecti lyase, pectinase, amylase, tannase, catalases are all produced from Aspergillus sp. Other sources include Saccharomyces sp., Kluveromyces sp., Candida sp. and Penicillium sp. Most of these enzymes are produced extracellularly into the medium with the exception of catalase, glucose oxidase and raffinase which are produced intracellularly and require cell disruption. It is also evident that most of the strains used have been employed by the food industry, especially for making dairy and milk products, making beverages and confectionaries as well as dietary foods.

Bacterial Enzymes

These are enzymes which are obtained by bacterial fermentation. Some of the industrially important enzymes which are sourced from bacterial sources are given in table below.

Table: Industrial enzymes isolated from bacteria.

Enzyme	Source	Areas of industrial application
α-Amylase	Bacillus	Starch hydrolysis, Beverages, Bakery
β-Amylase	Bacillus	Starch hydrolysis, Beverages
Endo-βglucanase	Bacillus	Beverages

Glucose isomerase	Bacillus	Fructose syrup Starch hydrolysis, Beverages
Hemicellulase	Bacillus	Baking, fruit juice/beverage making, animal feed
Protease	Bacillus	Meat tenderizer, protein hydrolysis etc.
Pullulanase	Klebsiella	Starch hydrolysis, Beverages
Penicillin amidase	Bacillus	Pharmaceutical
Asparaginase	E. coli	Medical uses

Table above gives a broad overview of the various enzymes obtained from bacterial sources. The primary and most important bacterial source of enzymes is Bacillus sp. from which numerous enzymes like amylase, protease, glucanase, hemicellulase, glucose isomerase have been derived. Like fungal enzymes, vast application of bacterial enzymes is in food industry. Production of high fructose syrup using glucose isomerase is a noteworthy example. Hydrolytic enzymes like amylases and proteases are produced extracellularly by bacillus sp. in substantial amounts. On the other hand, enzymes like penicillin amidase and asparaginase which have been shown to have clinical/ pharmaceutical applications are produced intracellularly and in small amounts, which need to be scaled up further.

As evident, microorganisms present an attractive source for enzymes due to the fact that they can be cultured easily and that too, over a very short period of time by established fermentation strategies. Amongst microbes, it is noteworthy that a majority of microbial enzymes originate from a very limited number of genera. In case of intracellular enzyme production, sometimes levels are low and require optimization of medium composition and production conditions. Sometimes the microbial cells themselves can be utilized directly as an enzyme source.

The most essential feature for microbes secreting industrially important enzymes is that they are GRAS (Generally Regarded As Safe) listed, implying that they must be non-pathogenic, non-toxic and should not produce antibiotics. Industry usually favours high-yielding microbial strains of microbes which secrete enzymes extracellularly into the growth medium in the shortest possible time to lower manufacturing costs involved. Sometimes the production strains are genetically manipulated to increase production levels. Furthermore, it is also desirable that the enzyme preparations must have a long shelf life and have stability/activity over extended periods. It is often preferred that the enzyme is produced constitutively; else approaches for enzyme induction need to be rapid and inexpensive.

Screening for Novel Enzyme Sources

With industrial demands for stable and novel enzymes, being continuously on the rise, it appears necessary that new microbial strains capable of performing catalysis under harsh industrial conditions be screened. Search for new enzymes endowed with novel activities and enhanced stability thus continues to be a desirable pursuit in enzyme research. This is fuelled by the necessity of enzymatic interventions in new and challenging bioprocesses.

- Genetic and Protein Engineering: The upsurge in molecular biology tools has enabled engineering of enzymes to impart stability by variety of techniques such as random or site directed mutagenesis, gene shuffling, directed evolution, chemical modifications and immobilization. Use of molecular cloning techniques to genetically engineer bacteria/fungus for heterologous expression of almost any enzyme in sufficient amounts has also been an alternative proposition. Industrial enzymes produced recombinantly have added advantages of higher expression levels and purity, can be engineered to possess desired properties and facilitate large scale production. For example, rennin and chymosin has been expressed from genetically engineered yeast.

- Bio-prospecting: This pertains to search for new and promising enzymes with improved properties and catalytic abilities. Organisms living in extreme environments are naturally attuned to withstand extreme survival conditions. Their enzymes are naturally stable and exhibit optimum activities even under harsh conditions. Extremophiles, as they are referred to, are therefore perceived to be one excellent source of stable enzymes. They are endowed with inherent ability to function under extreme conditions like extremes of temperature, salt, pressure, radiation etc. Extremophiles include:
    - Thermophile (these are microbes growing hot springs); Example: Pyrococcus sp. Sulfolobus sp.
    - Psychrophile (these are microbes growing under extreme cold); Example: Arthrobacter sp., Psychrobacter sp.
    - Barophile or Piezophile (these are microbes growing under extreme pressure); Example: Halomonas sp.
    - Radiophiles (these are microbes growing under extreme radiation); Example: Deinococcus radiodurans.
    - Acidophile (these are microbes surviving under low pH conditions); Example: Thiobacillus sp., Acetobacter aceti.
    - Alkaliphile(these are microbes surviving under high pH conditions); Example: Halorhodospira halochloris, Natronomonas pharaonis.
    - Metallophile (these are microbes able to withstand high metal ions concentrations in its surrounding envrionment); Example: Ferroplasma sp.
    - Halophile (these are microbes able to survive under extreme salinity); Example: Salinibacter ruber, Halobacterium sp.
    - Osmophile (these are microbes growth in environments with a high sugar concentration); Example: Saccharomyces sp.
    - Xerophile (these are microbes that grow in extremely dry, desiccating conditions); Example: Trichosporonoides nigrescens.
    - Solvent tolerant (these are microbes growing in presence of solvent/non-aqueous conditions); Example: Pseudomonas sp.

For inhabiting such environments, these organisms suitably adapt themselves wherein important biochemical reactions, cellular activities and physiological functions take place. The various biomolecules, including enzymes which are involved in the cellular machinery of above microorganisms are attuned to function under such harsh and extreme conditions in which these microbes thrive. Thus, extremophiles represent a potentially valuable resource for novel and stable enzymes which can be explored for catalysis of industrial process and development of novel biotechnological processes.

Enzyme Production

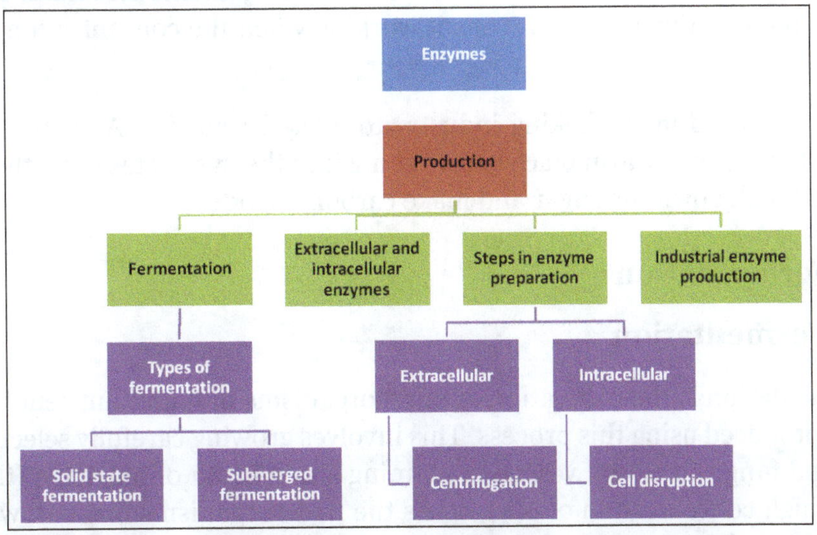

Industrial enzyme production has expanded rapidly in both volume and quantity, owing to the expanding market and increasing demand for novel and stable biocatalysts, which have established their potential applications in various industrial sectors.

Although microorganisms constitute the largest and most dominating source of enzymes, several enzymes have also been obtained from plant and animal sources.

Production of a microbial enzyme initially involves screening of microorganisms with desirable and useful property. The level of natural enzyme production from an organism is often very low and therefore, it is essential that production levels are elevated to meet the industrial demand. This is often achieved by optimization of medium and culture conditions or by causing mutation in the organism. The microorganisms used for enzyme production are grown in bioreactors or fermenters under optimized medium and culture conditions. Both solid state and submerged fermentation methods are used industrially. The enzymes are separated from the microbial culture using centrifugation or filtration and then stabilized to maintain suitable shelf lives.

Fermentation

Enzymes have been used for thousands of years to produce food and beverages, such as cheese, yoghurt, beer and wine. Yeast is a fungus whose enzymes aid the breakdown of glucose into ethanol and carbon dioxide anaerobically. The enzymes in yeast break down sugar (glucose) into alcohol (ethanol) and carbon dioxide gas:

$$\text{Glucose} \xrightarrow{\text{Yeast}} \text{Ethanol} + \text{Carbon dioxide}$$

$$C_6H_{12}O_6\ (aq) \xrightarrow{\text{Yeast}} 2\ C_2H_5OH\ (aq) + 2\ CO_2\ (g)$$

This reaction, which takes place in the absence of oxygen, is called fermentation.

Fermentation works best when the yeast and glucose solution is kept warm. Enzymes will also become ineffective if the temperature becomes too high. Fermentation is used in all production of alcoholic drinks. For stronger alcohol, such as whiskey and vodka, these need to be distilled after fermentation to increase the concentration of ethanol in the fermented mixture. This is due to the fact that ethanol poisons the yeast and stops it working when the concentration builds up about 18% by volume.

Fermentation is also used in the baking industry to make bread rise. After the dough has been prepared, it is left to rest in a warm place before going into the oven. This gives the enzymes in the yeast a chance to break down the sugar and make carbon dioxide.

Methods of Fermentation

Submerged Fermentation

Submerged fermentation is the cultivation of microorganisms in liquid nutrient broth. Industrial enzymes can be produced using this process. This involves growing carefully selected micro organisms (bacteria and fungi) in closed vessels containing a rich broth of nutrients (the fermentation medium) and a high concentration of oxygen. As the microorganisms break down the nutrients, they release the desired enzymes into solution.

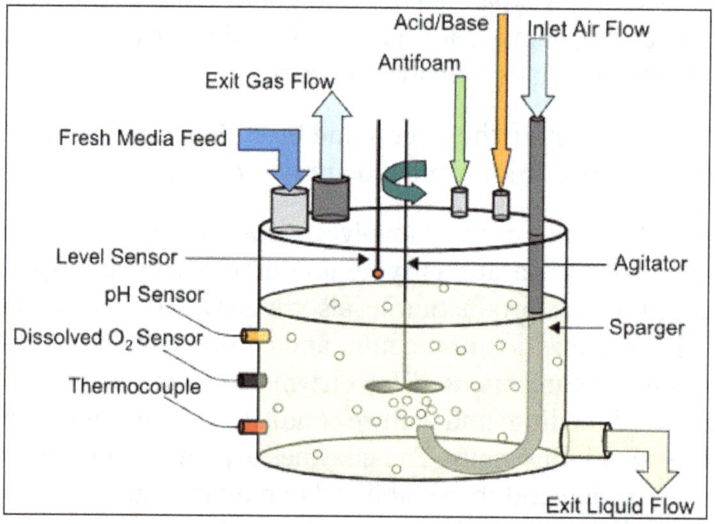

Figure: Typical fermenter.

Due to the development of large-scale fermentation technologies, the production of microbial enzymes accounts for a significant proportion of the biotechnology industries total output. Fermentation takes place in large vessels (fermenter) with volumes of up to 1,000 cubic metres.

The fermentation media sterilises nutrients based on renewable raw materials like maize, sugars and soya. Most industrial enzymes are secreted by microorganisms into the fermentation medium in order to break down the carbon and nitrogen sources. Batch-fed and continuous fermentation processes are common. In the batch-fed process, sterilised nutrients are added to the fermenter during the growth of the biomass. In the continuous process, sterilised liquid nutrients are fed into the fermenter at the same flow rate as the fermentation broth leaving the system.

This will achieve a steady-state production. Parameters like temperature, pH, oxygen consumption and carbon dioxide formation are measured and controlled to optimise the fermentation process.

Firstly, in harvesting enzymes from the fermentation medium one must remove insoluble products, e.g. microbial cells. This is normally done by centrifugation. As most industrial enzymes are extracellular (secreted by cells into the external environment), they remain in the fermented broth after the biomass has been removed. The biomass can be recycled as a fertiliser, but first it must be treated with lime to inactivate the microorganisms and stabilise it during storage. The enzymes in the remaining broth are then concentrated by evaporation, membrane filtration or crystallization depending on their intended application. If pure enzyme preparations are required, they are usually isolated by gel or ion exchange chromatography.

Certain applications require solid enzyme products, so the crude powder enzymes are made into granules to make them more convenient to use. Sometimes liquid formulations are preferred because they are easier to handle and dose along with other liquid ingredients. Enzymes used in starch conversion to convert glucose into fructose are immobilised, typically on the surfaces of inert granules held in reaction columns or towers. This is carried out to prolong their working life as these enzymes normally go on working for over a year.

Solid State Fermentation

Solid-state fermentation (SSF) is another method used for the production of enzymes. Solid-state fermentation involves the cultivation of microorganisms on a solid substrate, such as grains, rice and wheat bran. This method is an alternative to the production of enzymes in liquid by submerged fermentation. SSF has many advantages over submerged fermentation. These include, high volumetric productivity, relatively high concentration of product, less effluent generated and simple fermentation equipment.

There are many substrates that can be utilized for the production of enzymes by SSF. These include wheat bran, rice bran, sugar beet pulp and wheat and corn flour. The selection of substrate depends on many factors, which is mainly related to the cost and the availability of the substrate. Other factors include particle size and the level of moisture. Smaller substrate particles have a larger surface area for the proliferation of the microorganisms, but if too small the efficiency of respiration will be impeded and poor growth and hence poor production of enzymes will result. Larger particles provide more efficient aeration and respiration, but there is a reduction in the surface area. A compromise must be reached, regarding the particle size of the substrate for a particular process. SSF requires moisture to be present on the substrate, for the microorganisms to produce enzymes. As a consequence the water content of the substrate must also be optimized, as a higher or lower presence of water may adversely affect the microbial activity. Water also has implications for the physicochemical properties of the solid substrate. Enzymes of industrial importance have been produced by SSF. Some examples are proteases, pectinases, glucoamylases and cellulases.

Types of Fermentation Process

Fermentation in liquid media is of two types depending upon the mode of operation:

- Batch fermentation
- Continuous fermentation

Batch reactors are simplest type of mode of reactor operation. In this mode, the reactor is filled with medium and the fermentation is allowed to proceed. When the fermentation has finished the contents are emptied for downstream processing. The reactor is then cleaned, re-filled, re-inoculated and the fermentation process starts again.

Continuous reactors: Fresh media is continuously added and bioreactor fluid is continuously removed. As a result, cells continuously receive fresh medium and products and waste products and cells are continuously removed for processing. The reactor can thus be operated for long periods of time without having to be shut down. Continuous reactors can be many times more productive than batch reactors. This is partly due to the fact that the reactor does not have to be shut down as regularly and also due to the fact that the growth rate of the bacteria in the reactor can be more easily controlled and optimized.

In addition, cells can also be immobilized in continuous reactors, to prevent their removal and thus further increase the productivity of these reactors. Continuous reactors are as yet not widely used in industry but do find major application in wastewater treatment. Fed batch reactor is the most common type of reactor used in industry. In this reactor, fresh media is continuous or sometimes periodically added to the bioreactor but unlike a continuous reactor, there is no continuous removal. The fermenter is emptied or partially emptied when reactor is full or fermentation is finished. As with the continuous reactor, it is possible to achieve high productivities due to the fact that the growth rate of the cells can be optimized by controlling the flow rate of the feed entering the reactor.

Factors Influencing Fermentation

A fermentation is influenced by numerous factors, including temperature, pH, nature and composition of the medium, dissolved O_2, dissolved CO_2, operational system (e.g. batch, fed-batch, continuous), feeding with precursors, mixing (cycling through varying environments), and shear rates in the fermenter. Variation in these factors may affect: the rate of fermentation; the product spectrum and yield; the organoleptic properties of the product (appearances, taste, smell, and texture), the generation of toxins; nutritional quality; and other physic-chemical properties.

- Effect of Time: The effect of incubation period on enzyme production was investigated by checking the enzyme activity on 4th, 5th, 6th, 7th and 8th days of incubation in the different substrates at pH 7 and at room temperature.
- Effect of Temperature: The effect of temperature on enzyme production was investigated by fermentation in different substrates and incubated at 30°C, 37°C, 40°C, 45°C and 50°C at pH 7 for 6 days.

- Effect of pH: Fermentation investigated the effect of pH on enzyme production in different substrates by adjusting the pH of basal salt solutions to 4, 5.5, 6.5, 7, 7.5, 8 and 9. The substrates were then incubated for 6 days at room temperature.
- Effect of Carbon Source: The effect of carbon sources on enzyme production was investigated by supplementing the basal salt solution, pH 7, with 2% of different carbon sources such as glucose, maltose, lactose, sucrose and starch. The substrates were then incubated for 6 days at room temperature.
- Effect of Nitrogen Source: The effect of nitrogen source on enzyme production was studied by replacing the nitrogen source in basal salt solution, pH 7, with 2% of $NaNO_3$, $(NH_4)2SO_4$, NH_4Cl, NH_4NO_3 and KNO_3, and incubated at room temperature for 6 days.

Intracellular and Extracellular Enzymes

Enzyme produced by fermentation can be intracellular or extracellular:

- Intracellular Enzymes: These are enzymes which are produced internally within the cell for use inside the cell itself. For example, respiratory enzymes involved in the mitochondria of a cell are produced intracellularly.
- Extracellular Enzymes: These are enzymes which are produced inside the cell but released externally for biocatalysis outside the cell. For example, digestive hydrolytic enzymes like protease, amylase, and lipase are released by microbes into fermentation medium for hydrolysing large protein, carbohydrate or fat molecules into smaller fragments for easy absorption into the cell.

Preparation of Enzymes

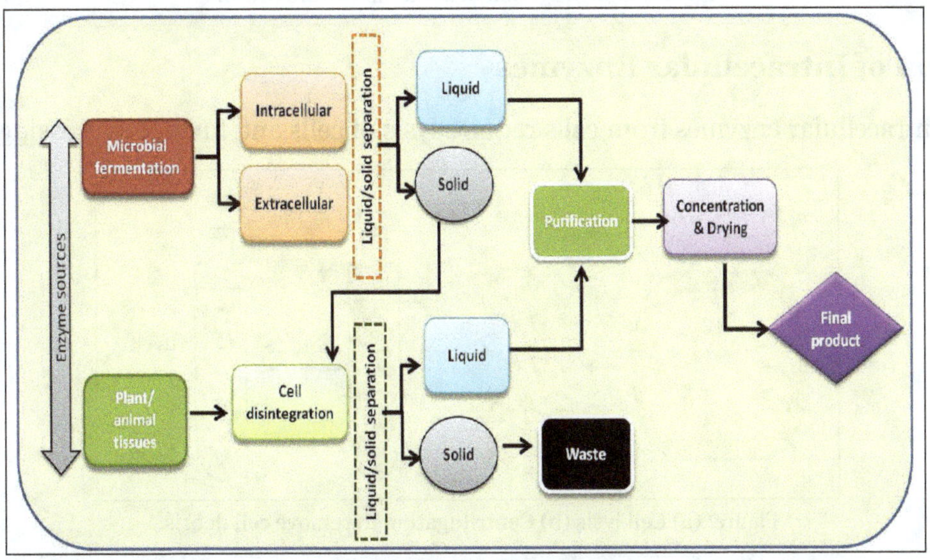

Figure: Schematic representation for preparing enzymes after its production.

Enzymes may be produced extracellularly/intracellularly by the microbe or are released from the cells/tissues (animal/plant) into solution. Figure outlines a schematic overview of the various stages for obtaining an enzyme preparation after being produced.

Post the microbial fermentation as well as the plant/animal tissue disintegration, the next and primary step is the solid/liquid separation which separates the cell mass from the broth or removes the cell debris that are produced after cell lysis. The filtrate containing soluble enzyme is collected by filtration or centrifugation. The filtrate is further processed or purified to obtain the desired enzyme product.

Steps in Enzyme Production

Production of Extracellular Enzymes

Centrifugation separates the liquid and solid phases on the basis of the particle size and density. For enzymes produced extracellularly, the culture broth is centrifuged at high sprred (for e.g, 10000-12000 rpm or roations per minute at 4°C) to obtain the filtrate which contains the enzyme of interest. The temperature is kept low during centrifugation to prevent the enzymes from getting denatured. The microbial pellet so obtained after centrifugation may be further disintegrated to obtain any intracellular enzyme of interest. Generally, centrifugation is the most preferred method for recovering extracellular enzyme.

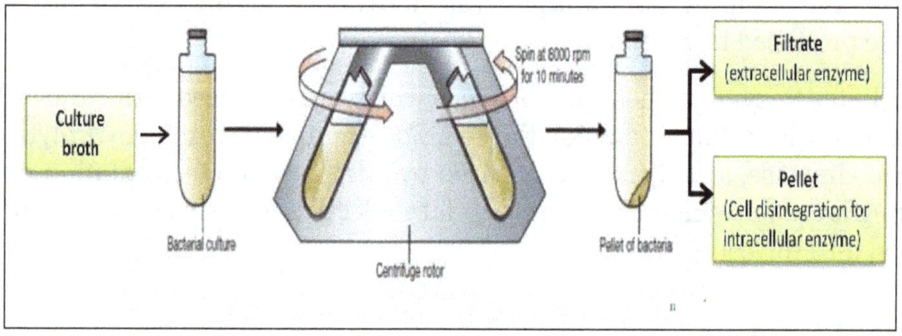

Figure: Centrifugation of bacterial culture broth.

Production of Intracellular Enzymes

Release of intracellular enzymes from cells require lysis of cells and further processing.

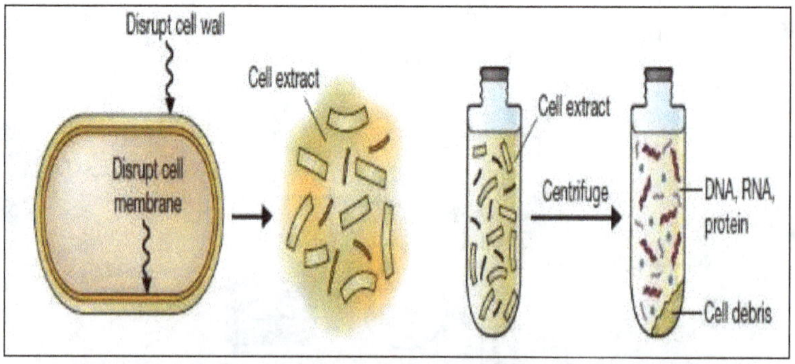

Figure: (a) Cell lysis (b) Centrifugation to remove cell debris.

Some types of cell (for e.g. some Gram-negative bacteria or animal cells) are readily lysed by gentle treatment such as osmotic shock, while some others (for e.g. yeasts, some Gram-positive bacteria) are able to resist cell breakage. Consequently a variety of cell disruption techniques have been developed. Figure below gives an overview of various techniques of cell disruption.

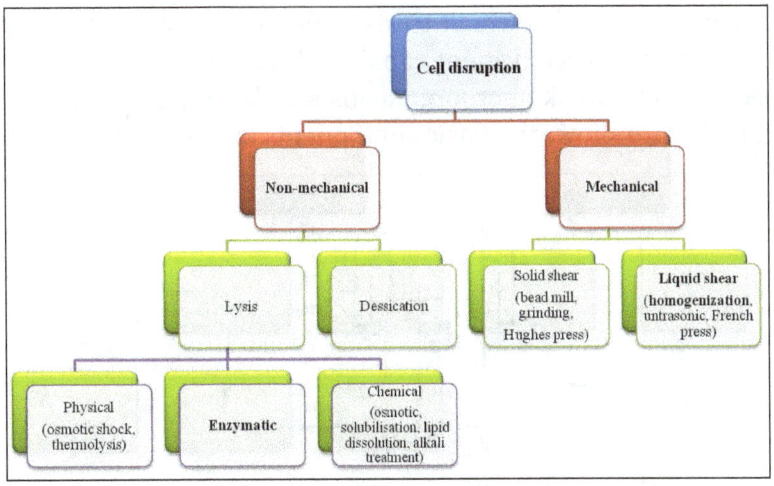

Figure: Techniques of cell disruption.

Mechanical Methods

Sonication

Sonication is a widely used and simple small-scale method used for cell disruption. This involves the treatment of microbial cells in suspension with high frequency ultrasound waves to disrupt cell walls and membranes which results in their disruption.

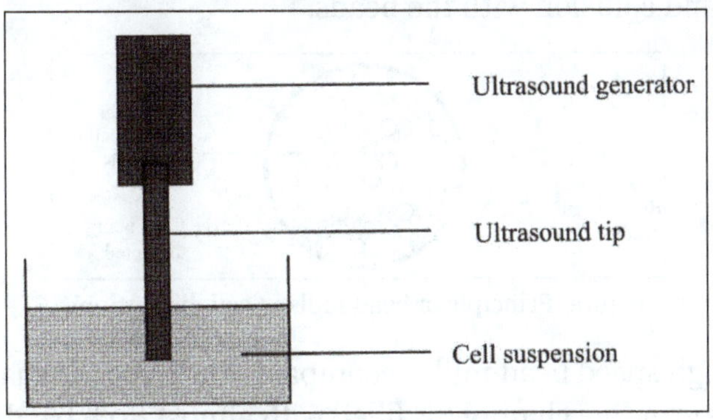

Figure: Cell breakage by ultrasonication.

The sinusoidal movement of the probe (that is immersed within the culture suspension) caused by ultrasonication leads to perforations or cavitation within the cells. However, a lot of heat is generated during cell-sonication which may often lead to damage of sensitive cellular biomolecules. Also, ultrasonication also leads to the production of very fine cell debris which can hinder further processing of the sample in the next steps. This method can be used as a continuous lysis method on a laboratory scale; however the large scale use of ultrasonication is often avoided due to high operating costs.

High Pressure Homogenizers

High pressure homogenizers are another well accepted technique for cell disruption, readily adapted to largescale and continuous operations. The principle of this technique lies in application of high pressure to microbial cells. The microbial cells are thus subjected to impact, shear and a severe pressure drop across the valve which tends to disrupt the cells as the pressure is released abruptly. The shear forces produced are not capable of damaging enzymes released in solution. This method has been used to break microorganisms like Aspergillus niger, Escherichia coli, and Bacillus megatherium. Figure shows the basic components of a high pressure homogenizer.

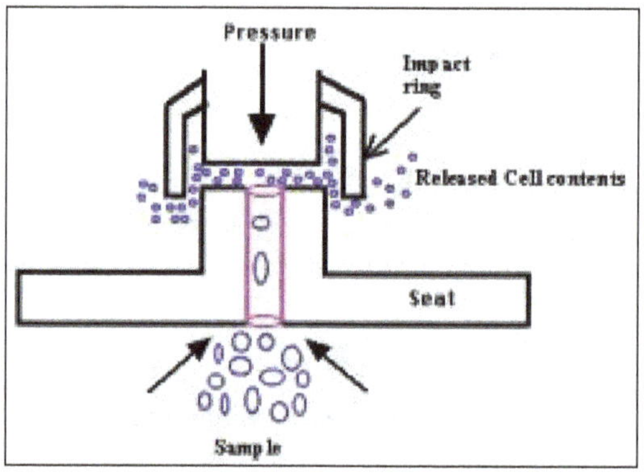

Figure: Cell breakage by high pressure homogenizer.

Use of Bead Mills

Another method for mechanical disruption of microbial cells is the use of bead mills. In this case, small steel, alumina, or glass (silica) beads (usually ranging in size 0.2 -.1.0 mm diameter) are used to agitate cell suspensions, during which they get broken/disrupted. The disruption is caused due to the high liquid shear and collision with the beads.

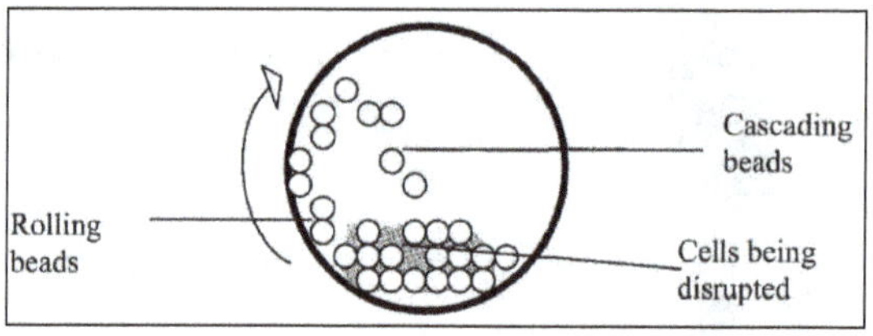

Figure: Principle of bead mills in cell disruption.

On an industrial scale, high speed bead mills accompanied with cooling jackets are used to agitate the cell suspensions. The grinding chamber is filled with almost 80% beads and a shaft with impellers is fixed within the chamber which rotates at high speed generating shear forces. The efficiency of cell disruption in a bead mill depends upon the amount and size of beads, the concentration of the cells in suspension as well as speed of rotation of the agitator. Bead mills are inexpensive and can be operated continuously on a large scale to lyse algae, bacteria and fungi.

Disadvantages of Mechanical Disruption

Mechanical disruption methods suffer from several drawbacks.

- Since, the cells are broken completely during mechanical disruption, all intracellular materials are released. Therefore, separation of the protein of interest from a complex mixture of cellular components becomes a complex task.
- The nucleic acids so released along with intracellular contents may increase the viscosity of the solution and interfere in subsequent processing steps.

- The cell debris, produced by mechanical lysis, often makes the solution difficult to clarify.
- Furthermore, cells are subjected to very harsh conditions and most enzymes will tend to get denatured by the heat generated in the instruments.

Non-Mechanical Methods

Physical Methods

- Osmotic shock: Changes in solute concentration around a cell cause a rapid change in the movement of water across the cell membrane. When concentration of solute inside the cell increases as compared to that outside the cell, water enters the cell in large amounts in order to establish osmotic balance, causing it to swell and burst. Use of this approach to cause cell lysis for releasing intracellular enzymes would indeed be a cheap, gentle and convenient method. However, this method has been used on a large scale.
- Thermolysis: Breaking down of cells by subjecting it to heat.

Chemical Methods

This refers to the use of chemicals to solubilize the components of the cell wall to release internal components of the cell. However, it is required that the internal cellular components, including enzymes are sufficiently robust, are insensitive to the chemicals being used and can be easily separated from the chemicals after lysis.

- Use of alkali-Alkali like NaOH can be used for extraction of some bacterial proteins.
- Use of organic solvents-Organic solvents like methanol, ethanol, isopropanol and butanol can be used to disrupt the cells.
- Use of detergents-Detergents can also be used to lyse cells. Detergents, such as Triton X-100 are used for releasing enzymes bound onto cellular membranes, but these are often difficult to be removed from the final product. Sometimes, detergents like trimethyl ammonium bromide (cationic detergent) or sodium lauryl sulphate (anionic detergent), which are also used for cell lysis can even denature cellular proteins and therefore, their removal after cell disruption is a must.

Enzymatic Methods

Use of Enzymatic Lysis Methods

Cell disruption using enzymatic lysis methods offers a gentler approach of releasing enzymes under milder conditions wherein the enzyme is not subjected to heat or shear. Enzymatic lysis is widely used in the laboratory scale but is less used for industrial purposes.

Lysozyme, sourced from hen egg-white, is often used to lyse the β-1, 4 glycosidic bonds present in the mucopeptide of bacterial walls. Gram-positive bacteria are more susceptible to enzymatic lysis than Gram negative bacteria. The major drawback to the use of lysozyme on a large scale is its cost, yet hen egg-white lysozyme is the only lytic enzyme available commercially as of now.

Drawbacks Associated with Cell Lysis Methods

Cell lysis approaches often damages internally released cellular components as shown in Table below.

Table: Hazards likely to damage enzymes during cell disruption.

Parameter	Effect on cell/ cellular components	Possible solution
Heat	All mechanical methods lead to generation of heat thus affecting enzyme activity and structure	Stabilizing agents like polyols may be used to stabilize the enzyme against deactivation by excess heat
Shear	Shear forces employed to disrupt cells may damage enzymes	
Proteases	Release of protein degrading enzymes, proteases upon cell lysis may affect other enzymes and lead to loss of enzyme activity.	Increased processing of sample under cooling. An alternative substrate for the proteases in the form of inexpensive protein or protease inhibitors can also be used in the extraction medium to prevent target intacellular enzymes from being degraded
pH	pH conditions at which the target intracellular enzyme is not stable may affect the enzyme activity/stability	Use of appropriate buffers to maintain steady pH
Chemical	Presence of detergent and/or solvents in the extracting medium may, at times, affect the enzyme conformation or even inhibit them	Adsorbents, such as polyvinylpyrrolidone, or ascorbic acid may be used
Heavymetal toxicity	Heavy metal ions may leach into the samples from the homogenization apparatus which may interfere and inactivate/destabilize the enzyme	Chelating reagents like EDTA may be used to protect the enzyme from metal ion promoted deactivation

Clarification of Culture Broth/Homogenates

The extract which is obtained after cell lysis is clarified. Clarification refers to the removal of any particulate matter, remaining cellular debris or other suspended contaminants in the extract, which are not compatible with chromatography and needs to be removed prior to downstream processing. The clarified sample of soluble protein thus forms the starting material for further purification.

Microbial Enzymes

Enzymes are biocatalysts that play an important role in metabolic and biochemical reactions. Microorganisms are the primary source of enzymes, because they are cultured in large quantities in short span of time and genetic manipulations can be done on bacterial cells to enhance the enzyme production. In addition, the microbial enzymes have been paid more attention due to their active and stable nature than enzymes from plant and animal. Most of the microorganisms are unable to grow and produce enzyme under harsh environments that cause toxicity to microorganisms. However, some microorganisms have undergone various adaptations enabling them to grow and produce enzymes under harsh conditions.

Enzymes are large macromolecules composed of polymers of amino acids connected by amide bonds, ranging from kilodalton (insulin) to megadalton (ribosome) in molecular mass. Catalytic site of these macromolecules is often buried deep within hydrophobic pockets, which determines the specificity for their substrate. This specificity of enzyme to catalyze reactions between one types of chemical compound over the other provides the basis of its classification and name. With the great advancement achieved in the area of biochemistry after 1940, a large number of enzymes were isolated and characterized, and therefore, it was necessary to regulate the enzyme nomenclature. According to the type of reaction catalyzed, the enzyme commission has classified the enzymes into six main classes, mentioned in Table below.

Table: Enzyme classes, types of reactions and corresponding enzymes.

S. no.	Class	Reactions	Enzymes
1	Oxidoreductases	Transfer of hydrogen or oxygen or electrons between molecules	Dehydrogenases, oxidases, oxygenases, peroxidases
2	Transferases	Transfer of groups of atoms from one molecule to another	Fructosyltransferases, transketolases, acyltransferases, transaminases
3	Hydrolases	Hydrolytic cleavage of bonds	Proteases, amylases, acylases, lipases, phosphatases, cutinases
4	Lyases	Non-hydrolytic cleavage by elimination or addition reactions	Pectate lyases, hydratases, dehydratases, decarboxylases, fumarase, argino succinase
5	Isomerases	Transfer of group from one position to another within one molecule	Isomerases, epimerases, racemases
6	Ligases	Covalent joining of two molecules coupled with the hydrolysis of an energy rich bond in ATP or similar triphosphates	Synthetases, ligases

Microorganisms are favored sources for industrial enzymes due to easy availability, and fast growth rate. Genetic changes using recombinant DNA technology can easily be done on microbial cells for elevated enzyme production and scientific development. Production of microbial enzymes is a necessary event in the industrial sectors, due to the high and superior performances of enzymes from different microbes, which work well under a wide range of varied physical and chemical conditions. Further, microbial enzymes are used in the treatment of health disorders associated with deficiency of human enzymes caused by genetic problems. For instance, patients with inherited congenital sucrase-isomaltase deficiency are unable to digest sucrose, and therefore, sacrosidase (β-fructofuranoside fructohydrolase, EC 3.2.1.26) enzyme is given orally to facilitate digestion of sucrose. In addition, phenylalanine ammonia lyase (EC 4.3.1.24) is used to degrade phenylalanine in genetic phenylketonuria disorder.

The extensive application of microbes in different bioprocess is used to deliver a variety of products in applied industries. Table summarizes several applications of microorganisms to deliver a variety of products. The Schematic representation of industrial production of microbial enzymes has been shown in Figure below.

Table: Industrial applications of microbial enzymes.

Industry	Enzyme	Function	Microorganisms
Dairy	Acid proteinase	Milk coagulation	Aspergillus sp.
	Neutral proteinase	Faster cheese ripening, debittering	Bacillus subtilis, A. oryzae
	Lipase	Faster cheese ripening, flavor customized cheese,	Aspergillus niger, A. oryzae

	Lactase (β-galacto-sidase)	Lactose reduced milk and whey products	Escherichia coli, Kluyveromyces sp.
	Aminopeptidase	Faster cheese ripening	Lactobacillus sp.
	catalase	Cheese processing	Aspergillus niger
	Transglutaminase	Protein cross linking	Streptomyces sp.
Baking	Amylase	Flour adjustment, bread softness	Aspergillus sp., Bacillus sp.
	Maltogenic α-Amylase	Enhance shelf life of breads	Bacillus stearothermophilus
	Xylanase	Dough conditioning	Aspergillus niger
	Lipase	Dough stability and conditioning	Aspergillus niger
	Glucose oxidase	Dough strengthening	Aspergillus niger, Penicillium chrysogenum
	Transglutaminase	Laminated dough strength	Streptoverticillium sp., streptomyces sp.
Beverage	Pectinase	Depectinization	Aspergillus oryzae, Penicillium funiculosum
	Glucose oxidase	Oxygen removal from beer	Aspergillus niger
	Cellulase	Fruit liquefaction	Aspergillus niger, Trichoderma atroviride
	α-Amylase	Starch hydrolysis	Bacillus, Aspergillus
	β-Amylase	Starch hydrolysis	Bacillus, Streptomyces, Rhizopus
	β-Glucanase	Restrict haze formation	Bacillus subtilis, Aspergillus spp.
	protease	Restrict haze formation	Aspergillus niger
	Pullulanase	Starch saccharification	Bacillus sp., Klebsiella sp.
	Naringinase	Debittering	Aspergillus niger
	limoninase	Debittering	Aspergillus niger, A. oryzae
	Aminopeptidases	Protein breakdown during mashing	Lactobacillus brevis, L. plantarum
Animal feed	Phytase	Hydrolyze phytic acid to release phosphorous	Aspergillus niger
	Xylanase	Enhanced digestibility of starch	Aspergillus sp., Bacillus sp.
	β-glucanase	Digestive aid	Aspergillus niger
Pulp and paper	Lipase	Pitch control	Candida Antarctica
	Protease	Biofilm removal	Bacillus subtilis
	Amylase	Deinking, drainage improvement	Bacillus licheniformis
	Xylanase	Bleach boosting	Trichoderma reesei, Thermomyces lanuginosus, Aureobasidium pullulans
	Laccase	Non-chlorine bleaching, delignification	Bacillus subtilis
	Cellulase	Deinking, drainage improvement	Bacillus sp., Aspergillus niger
Polymer	Lipase	Polycondensation, ring-opening polymerization of lactones, carbonates	Candida Antarctica
	Laccase	Polymerization of bisphenol A	Trametes hirsuta
	Glucose oxidase	Polymerization of anilines	Aspergillus niger, Penicillium chrysogenum
	Transglutaminase	Crosslinking of protein	Streptomyces mobaraensis
	Tyrosinase	Polymerization of lignin and chitosan	Trichoderma reesei
Detergent	Amylase	Carbohydrate stain removal	Aspergillus sp., Bacillus subtilis
	Lipase	Fat stain elimination	Aspergillus oryzae, A. flavus,

	Protease	Protein stain removal	Aspergillus oryzae, Bacillus subtilis
	Cellulase	Color clarification	Aspergillus niger, Bacillus sp.
	Cutinase	Triglyceride removal	Fusarium solani f. pisi
	Mannanase	Mannan spot removal	Bacillus sp.
Leather	Alkaline protease	Dehairing, bating	Alcaligenes faecalis
	Neutral Protease	Dehairing, soaking	Aspergillus niger, A. flavus, Bacillus subtilis
	Lipase	Degreasing	Aspergillus oryzae, A. flavus,
	Amylase	Fiber splitting	Aspergillus sp., Bacillus subtilis
Cosmetics	Superoxide dismutase	Free radical scavenging, skin care	Corynebacterium Glutamicum, Lactobacillus plantarum
	Protease	Removal of dead skin	Aspergillus niger, A. flavus, Bacillus subtilis
	Endoglycosidase	Teeth and gum tissue care	Mucor hiemalis
	laccase	Hair dye	Bacillus subtilis, Trametes versicolor
	lipase	Skin care	Aspergillus oryzae, A. flavus
Organic synthesis	Lipase	Synthesis of pharmaceuticals, polymers, biodiesels, biosurfactants	Aspergillus oryzae, A. flavus
	Glycosyl tranferase	Synthesis of oligosaccharides	Bacillus sp.
	Nitrile hydratase	Synthesis of acrylamide, butyramide, nicotinamide	Rhodococcus rhodochrous PA-34, Bacillus sp. APB-6
	Glucose isomerase	Production of High fructose corn syrup	Corynebacterium sp., streptomyces murinus
	Acyltransferase	Synthesis of hydroxamic acids	Bacillus sp. APB-6
	Laccase	Production of textile dyes, cosmetic pigments, flavor agents, and pesticides	Trametes versicolor, Bacillus subtilis
Waste management	Amidase	Degradation of nitriles containing wastes	Rhodococcus erythropolis
	Amylase	Bioremediation of vegetables wastes	B. licheniformis, Aspergillus sp.
	Amyloglucosidase	Starch hydrolysis for bioremediation	Aspergillus niger
	Lipase	Degradation of crude oil hydrocarbons	Aspergillus oryzae, Candida tropicalis
	Nitrile hydratase	Degradation of nitriles containing wastes	Rhodococcus sp.
	Protease	Bioremediation of keratinic wastes	Chrysosporium keratinophilum
	Laccase	Degradation of waste containing olefin unit, polyurethane and phenolic compounds	Trametes versicolor
	Cutinase	Degradation of plastics, Polycaprolactone	Fusarium solani f. pisi
	Manganese peroxidase	Degradation of phenolic compounds	Phanerochaete chrysosporium, Coprinus cinereus
	Lignin peroxidase	Degradation of phenolic compounds	Phanerochaete chrysosporium, Coprinus cinereus
	Oxygenase	Degradation of halogenated contaminants	Pseudomonas sp., Rhodococcus sp.

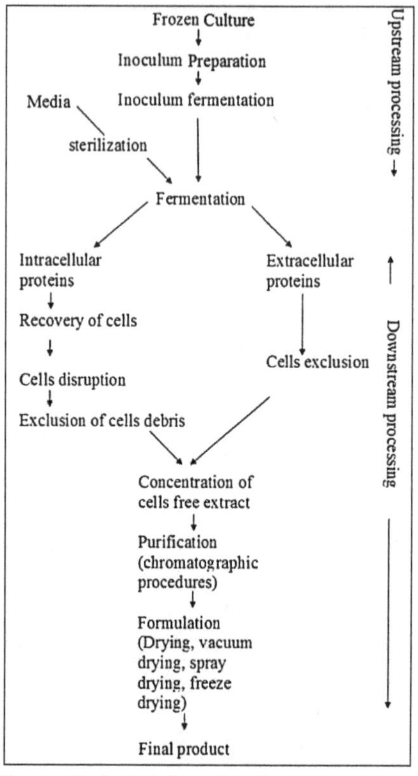

Figure: Industrial production of enzymes.

Enzymes, particularly of microbial origin, can be cultured largely by gene manipulations, as per the need for industrial applications. Applications of microbial enzymes in food, pharmaceutical, textile, paper, leather, and other industries are numerous and increasing rapidly over conventional methods due to less harm to the environment, greater efficiency, and the higher quality products.

Here the attentions are given to present a succinct role of all microbial enzymes involved in various fields of technical applications, such as feed industry, food processing and cosmetics. Furthermore, efforts are made to draw a simple and clear scenario about the industrial structure of global enzyme market.

The use of Enzymes in Industrial Processes

Enzymes are used in industrial processes, such as baking, brewing, detergents, fermented products, pharmaceuticals, textiles, leather processing. Here are a range of processes showing how enzymes are used.

Pharmaceutical and Analytical Industry

Enzymes have many significant and vital roles in the pharmaceutical and diagnostic industries. These are extensively used as therapeutic drugs in health issues associated with enzymatic deficiency and digestive disorders, and in diagnostic procedures such as ELISA and diabetes testing kits.

Enzyme applications in medicine are as extensive as in industry and are growing rapidly. At present, most prominent medical uses of microbial enzymes are removal of dead skin, and burns by proteolytic enzymes, and clot busting by fibrinolytic enzymes. Nattokinase (EC 3.4.21.62), a potent fibrinolytic enzyme, is a promising agent for thrombosis therapy. Acid protease, dextranase (EC 2.4.1.2) and rhodanase (EC 2.8.1.1) may be used to treat alimentary dyspepsia, tooth decay and cyanide poisoning, respectively. Lipases (EC 3.1.1.3) are the most frequently used enzymes in the organic synthesis and are used in the synthesis of optically active alcohols, acids, esters, and lactones. Microbial lipases and polyphenol oxidases (EC 1.10.3.2) are involved in the synthesis of (2R,3S)-3-(4-methoxyphenyl) methyl glycidate (an intermediate for diltiazem) and 3, 4-dihydroxylphenyl alanine (DOPA, for treatment of Parkinson's disease), respectively. Tyrosinase (EC 1.14.18.1), an important oxidase enzyme, is involved in melanogenesis and in the production of l-Dihydroxy phenylalanine (L-DOPA). L-DOPA is used as a precursor for the production of dopamine which is a potent drug for the treatment of Parkinson's disease and to control the myocardium neurogenic injury. Chitosanase (EC 3.2.1.132) catalyze hydrolysis of chitosan to biologically active chitosan oligosaccharides (COS), which is used as antimicrobial, antioxidant, lowering of blood cholesterol and high blood pressure, controlling arthritis, protective effects against infections and improving antitumor properties. Applications of microbial enzymes for different health problems are illustrated in Table below.

Table: Some therapeutic applications of microbial enzymes.

Treatment	Enzymes	Microorganisms
Antitumor	l-Asparaginase, l-glutaminase, l-tyrosinase, galactosidase	Escherichia coli, Pseudomonas acidovorans, Beauveria bassiana, Acinetobacter
Antiinflammatory	Superoxide dismutase, Serrapeptase	Lactobacillus plantarum, Nocardia sp., Mycobacterium sp., Corynebacterium Glutamicum
Anticoagulants	Streptokinase, urokinase	Streptococci sp., Bacillus subtilis
Antibiotic synthesis	Penicillin oxidase, rifamycin B oxidase	Penicillium sp.
Antioxidants	Superoxide dismutases, glutathione peroxidases, catalase	Lactobacillus plantarum, Corynebacterium glutamicum
Skin ulcers	Collagenase	Clostridium perfringens
Detoxification	Laccase, rhodanese	Pseudomonas aeruginosa
Antibiotic resistance	β-Lactamase	Klebsiella pneumonia, Citrobacter freundii, Serratia marcescens
Antiviral	Ribonuclease, Serrapeptase	Saccharomyces cerevisiae
Gout	Uricase	Aspergillus flavus
Digestive disorders	α-Amylase, lipase	Bacillus spp., Candida lipolytica, A. oryzae
Cyanide poisoning	Rhodanase	Sulfobacillus sibiricus

The extensive utilization of enzymes for scientific and analytical purposes is used to estimate the concentration of substrates and to determine the catalytic activity of enzymes present in biological samples. Advances in the enzyme technology have replaced or minimized the use to harmful radioactive elements in different immunoassays, which are used for the determination of a variety of proteins and hormones.

Furthermore, enzymes are used in clinical diagnostic for the quantitative determination of diabetes and other health disorders, for example, glucose oxidase (EC 1.1.3.4) for glucose; urease (EC 3.5.1.5) and glutamate dehydrogenase (EC 1.4.1.2) for urea; lipase, carboxyl esterase, and glycerol kinase for triglycerides; urate oxidase (EC 1.7.3.3) for uric acid; creatinase (EC 3.5.3.3) and sarcosine oxidases

(EC 1.5.3.1) for creatinine. Cholesterol oxidase (EC 1.1.3.6) has also been reported for useful biotechnological applications in the detection and conversion of cholesterol. Putrescine oxidase (EC 1.4.3.10) is used to detect biogenic amines, such as putrescine, a marker for food spoilage.

Enzymes are indispensable in nucleic acid manipulation for research and development in the field of genetic engineering, such as restriction endonucleases are used for site specific cleavage of DNA for molecular cloning and DNA polymerases for the DNA amplification by polymerase chain reaction (PCR).

Food Industry

Quality food supply issue can be addressed by the application of enzymes in the food industry. These biomolecules are efficiently involved in improving food production and components, such as flavor, aroma, color, texture, appearance and nutritive value. The profound understanding of the role of enzymes in the food manufacturing and ingredients industry have improved the basic processes to provide better markets with safer and higher quality products. Furthermore, the enzymes gained interest in new areas such as fat modification and sweetener technology. Currently, the food and beverage segment dominated the industrial enzyme market and it is projected to reach a value of $2.3 billion by 2020. In beverage and food industry, enzymes are added to control the brewing process and produce consistent, high-quality beer; to enhance the functional and nutritional properties of animal and vegetables proteins by the enzymatic hydrolysis of proteins, for higher juice yield with improved color and aroma.

The application of enzymes in food industry is segmented into different sectors, such as baking, dairy, juice production and brewing. Worldwide, microbial enzymes are efficiently utilized in bakery—the principal application market in food industry—to improve dough stability, crumb softness and structure, and shelf life of products. Increased uses of microbial enzymes in cheese processing are largely responsible for the use of enzymes in dairy industry, which is the next largest application industry followed by the beverages industry.

Baking Industry

Baking enzymes are used for providing flour enhancement, dough stability, improving texture, volume and color, prolonging crumb softness, uniform crumb structure and prolonging freshness of bread. To meet rising demand for quality, enzymes are seen as natural solutions in today's baking market. The baking enzymes industry is expected to reach $695.1 million by 2019 growing at a CAGR 8.2 % from 2013 to 2019.

Bread making is one of the most common food processing techniques globally. The use of enzymes in bread manufacturing shows their value in quality control and efficiency of production. Amylase, alone or in combination with other enzymes, is added to the bread flour for retaining the moisture more efficiently to increase softness, freshness and shelf life. Additionally, lipase and xylanase (EC 3.2.1.8) are used for dough stability and conditioning while glucose oxidase and lipoxygenase added to improve dough strengthening and whiteness. Transglutaminase (EC 2.3.2.13) is used in baking industry to enhance the quality of flour, the amount and texture of bread and the texture of cooked pasta. Lipases are also used to improve the flavor content of bakery products by liberating short-chain fatty acids through esterification and to prolong the shelf life of the bakery products.

Dairy Industry

Dairy enzymes, an important segment of food enzyme industry, are used for the development and enhancing organoleptic characteristics (aroma, flavor and color) and higher yield of milk products. The use of enzymes (proteases, lipases, esterases, lactase, aminopeptidase, lysozyme, lactoperoxidase, transglutaminase, catalase, etc.) in dairy market is well recognized and varies from coagulant to bio-protective enzyme to enhance the shelf life and safety of dairy products. Dairy enzymes are used for the production of cheese, yogurt and other milk products.

Rennet, a combination of chymosin and pepsin, is used for coagulation of milk into solid curds for cheese production and liquid whey. Currently, approximately 33 % of global demand of cheese produced using microbial rennet. Other proteases find applications for accelerated cheese processing and in reduction of allergenic properties of milk products. Currently, lipases are involved in flavor improvement, faster cheese preparation, production of customized milk products, and lipolysis of milk fat. Transglutaminase catalyzes polymerization of milk proteins and improves the functional properties of dairy products.

Lactose intolerance is the lack of ability of human being to digest lactose due to deficiency of lactase enzyme. Lactase (β-galactosidase, EC 3.2.1.23) catalyzes hydrolysis of lactose to glucose and galactose, and therefore, is used as a digestive aid and to enhance the solubility and sweetness in milk products. It is required to minimize or removal of lactose content of milk products for lactose-intolerant people to prevent severe tissue dehydration, diarrhea, and sometimes fatal consequences.

Beverages Industry

The beverage industry is divided into two major groups and eight sub-groups. The nonalcoholic group contains soft drink and syrup, packaged water, fruit juices along with tea and coffee industry. Alcoholic group comprised distilled spirits, wine and beer. Industrial enzymes are used in breweries as processing aids and to produce consistent and high-quality products. In the brewing industries, microbial enzymes are used to digest cell wall during extraction of plant material to provide improved yield, color, and aroma and clearer products.

The enzyme applications are an integrating ingredient of the current fruit and vegetable juice industry. Enzymes are used in fruit and vegetable juice industry as processing aids to increase the efficiency of operation, for instance, peeling, juicing, clarification, extraction and improve the product quality. Application of cellulases, amylases, and pectinases during fruit juice processing for maceration, liquefaction, and clarification, improve yield and cost effectiveness. The quality and stability of juices manufactured are enhanced by the addition of enzymes. Enzymes digest pectin, starch, proteins and cellulose of fruits and vegetables and facilitate improved yields, shortening of processing time and enhancing sensory characteristics. Amylases are used for clarification of juices to maximize the production of clear or cloudy juice. Cellulases and pectinases are used to improve extraction, yield, cloud stability and texture in juices. Naringinase (EC 3.2.1.40) and liminonase, debittering enzymes, hydrolyze bitter components and improves the quality attributes of citrus juices. Pectin, a structural heteropolysaccharide, present in nearly all fruits is required to be maintained to regulate cloudiness of juices by polygalacturonase (EC 3.2.1.15), pectin esterases (EC 3.1.1.11), pectin lyase (EC 4.2.2.10) and various arabanases.

Microbial amylases may be utilized in the distilled alcoholic beverages to hydrolyze starch to sugars prior to fermentation and to minimize or remove turbidities due to starch. The application of enzymes to hydrolyze unmalted barley and other starchy adjuncts facilitate in cost reduction of beer brewing. In brewing, development of chill-hazes in beer may be control by the addition of proteases.

Feed Industry

To meet the continuously increasing worldwide demand of milk and meat consumption, growth of feed enzymes occurred steadily. The use of enzymes in animal diets initiated in the 1980s and exploded in the 1990s. Feed enzymes are gaining importance as they can increase the digestibility of nutrients and higher feed utilization by animals. The global market for feed enzymes was estimated $899.19 million in 2014 and expected to reach nearly $1.3 billion by 2020, at a CAGR of 7.3 % from 2015 to 2020.

Feed enzymes may be used in animal diet formulation. For instance, these are added to degrade specific feed components which are otherwise harmful or no nutritional value. In addition, the protein dietary value of feeds available for poultry may also be enhanced by the application of feed enzymes. Feed enzymes mainly used for poultry are phytases, proteases, α-galactosidases, glucanases, xylanases, α-amylases, and polygalacturonases. The phytase, largest enzyme segment in the feed industry, is used to utilize natural phosphorous bound in phytic acid in cereal-based feed. Monogastric animals are unable to digest plant based feeds containing high amount of cellulose and hemicelluloses. Xylanase and β-glucanase are added to their feeds as these enzymes fully degrade and digest high amount of starch. Proteases are also used in animal feeds to overcome anti-nutritional factors by degrading proteins into their constituent amino acids. Apart from improving the nutritional value of feed for better feed conversion by the animals, these feed enzymes are gaining importance for their role in feed cost reduction and meat quality improvement.

Polymer Industry

To meet the increased consumption of polymers and the growing concern for human health and environmental safety has led to the utilization of microbial enzymes for synthesis of biodegradable polymer. In vitro enzyme catalyzed synthesis of polymer is an environmental safe process having several advantages over conventional chemical methods. Biopolymers are environmentally friendly materials as these are synthesized from renewable carbon sources via biological processes, degrade biologically after use and return to the natural environment as renewable resources, such as CO_2 and biomass. Biopolymers, such as polyesters, polycarbonates and polyphosphates are used in various biomedical applications, e.g., orthopedic devices, tissue engineering, adhesion barriers, control drug delivery, etc.

The biopolymers market is growing at a CAGR of 14.5 % due to high penetration of materials in industries like medical, packaging, appliances, automotive, electronics, and furniture and the market is expected to reach nearly $3.6 billion by 2018. Increasing demand of packaging materials and environmental safety can be addressed by the biodegradable polymers. Lipases catalyze the polymerization of lactones, cyclic diesters and cyclic carbonates to produce polyesters or polycarbonates. Lipase catalyzed polymerization is an eco-friendly technique for the preparation of useful polyesters by polycondensation as well as poly-addition reactions. The other enzymes involved in biopolymer industries are laccase (EC 1.10.3.2), peroxidase and transglutaminase for forming cross-links in biopolymers to produce materials in situ by means of polymerization processes.

Paper and Pulp Industry

With increasing awareness of sustainability issues, uses of microbial enzymes in paper and pulp industry have grown steadily to reduce adverse effect on ecosystem. The utilization of enzymes reduce processing time, energy consumption and amount of chemicals in processing. Enzymes are also used to enhance deinking, and bleach in paper and pulp industry and waste treatment by increasing biological oxygen demand (BOD) and chemical oxygen demand (COD). Xylanases and ligninases are used in paper and pulp industries to augment the value of the pulp by removing lignin and hemicelluloses. In these industries, amylases uses include starch coating, deinking, improving paper cleanliness and drainage improvement. Lipases are employed for deinking and enhancing pitch control while cellulases are used for deinking, improving softness and drainage improvement. Cellulase has also been used for the development of the bioprocess for recycling of used printed papers. The application of laccase is an alternative to usage and requirement of large amount of chlorine in chemical pulping process; subsequently, reduce the waste quantity that causes ozone depletion and acidification. Moreover, mannases are used for degrading glucomannan to improve brightness in paper industry.

Leather Industry

The leather industry is more customary, and therefore, discharges and waste disposed from different stages of leather processing are causing severe health hazards and environmental problems. The biodegradable enzymes are efficient alternative to improve the quality of leather and help to shrink waste. The initial attempt for application of enzyme in leather industry was made for dehairing process, the largest process in leather preparation which require bulk amount of enzymes like proteases, lipases and amylases. Enzymatic dehairing applications are attractive because it can preserve the hair and contribute to fall in the organic load released into the effluent. Enzymatic dehairing processes minimize or eliminate the dependence on harmful chemicals, such as sulfide, lime and amines.

Enzymes are required for facilitating procedure and enhancing leather quality during different stages in leather processing, such as, curing, soaking, liming, dehairing, bating, picking, degreasing and tanning. The enzymes used in leather industries are alkaline proteases, neutral proteases, and lipases. Alkaline proteases are used to remove non fibrillar proteins during soaking, in bating to make leather soft, supple and pliable. Neutral and alkaline proteases, both are used in dehairing to reduce water wastage. In addition to this, lipases are used during degreasing to remove fats. The advantages of using enzymes instead of chemicals in liming are stainless pelt, reduced odor, low BOD and COD in effluents, and improved hair recovery.

Textile Industry

The textile industry is responsible for vast generation of waste from desizing of fabrics, bleaching chemicals and dye is one of the largest contributors to environmental pollution. In such industries, enzymes are used to allow the development of environmentally friendly technologies in fiber processing and strategies to improve the final product quality. The main classes of enzymes involved in cotton pre-treatment and finishing processes are hydrolase and oxidoreductase. The group of hydrolase includes amylase, cellulase, cutinase, protease, pectinase and lipase/esterase, which are involved in the bio-polishing and bio-scouring of fabric, anti-felting of wool, cotton softening, denim finishing, de-sizing, wool finishing, modification of synthetic fibers, etc. Oxidoreductase, other group of enzyme, includes catalase, laccase, peroxidase, and ligninase, which are involved in bio-bleaching, bleach termination, dye de-colorization, fabric, wool finishing, etc. A brief detail of applications of enzymes in textiles industries are shown in Table below.

Table: Uses of enzymes in textile industry.

Enzyme	Use	Microorganisms
Amylase	Desizing	Bacillus sp., B. licheniformis
Cellulose	Cotton softening, denim finishing	Aspergillus niger, Penicillium funiculosum
Catalase	Bleach termination	Aspergillus sp.
Laccase	Non-chlorine Bleaching, fabric dyeing	Bacillus subtilis
Pectate lyase	Bioscouring	Bacillus sp., Pseudomonas sp.
Amylase	Desizing	Bacillus sp., B. licheniformis
Cellulose	Cotton softening, denim finishing	Aspergillus niger, Penicillium funiculosum
Protease	Removal of wool fiber scales, degumming of silk	Aspergillus niger, B. subtilis
Lipase	Removal of size lubricants, denim finishing,	Candida Antarctica
Ligninase	Wool finishing	Trametes versicolor, Phlebia radiata
Collagenase	Wool finishing	Clostridium histolyticum
Cutinase	Cotton scouring, synthetic fiber modification	Pseudomonas mendocina, Fusarium solani pisi, Thermomonospora fusca

Enzymes in Cosmetics

The applications of enzymes in cosmetics have been continuously increased. Enzymes are used as free radical scavengers in sunscreen cream, toothpaste, mouthwashes, hair waving and dyeing. The superoxide dismutase (SOD, EC 1.15.1.1) is used to arrest free radicals and to control damage to skin caused by air and water pollutions, microbes and other harmful factors. SOD and peroxidases are used in combination in sunscreen cream as free radical scavengers to reduce erythema. Proteases are used in skin creams to clean and smoothen the skin through peeling off dead or damaged skin.

Other widely used enzymes in toothpaste and mouthwash are endoglycosidase and papain, which are used to whiten teeth, to remove plaque and to remove odor-causing deposits on teeth and gum tissue. Laccase, oxidases, peroxidases, and polyphenol oxidases are used in hair dyeing, lipase, catalase, papain, bromelain and subtilisin in skin care; and protein disulfide isomerase, glutathione sulfhydryl oxidase and transglutaminase in hair waving. Additionally, enzymes are also used in contact lens cleaners to remove protein films.

Enzymes in Detergents

Detergents represent the largest industrial application of enzymes amounting to 25–30 % of the total sales of enzymes and expected to grow faster at a CAGR of about 11.5 % from 2015 to 2020. Enzymes have contributed significantly to the growth and development of industrial detergents, which is a prime application area for enzymes today. Detergents are used in miscellaneous applications as dishwashing, laundering, domestic, industrial and institutional cleaning. The enzymes in detergent products are used to remove protein, starch, oil and fats based stains and to increase the effectiveness of detergents. The enzymes in laundry detergents are weight efficient, cleave off damaged cotton fibers, improve whiteness, color and fabric care. Enzymes mainly used in detergent products are of hydrolase group and currently, most commonly used enzymes are amylase and protease. Sometimes a combination of enzymes, including proteases, amylases, pectinases, cellulases and lipases used to increase efficiency on stain cleaning and fabric care.

Amylases and lipases are effective on removing starchy food deposits and stains resulting from fatty products, respectively. Cutinase (EC 3.1.1.74), a hydrolytic enzyme, is used as a lipolytic enzyme in dishwashing and laundry detergents. Protease digests on organic stains, such as grass, blood, egg and human sweat, whereas cellulases are used to brighten colors, soften fabrics and to eliminate small fibers from the fabric without damaging the major fibers of the fabric. Protease and amylase are used particularly in dishwasher detergents to remove protein and carbohydrate containing food particles. The application of enzymes in detergent products is advantageous as these products contain less bleaching agents, phosphates, and consequently have beneficial effects on public and environmental health.

Organic Synthesis Industry

Enzyme based processes for production of fine chemicals are rapidly gaining practical significance owing to more economical high purity products in an eco-environmentally acceptable manner. Enzymes are preferred in industrial chemical synthesis over conventional methods for their high selectivity, i.e., chiral, positional and functional group specific. Such high selectivity is extremely advantageous in chemical synthesis as it may offer several benefits such as minimal or no by-product formation, easier separation, and less environmental problems. Besides, high catalytic efficiency and mild operational conditions are advantages of enzyme mediated commercial applications. Catalytic potential of microorganisms have been employed for hundreds of years in the production of alcohol, and cheese for industrial synthetic chemistry. Among the enzymes in organic synthesis, lipases are the most frequently used, particularly, in the formation of a wide range of optically active alcohols, acids, esters, and lactones. Lipases are used for the production of (S, R)-2, 3-*p*-ethoxyphenylglycyclic acid, an intermediate for diltiazem. Oxidoreductases, such as polyphenol oxidase is involved in the synthesis of 3,4-dihydroxylphenyl alanine (DOPA), a chemical used in the treatment of Parkinson's disease. Oligosaccharides and polysaccharides, play vital roles in cellular recognition and communication processes, are synthesized industrially using high regio- and stereoselectivity of glycosyltransferases. Lyases are involved in organic synthesis of cyanohydrins from ketones, acrylamide from acrylonitrile, malic acid from fumaric acid. The nitrile hydratase mediated process for the production of acrylamide is carried out by the Nitto Chemical Company of Japan at a scale of more than 40,000 tons per year. A multi-million ton of high fructose corn syrup (HFCS), an alternative sweetener to sucrose in the food and beverage industry, is produced every year commercially using glucose isomerase.

Waste Treatment

The use of enzyme for waste management is extensive and a number of enzymes are involved in the degradation of toxic pollutants. The industrial effluents as well as domestic waste contain many chemical commodities, which are hazardous or toxic to the living being and ecosystem. Microbial enzyme(s), alone or in combinations, are used for the treatment of industrial effluents containing phenols, aromatic amines, nitriles, etc., by degradation or bioconversion of toxic chemical compound(s) to innocuous products. A number of enzymes employed for waste treatment are amidases, amylases, amyloglucosidases, cellulases, glucoamylases, lipases, nitrile hydratases, pectinases and proteases. The detoxification of toxic organic compounds through oxidative coupling is mediated with oxidoreductases. These enzymes, like laccase, manganese peroxidase, lignin peroxidase and tyrosinase catalyze the removal of chlorinated phenolic compounds from industrial effluents. The microbial enzymes are also utilized to recycle the waste for reuse, e.g., to recover additional oil from oil seeds, to convert starch to sugar, to convert whey to various useful products. Microbial oxygenases, such as monooxygenases and dioxygenases have a broad substrate range, and are active against a wide range of compounds, including the chlorinated aliphatics. These are used in the degradation of halogenated organic compounds containing pollutants, like herbicides, insecticides, fungicides, hydraulic and heat transfer fluids, plasticizers, and intermediates for chemical synthesis.

Enzyme Purification

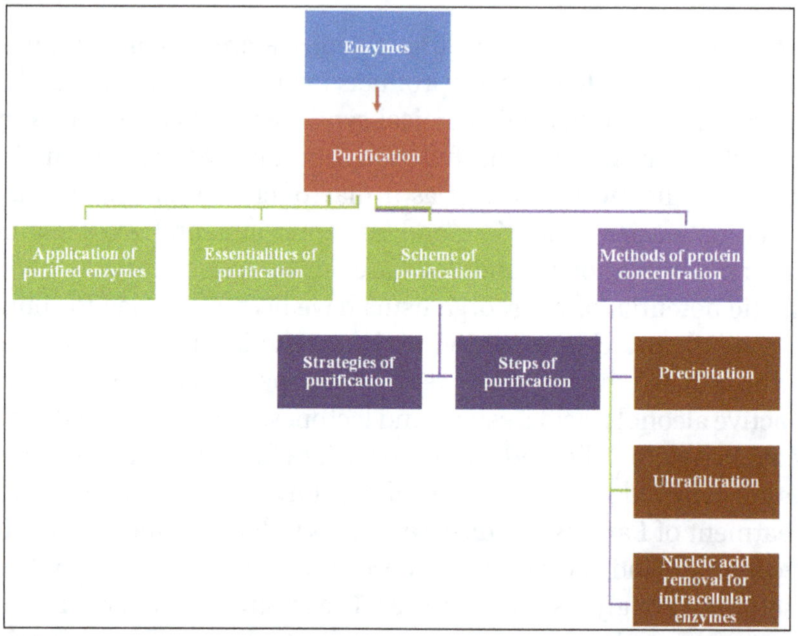

Protein purification is an essential component for the characterization of protein properties, structure, functions and related interactions, for example conformational alterations, substrate specificities, specific activities as well as interaction with other ligands. The study of proteins and their function is imperative from the perspective of understanding of both cells and organisms. The importance of proteins in emphasized by their role as biocatalysts for cellular metabolic processes, as integral cellular structural components both within and outside the cell, as cellular receptors for conveying relevant and necessary information to the cell, as intracellular signaling components or as components of the genetic machinery of the cell. Thus, it is often reasonable to purify an isolated protein so that one is able to study the protein in isolation. Additionally, purified proteins also find applications in various areas as shown in Figure below.

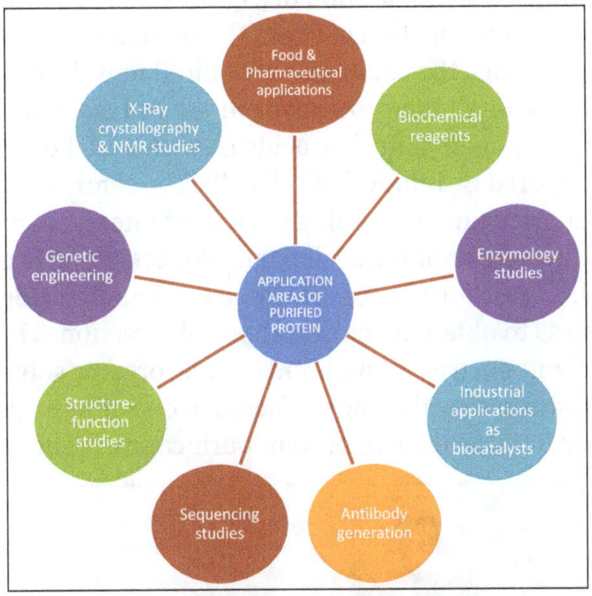

Figure: Areas where pure protein is required.

The degree of protein purity required depends to a great extent on the end use of the protein. For example, in some applications such as an enzyme additive in washing powder, a relatively impure sample is sufficient. However, for other applications, for example in the food and pharmaceutical industry, a high level of protein purity is essential and the desired protein must be purified over a number of steps. This is thus achieved through several protein purification methods, carried out in a sequence of purification steps.

Purification procedure is not restricted to any specific number of steps or stages. Each step is targeted to bringing about some fold of purification in the protein to be purified. This is also accompanied by some extent of product loss most of the times. An ideal protein purification strategy would be the one which leads to the highest level of purification in fewer numbers of steps and also simultaneously ensures that there is minimum product loss. The selection of steps in the protein purification protocol depends largely on the properties of the target protein such as size, charge, solubility etc.

Essentialities of Purification

- Sources: One of the key essentialities of protein purification is that there must exist a reasonable source from which the protein could be derived. The source should be both cheap and readily available.
- Assays: An assay is essential for quantitative determination of a particular activity present in the protein within a particular fraction. Assays can be diverse ranging from colorimetric assays (change in absorbance) or a change in molecular weight of a protein or ability to stimulate cell differentiation, cell proliferation or cell death. Since an assay is required to be repeated many times, it is essential that the assay procedure is simple and not much time consuming.

Scheme of Protein Purification

The first step in purifying proteins is preparation of a crude extract. Crude preparations of extracellular proteins are obtained by centrifugation wherein the cellular biomass is removed. In case of intracellular (inside the cell) proteins, the cells will be required to be lysed and the lysate so obtained will contain a complex mixture of all the proteins from the cell cytoplasm, additional macromolecules and internal cellular components, nucleic acids as well as cell debris etc. Intracellular crude protein extracts are then prepared by removal of cell debris by centrifugation and removal of nucleic acid, following which a supernatant containing protein of interest is recovered. The soluble protein containing fraction (extracellular or intracellular) is concentrated to obtain an enriched protein fraction. The protein extract can be further concentrated before subjecting it to any chromatographic purification procedure so as to obtain an enriched protein fraction. This can be achieved by adding various types of precipitants such as ammonium sulfate, organic solvents or polyethylene glycol (PEG). The partially pure fraction may then be subjected to one or more purification strategies until purification is achieved. The basic scheme of protein purification is shown in figure below.

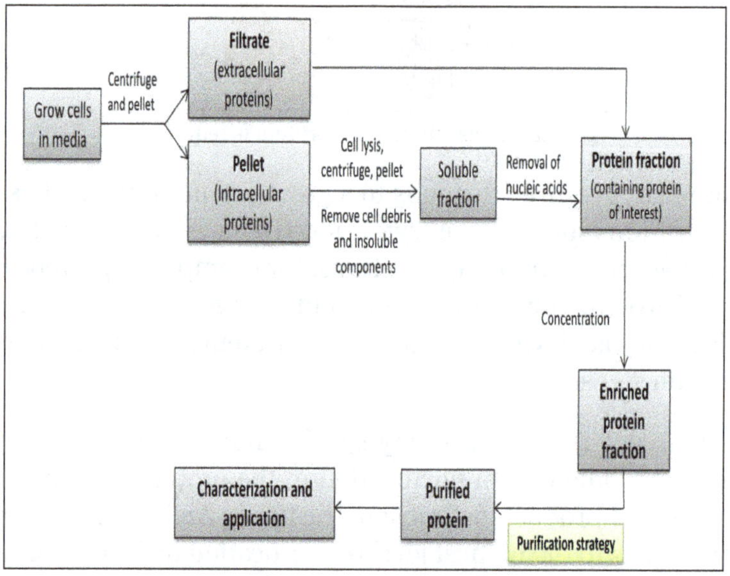

Figure: Scheme of protein purification.

Purification Strategy

To enable simple planning and execution of protein/enzyme purification protocols, a three step protein purification strategy has been developed. This is known as the CIPP strategy-Capture, Intermediate Purification and Polishing. This strategy is used widely both on laboratory and industrial scale to ensure faster process, shorter time required and a good economy. Together, this enables a cost effective and robust production of target production.

- The Capture Phase- In this phase, isolation, concentration and stabilization of the target protein is achieved.
- The Intermediate Purification Phase-In this phase, bulk impurities like removal of other unwanted proteins, nucleic acids, endotoxins which are closely related to the protein are removed.

- The Polishing Phase-In this phase, most of the major impurities have already been removed. This stage removes any trace amounts of specified impurities or any wanted forms of the target protein that may have been formed during purification.

The basic objectives of each phase in the CIPP purification strategy is shown in Figure below.

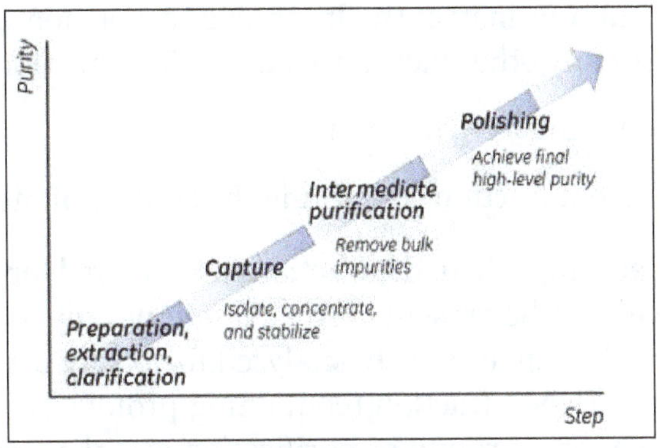

Figure: The CIPP strategy.

Although this strategy consists of three broad phases, it does not imply that every purification process will contain three steps. Each phase may include one or more purification steps wherein each step may be used to address a specific objective within a purification process. The number of steps will primarily depend on the intended use of the protein. It should however be kept in mind that as the number of purification steps are increased, the overall protein recovery and activity will decrease, and purification time and cost involved will go up. Therefore, efforts should be made to optimize a purification strategy that gives best results with maximum yield of the purified protein and fold purification.

A multistep Procedure

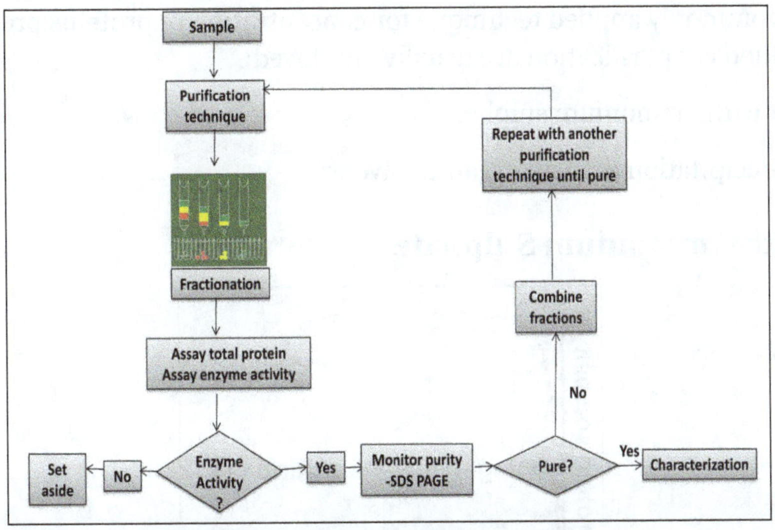

Figure: A multistep purification procedure.

Purification is a multistep procedure. The various steps in the purification process is aimed removing non-protein components and impurities from the protein of interest, freeing the protein from the matrix that confines it, and finally separating out the desired protein in its pure form. Figure gives outlines the various steps in a purification process.

The soluble fraction containing target protein is concentrated by precipitation with salt, following which, it is subjected to fractionation by passing them through chromatographic columns for further purification. This could include separation based on charge on protein surface (Ion-exchange chromatography), separation based on binding to affinity columns (affinity chromatography), on size of protein molecule (gel filtration chromatography), separation based on hydrophobic interactions of protein with the column matrix (hydrophobic interaction chromatography), fractionation by isoelectric focusing or any other method available. The fractionation serves two purposes.

- Removal of added contaminating material.
- Enrichment of the protein fraction containing the activity of interest.

Once such a fractionation is accomplished, the fractions are assayed for activity to detect the activity of interest. This activity will only be present in some fractions only, which may contain the target proteins. These fractions are then qualitatively analyzed for purity using methods like SDS Page. If not completely pure, this enriched fraction (containing protein of interest but not completely pure) is further put through another round of purification and the process can be repeated till a pure protein is achieved.

Concentration of Proteins Prior to Purification

Concentration of the soluble fraction containing desired protein to be purified can be achieved through:

- Precipitation
- Ultrafiltration

Precipitation

Precipitation is a commonly applied technique for concentration of proteins present in an extract. The following methods of purification are usually employed:

- Salting out with ammonium sulphate.
- Selective precipitation with an organic solvent.

Salting Out with Ammonium Sulphate

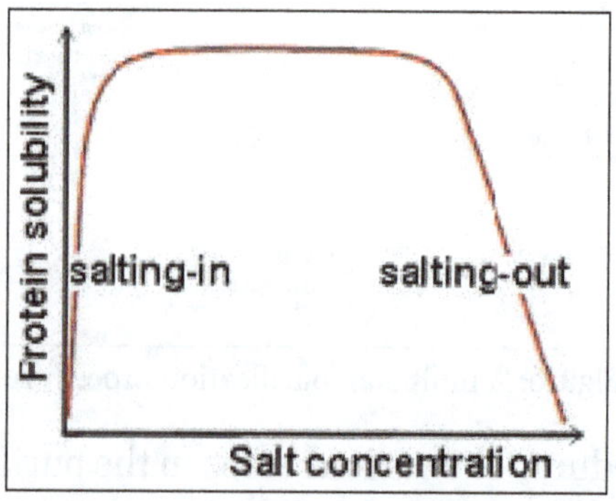

Figure: Salting in and salting out.

Salting out is an effective means for purification which explores the reduced solubility of proteins present in a solution of very high ionic strength causing certain proteins to precipitate. Figure below shows salting in and salting out processes.

Each protein molecule in solution is uniformly layered by an essential "layer of hydration" by water molecules which enable the molecule to repel each other and stay in solution. As more and more salt is added to the protein, the solubility of the salt added gradually tends to become higher than protein. Owing to the increased affinity of salt molecules for water over protein molecules, the hydration shell around the protein molecule is thus gradually displaced by the increasing ionic concentration in the solvent. In other words, the protein molecules are thus "stripped off" their hydration layer, allowing hydrophobic interaction between proteins (interactions between hydrophobic patches on protein surface) to predominate which leads to aggregation of the protein molecules and precipitation. It is important to note that salting out occurs at high salt concentration. Use of salt at very high concentrations also cause a further increase in surface tension, inducing the protein to aggregate, resulting in salt-precipitation.

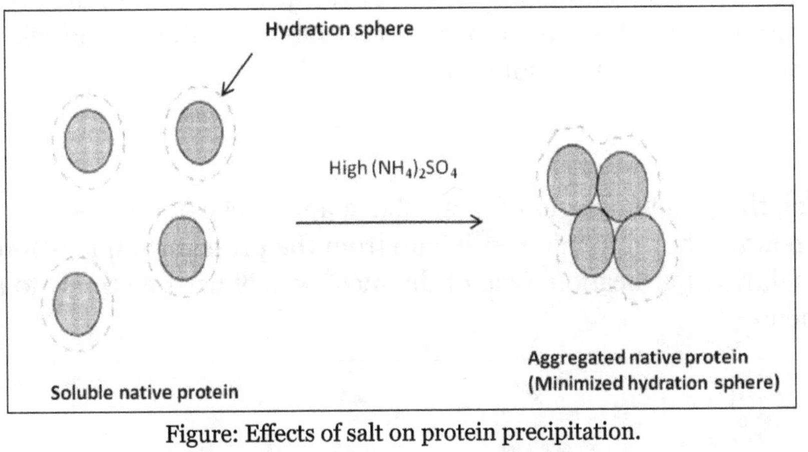

Figure: Effects of salt on protein precipitation.

Precipitation of proteins by salt does not usually lead to a highly purified protein, but it can assist in removal of some unwanted bulk proteins in a mixture and also in concentrating the sample. Ammonium sulphate is convenient and effective because it is highly soluble, cheap, less toxics and stabilizes most proteins/enzymes. Fractionation of protein mixtures by the stepwise increase in the ionic strength of the salt being used for protein precipitation can prove to be an effective strategy of obtaining partially purified enzymes.

Hofmeister Series

The effectiveness of the different ions towards protein precipitation was established by Franz Hofmeister in 1888 and the ordering of cations and anions arranged in order of their effectiveness is called Hofmeister series.

Cations: $N(CH_3)_3^+ > NH_4^+ > K^+ > Li^+ > Mg^{2+} > Ca^{2+} > Al^{3+} >$ guanidinium

Anions: $SO_4^{2-} > HPO_4^{2-} > CH_3COO^- >$ citrate > tartrate $> F^- > Cl^- > Br^- > I^- > NO_3^- > ClO_4^- >$ SCN

Between cations and anions, the anions have the greatest effect on protein precipitation.

The starting molecules decreasing solubility of the non-polar molecules and strengthen hydrophobic interactions, thus salting out the system. Contrarily, the latter molecules tend to form strong ionic interactions with the protein that disrupt hydrogen bonding, thus contributing to the denaturation of the protein.

Precipitation by Organic Solvents

Sometimes polar organic solvents such as methanol, ethanol, propan-2-ol and acetone are used for precipitating some enzymes/proteins.

The mechanism of precipitation is almost similar to that of salting out, as explained earlier. Use of solvents such as ethanol or methanol (which are miscible in nature) to a solution may cause proteins to precipitate. This is because the organic solvent molecules gradually displaces water from the protein surface thus reducing the solvation layer around the protein and binding the water molecules, so displaced, as hydration layers around the organic solvent molecules. The protein molecules, with reduced/no hydration layers thus tend to aggregate due to predominating protein-protein hydrophobic interactions as well as electrostatic forces. Miscible organic solvents also act by reducing the dielectric of the medium and consequently reducing the solubility of proteins, leading them to precipitate out of the solution.

Dialysis

After precipitating the protein fraction (using above approaches) and re-dissolving it in buffer, it is essential to remove the ammonium sulphate from the protein sample before subjecting it to subsequent steps during purification. One of the most widely used methods to achieve this is to dialyse the solution.

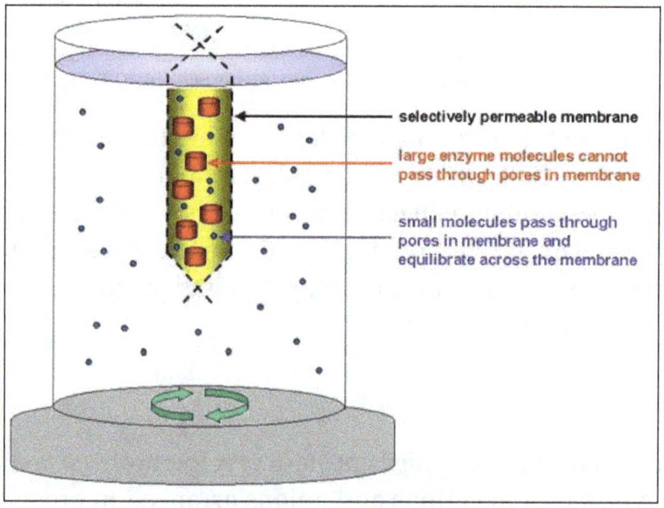

Figure: Dialysis of proteins.

Dialysis leads to the separation of protein molecules from other small molecules, such as salt, by using a semi-permeable membrane (for e.g., cellulose membrane) with pores. The principle behing dialysis is usually diffusion, which is explained further. This membrane contains micro pores through which smaller molecules and ions (present along with the protein sample) move out (from region of higher concentration) and emerge in the dialysate outside the bag (to their region of lower concentration), across a concentration gradient, till equilibrium is achieved. To balance this, water/buffer molecules traverse into the dialysis bag, across a concentration gradient. However, the protein molecules which have sizes that are significantly greater than the membrane pore diameter, therefore cannot traverse out of the dialysis tubing and remain retained inside the dialysis bag. Figure shows a typical set up for protein dialysis.

As shown in Figure, the protein/enzyme solution is placed in a dialysis bag and immersed in a large volume of buffer that is stirred and maintained at about 4°C. During dialysis, the salt molecules will tend to diffuse out of the dialysis bag. If the buffer is changed several times, more or less all the salt will be removed from the protein solution. It is important to note that dialysis will result in an increase the volume of the enzyme solution, because water molecules from the buffer enter into the bag. It is therefore necessary to leave some space/gap at the top of the membrane tube as shown in Figure, to prevent it from bursting.

Ultrafiltration

Ultrafiltration is another method for protein concentration in which water and low molecular weight materials are removed from the protein by passing it through a membrane in an inert atmosphere and under pressure. The protein is retained and does not pass through the membrane, leading to its concentration with little loss of enzyme activity. Membranes with molecular weight cut-offs from 1000 to 100,000 kDa are available which can be used under pressure up to 2 MPa.

Figure represents the simplest ultrafiltration cell. The protein solution is contained in a cylindrical stirred cell where the membrane is placed at the base of a cell and the cell is equipped with a magnetic stirrer. An inert gas, for example, nitrogen is passed continuously through the stirrer which can force the water and low molecular weight materials through the membrane. Ultrafiltraion procedures for concentration of laboratory column eluates are commonly used, however their large scale use is not much explored.

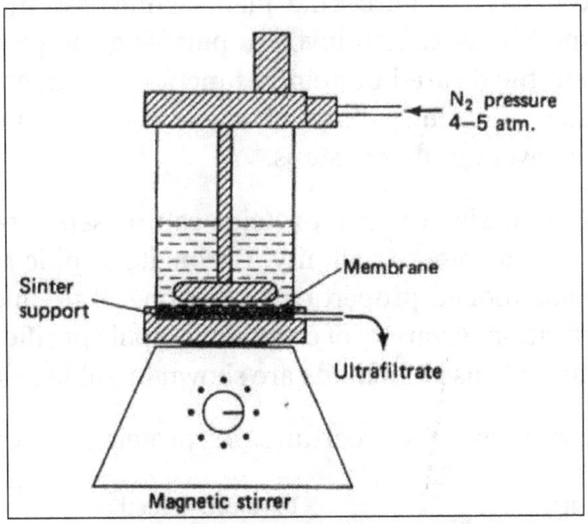

Figure: A stirred ultrafiltration cell.

Nucleic Acid Removal

The intracellular enzymes obtained after cell lysis often contain nucleic acids which give rise to increased viscosity and interfere with enzyme purification procedures. Thus removal of such nucleic acids before proceeding to next purification steps becomes pertinent.

Nucleic acids can be removed by precipitation or degraded by externally added nucleases. In some cases, sufficient nuclease activity is priorly present in the lysed cell extract which can naturally degrade the nucleic acids present.

Positively-charged materials like polyethyleneimine, cetyltrimethyl ammonium bromide (a cationic detergent), streptomycin sulphate and protamine sulphate, which form complexes with the negatively-charged phosphate residues of the nucleic acids are used as precipitating agents for removal of nucleic acids. The nucleic acids complex with these compounds and the precipitates can be removed via centrifugation.

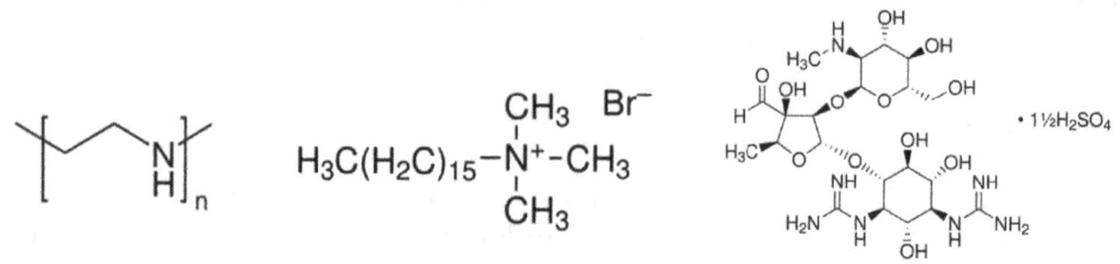

Polyethyleneimine

Cetyltrimethyl ammonium bromide

Streptomycin sulphate

Figure: Nucleic acid precipitants.

However, all of the above precipitants are expensive and sometimes toxic. Treatment with bovine pancreatic nucleases is another cost-effective and safer alternative for nucleic acid removal.

Protein Separation

Proteins can be obtained from natural sources like plants, animals or microbes or by their over-expression in a host organism. The next step involves purifying the protein to remove associated contaminants so as to obtain the desired protein in functional form and in substantial amounts. The purification scheme must be optimized in such a way that maximum yield of pure enzyme is achieved in less time and in fewer number of steps.

The various properties on the basis of which proteins can be separated include solubility, size, charge, hydrophobicity as well as binding affinity. Chromatographic separation of protein mixtures, based on these aforementioned properties, are among of the most widely and commonly used means of protein purification. A variety of chromatographic purification techniques are available. Some of the most commonly used methods are shown in Table below.

Table: Commonly used chromatographic procedures for protein purification.

Property	Purification method
Solubility	Precipitation with ammonium sulphate (salting out)
Size/shape	Size exclusion (gel filtration) chromatography
Charge	Ion-exchange chromatography
Binding affinity	Affinity chromatography
Hydrophobicity	Hydrophobic interaction chromatography

Column Chromatographic Procedures for Protein Purification

Column chromatography is based on the principle of separation of a one/more proteins from a protein mixture by passing the mixture through a column containing an appropriate matrix, to which the protein of interest may bind and can be later eluted out from the column using an eluant. The basic equipment required for chromatography is discussed further, although specialized and expensive equipments are also available.

Basic Set up for Column Chromatography

A schematic representation of protein purification set up is shown in Figure below.

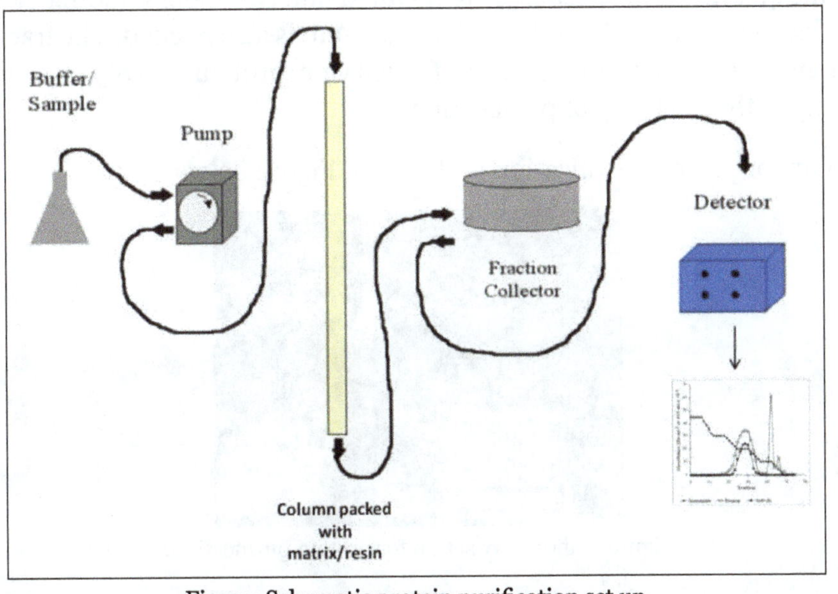

Figure: Schematic protein purification set up.

The basic components required for column chromatography are as follows:

- Column: A columnar glass reservoir (of various lengths and diameters) is cleaned and pre-packed with a matrix/resin (stationary phase) and attached to chromatographic set up.
- Stationary phase: The stationary phase usually consists of an inert matrix or a resin which can be directly packed into the column or pre-swelled/pre-equilibriated before column packing.
- Mobile phase/solvents: The mobile phase is constituted by various buffers which are used for column washing, equilibration, and elution of bound protein from the stationary phase. Different chromatographic methods require different mobile phases.
- Protein Sample: The protein sample to be purified is an essential component which should be present in sufficient concentration and loaded onto the chromatographic column.
- A pump: The chromatographic set up is usually connected to a peristaltic pump with variable and adjustable flow rate to push buffer through the column and prevent low pressure conditions or creation of air bubbles which may disrupt the column packing and affect purification process.

- Fraction collector: This is necessary for collecting the fractions as and when they are eluted out of the column. The fraction collector is sometimes automated to collect a specific volume of fraction based on number of drops or on time, after which, the fraction shifts to the next empty collecting tube placed in the collector.
- A detector: The fraction collector is usually connected to a detector which records protein absorbance at 280 nm. The absorbance of these individual fractions can either be read on a UV spectrophotometer or on a chart recorder, which is often directly connected to the detector.
- Standard assay to measure protein content and activity: After the proteins are detected in relevant fractions eluted from the column, there should be a standard assay to determine the specific quantity of a specific protein eluted out. An understanding of the total protein that was loaded and eluted out through the column is also needed. The fractions containing the protein of interest can be assayed for enzyme/protein activity after each step before proceeding to the next step of purification.

A laboratory set up for protein purification is shown in Figure below.

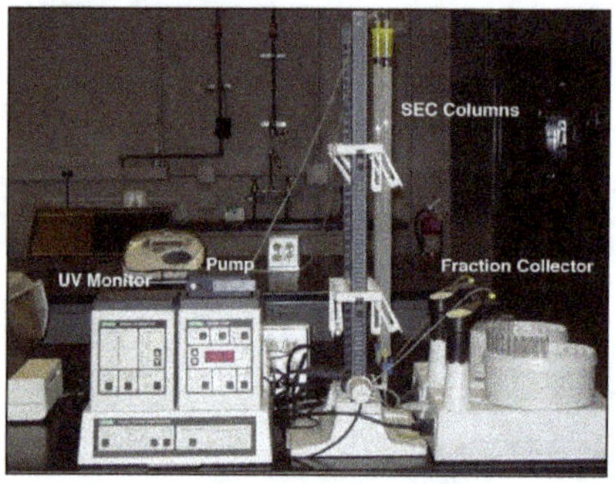

Figure: Laboratory set up for protein purification.

Systems for Column Chromatography

Column chromatography can be performed with automated systems (wherein a pump is used to pass solvent over a packed column as described above), or can be run by gravity flow (in which the solvent runs down the column due to force of gravity). There are advantages and disadvantages to each system are shown in table.

Table: Advantages and disadvantages of automated and gravity controlled column chromatography.

	Automated Column chromatography	Gravity Flow Column chromatography
Advantages	• Self-operated • Often coupled to an absorbance detector • Programmable • Easy to set up an elution gradient • Reproducible results produced	• Less expensive • The user has more control
Disadvantages	• Requires costly equipment • Possibility of back flow on column • Maximum flow rate dependent on pressure limit of column	• More labor intensive and time • Consuming • Reproducibility is difficult at • Times • Flow rate depends on gravity

Preparation of Resins

The steps for preparing resin are shown in Figure. Resins for protein purification are available in dry or pre-swollen form. In case dry, the resin needs to be hydrated before it is used for column packing. This is done by mixing the dry resin with buffer and leaving it for overnight incubation at room temperature (25°C). After the hydrated resin has settled, very fine particles, which remain suspended at the top, are carefully decanted and discarded. The next step is the equilibration of the hydrated resin in the purification buffer. Finally, the slurry is degassed (to minimize the formation of air bubbles) prior to packing the column.

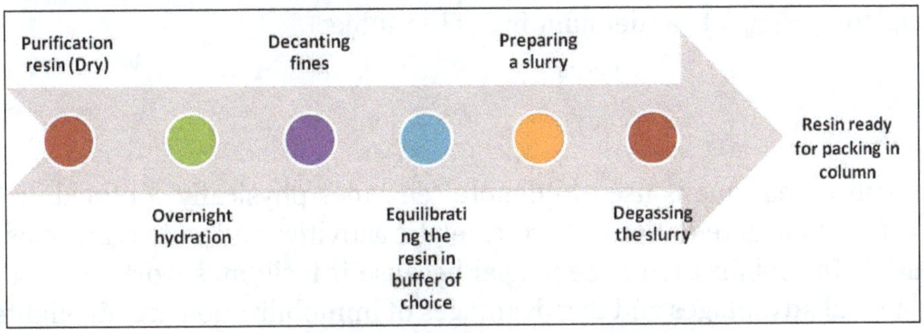

Figure: Preparation of resin for purification.

Key Features

- Excellent affinity chromatography media for protein and antibody purification.
- Fast and simple procedures for batch or gravity-flow operation.
- High binding capacity with ultra-low nonspecific binding.
- Compatible with a variety of sample formats and buffer additives.
- Boosted cost-effectiveness due to reusability for several times.

Applications

- Antibody purification.
- His and GST tagged protein purification.
- Purification of streptavidin, avidin or their conjugates.
- Purification of biotinylated protein, antibody, lectin etc.

References

- Enzyme-synthesis-by-fermentation-method--a-review: tsijournals.com, Retrieved 19, January 2020
- Protein-purification-resins: genscript.com, Retrieved 16, July 2020

Chapter 4

Enzyme Immobilization

The process of attaching an enzyme to an inert and insoluble material is known as enzyme immobilization. Some of the important topics associated with enzyme immobilization are support materials for enzyme immobilization, properties of immobilized enzymes, etc. These topics are crucial for a complete understanding of the subject.

The term immobilized enzyme is used to denote "enzymes physically confined or localized in a defined region of space with retention of their catalytic activities and which can be used repeatedly and continuously". Immobilization is beneficial because it facilitates work up product isolation. Some of the potential advantages and disadvantages of immobilization are highlighted below.

Soluble Enzyme + Substrate----------- Product (single time usage of enzyme)

Immobilized Enzyme + Substrate---------Product (Repeated usage of enzyme)

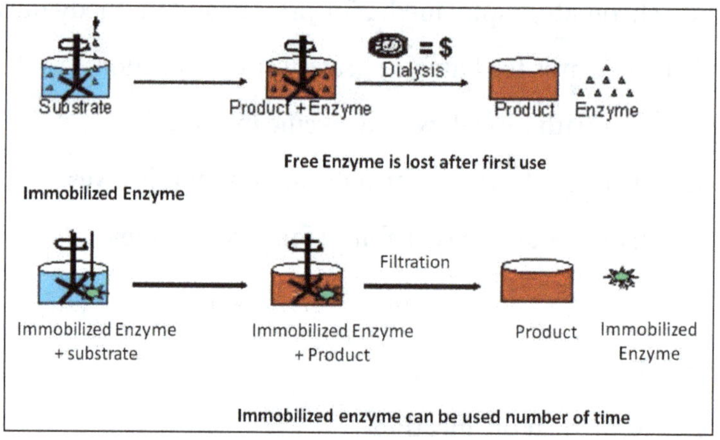

Figure: Enzyme Immobilization

Advantages

- Easy recovery of the product.
- Product is free from enzyme, so there is no cost of purification of enzyme.
- Enzyme can be used repeatedly.
- The enzyme generally gets stabilized after adsorption.

Disadvantages

- Loss of catalytic properties for some enzymes.

- Some enzymes become unstable.
- Additional cost of immobilization.
- Differential limitations.

Intensive study in the area of enzyme immobilization started in mid 1950 and has since continued. The first industrial use of immobilized enzymes was reported in 1967 by Chibata and co-workers who immobilized Aspergillus oryzae amino acylase for the resolution of synthetic , DL amino acids into the corresponding optically active enantiomers. At present, the use of immobilized enzyme is well established in various industries.

Number of important points should be kept in mind while immobilizing an enzyme:

- The biological activity of the enzyme should be retained.
- The enzyme should be more stable as compared its soluble counterpart.
- The cost of immobilization should not be too high.
- It should be used repeatedly.

Enzyme Immobilization Techniques

The various methods of enzyme immobilization are broadly classified as:

- Reversible immobilization:
    - Adsorption.
- Irreversible immobilization:
    - Covalent coupling.
    - Entrapment and microencapsulation.
    - Crosslinking.

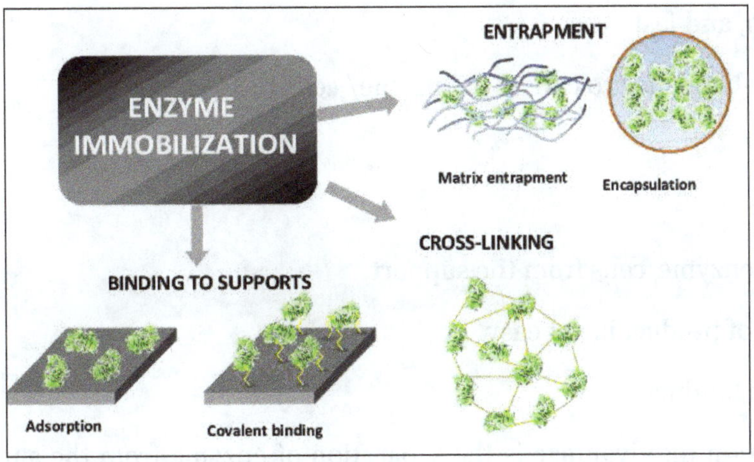

Figure: Diagrammatic representation of the various methods of immobilization.

The advantage of reversible immobilization over irreversible systems can be summarized as follows:

- No chemical modification of the enzyme is required.
- If the enzyme gets activated during use, it can be replaced in reversible immobilization.
- The immobilization of enzyme by adsorption of bio-affinity can be accomplished rapidly.

Methods of Enzyme Immobilization

Adsorption

Using adsorption as immobilization method is the easiest technique and includes reversible surface interaction between carrier and enzyme. The forces formed are weak force, mostly electrostatic, for example Van der Waals forces, ionic bond and hydrogen bonding interactions, although hydrophobic bonding can be significant, but although these forces are very weak, but sufficiently large in number to enable reasonable binding. This method done by mixing the enzyme(s) and a support material with each other in adsorption properties, at optimum pH, ionic strength, etc., for a time, after that collect immobilized enzyme and wash it to remove unbound enzymes.

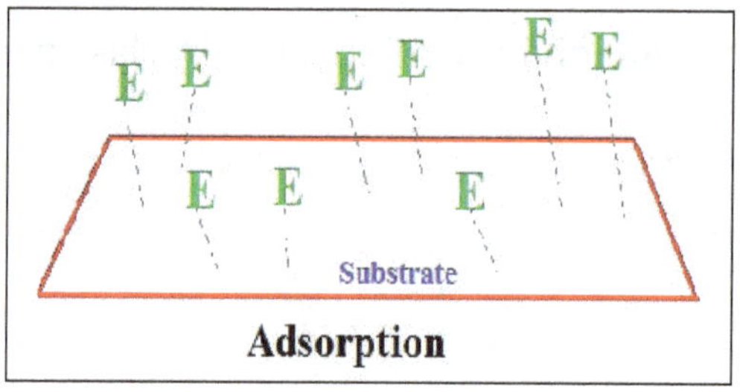

Figure: Immobilization by adsorption.

Advantages of this method are:

- Little or no damage to enzyme /cells.
- Easy, cheap, and fast.
- No changes happened to carrier or enzyme/ cells.
- Reversible.

Disadvantages are:

- Leakage of enzyme/cells from the support.
- Separation of product is not easy.
- Nonspecific binding.

The most signification disadvantage is the separation of enzyme from the support material, desorption may be happen under changes in pH, temperature, and also ionic strength.

Desorption may be happen as a result of physical factors, for example, flow rate, agitation, particle-particle Collisions.

Nonspecific binding may be lead to diffusion limitation and reaction kinetic problem, with consequent alteration in parameter Vmax and Km. Further, binding of protons to the support with consequent shift in pH optimum (1-2 pH), which may be important enzymes with precise pH optimum requirement. Unless carefully controlled, overloading the support can lead to low catalytic activity, and the presence of a suitable spacer between the enzyme molecule and the support can produce problems related to steric hindrance.

Covalent Binding

Covalent binding immobilization method consists of formation of a covalent bond, strong bond, between the enzyme/cell and a carrier. This covalent bond formed between the functional groups present on the surface of carrier and the surface functional groups of the enzyme.

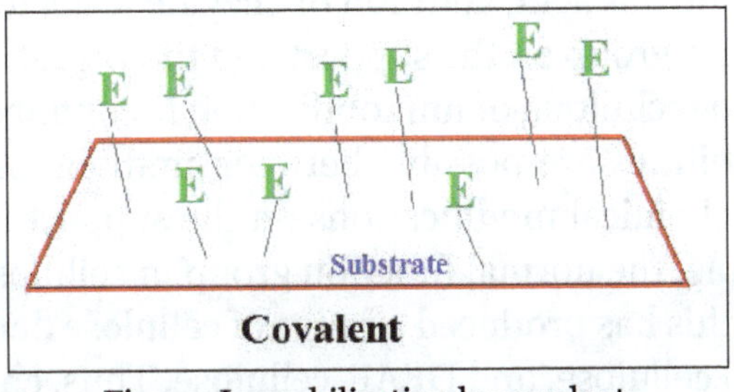

Figure: Immobilization by covalent.

These functional groups on the surface of the enzyme such as amino groups (NH_2) of arginine or lysine, carboxylic group (COOH) of glutamic acid or aspartic acid, hydroxyl group (OH) of threonine or serine, and sulfhydryl group (SH) of cysteine.

Many factors affect the choice of specific carrier, and research work has demonstrated that hydrophilicity is one of the most important factors for keeping up enzyme activity. Thus, hydrophilic carriers such as polysaccharide polymers are popular materials for enzyme immobilization. For example: cellulose, starch, dextran (sephadex), and agarose (sepharose). The sugar residues in these polymers contain ideal functional groups, hydroxyl groups, for covalent bond formation. Also, hydroxyl groups can form hydrogen bonds with water and create an aqueous (hydrophilic) environment in the support. The supports are usually used in bead form.

Other popular supports for immobilization of enzymes are porous silica and porous glass. Porous silica contains small spherical particles of silica fused together having micro cavities and small channels. The carrier is normally sold in bead form, and is very strong and durable. Sintered borosilicate glass has a system of uniform channels. Porous glass is also durable and resistant to microbial disintegration or solvent distortion. However, these two supports are procedures for coupling an enzyme and a carrier is a covalent bond. Most reactions may be one of the following categories:

- Isourea linkage Formation
- Diazo linkage Formation
- Peptide bond Formation

- An alkylation reaction

It is very important to choose a technique that no effect on the enzyme as it may be inactivate it by reacting with enzyme active site. Covalent binding consists of two steps. First one, activation of functional groups found on carrier surface by a specific reagent, and the second, adding enzyme to form covalent bond with activated surface of carrier. Normally the activation reaction is designed to make strong electrophilic (electron deficient) functional groups on the carrier. In the coupling reaction, these activated groups will react with strong electron donating nucleophiles, such as the amino group (NH_2), functional groups of certain amino acids on the surface of most enzymes, to form strong covalent bond.

Cyanogen bromide (CNBr) is usually used to activate the hydroxyl groups in polysaccharide materials. This method contains isourea linkage between enzyme and carrier. In the case of carbodiimide activation, the support materials should contain carboxyl group (CO_2H) then enzyme and support are combined by peptide bond. If the support material contains an aromatic amino group, it can be diazotized utilizing nitrous acid, addition of enzyme leads to the formation of diazo linkage between the activated diazo group on the support and the ring structure of an aromatic amino acid, for example tyrosine. No technique of immobilization is confined to a specific type of carrier, and large numbers of Probabilities are possible between immobilization technique and support material. This is possible by chemical modifications on the support material to produce different functional groups. For example, the normal function group in cellulose is the hydroxyl group, and the chemical modification of this has produced a range of cellulose derivative, such as AE-cellulose (amino ethyl), carboxymethyl cellulose, and DEAE-cellulose. Thus, chemical modification increases the number of immobilization methods that can be utilized for each support material.

Entrapment

One of the easiest techniques of immobilization is entrapment. In recent years, calcium alginate has attraction as an immobilization support material. It has been utilized for immobilization of variety of cell types, sub-cellular organelles, multi-component systems, and enzymes. The physicochemical characteristics of this matrix in gel form have an important effect on the reactions of entrapped biologically active material in the gel. Critical parameter in selecting a matrix is pore size. The difference between entrapment technique and adsorption and covalent binding is that however the enzyme is restricted in movement by the structure of a gel lattice but it is free in solution. The pore size of a gel lattice is controlled to ensure that the structure become tight enough to prevent loose of enzyme or cells, it also allow free movement of the substrate and product. The support acts as a barrier to mass transfer, and although this have serious reaction kinetics implications, but it can prevent interaction between harmful cell, proteins, and enzymes and immobilized biocatalyst.

There are several major methods of entrapment:

- Ionotropic gelation of macromolecules with multivalent cations (e.g. alginate)
- Temperature-induced gelation (e.g. agarose, gelatin)
- Organic polymerization reaction by chemical/photochemical (e.g.Polyacrylamide)
- Precipitation from an immiscible solvent (e.g. polystyrene)

Entrapment can be accomplished by cross linking the polyionic polymer material with multivalent cations in an ion-exchange reaction after mixing with enzyme to form a structure that traps the enzymes/cells (ionotropic gelation). Change in temperature is a simple way of gelation by phase transition utilizing 1-4% solutions of gelation. κ-carrageen a polymers that can easy form gels by ionotropic gelation and by temperature-induced phase transition, which has form a greater degree of flexibility in gelation frameworks for immobilization.

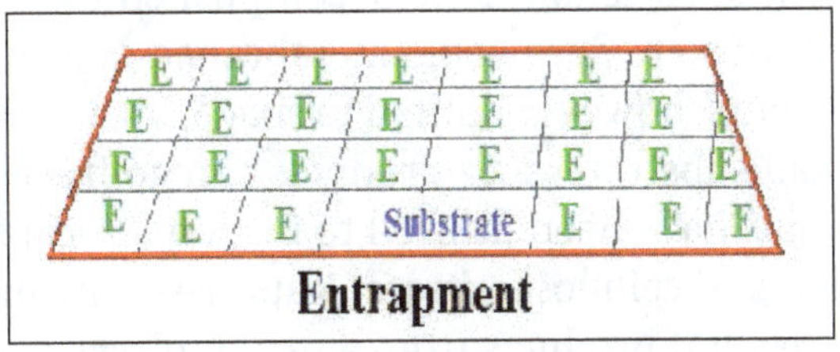

Figure: Immobilization by Entrapment.

On the other hand, it is possible to mix the enzyme with material that is then polymerized to frame a cross-linked polymeric system, trapping the enzyme in the internal spaces of the lattice. The last method is more widely utilized, and various acrylic materials are available for the formation of hydrophilic copolymers. For example, acrylamide monomer is polymerized to form polyacrylamide and methylacrylate is polymerized to form polymethacrylate. In addition to the monomer, a cross-linking agent is added during polymerization to form cross linkage between the polymer chains and help to create a three-dimensional network lattice. The formed polymer may be broken up into particles of a desired size, or polymerization can be arranged to form beads of defined size. Precipitation occurs by phase separation rather than by chemical reaction, but does bring the enzymes/cells into contact with a water-miscible organic solvent, and most enzymes/cells are not tolerant of such solvents. Thus, this method is limited to highly stable/previously stabilized enzymes or non-living cells.

This method is depending on localization of an enzyme inside polymer network or membrane lattice. Entrapment has been advanced and broadly utilized for the immobilization of cells more than for enzymes. It is limited for enzymes immobilization as it may be lost during repeatedly using because of the small molecular size of enzyme compared to the cells. Diffusion limitations are also disadvantages for this method. This method may be classified into five categories: lattice, microcapsule, liposome, membrane, and reverse micelle. The most widely one is the lattice method, in this type enzyme is entrapped in the lattice of the different natural or synthetic polymers. Alginate which is naturally occurring polysaccharide that has the ability to form gels by ionotropic gelation, is the most popular one. Another type, microcapsule, involves entrapment to porous polymer. The preparation of micro capsules containing enzyme requires highly controlled conditions. Taqieddin and Amiji24 developed a new method for encapsulation in which the alginate-chitosan core-shell microcapsules were prepared to immobilize β-galactosidase. The enzyme was confined and protected in the inner core, alginate, while the outer shell, chitosan, manages the transport properties. Utilizing Ca^{2+} ions for alginate crosslinking, microcapsules with liquid core were developed with 60% loading efficiency. And using Ba^{2+} ions, microcapsules with solid core were produced and 100% loading efficiency was obtained. The entrapment in liposome is increasingly recognized as a technique of protecting biocatalysts from inactivation by proteolytic enzymes. Also, this enzyme, liposome, offers a noticeable increment in thermal protection. In the third type, reversed micelle, entrapping within the reversed micelles, it is formed by adding a surfactant with an organic solvent, such as aerosol OT/isooctane reverse micelles26 and used to entrap βgalactosidase. While in the membrane type, the enzyme is isolated from the reaction solution microfiltration membrane or a hollow fiber.

Encapsulation

Encapsulation of enzymes as well as cells can be accomplished by wrapping the biological components inside different forms of semi permeable membranes. It is as entrapment in that the enzymes/cells are free in movements, however limited in space. Vast proteins or enzymes cannot out or inter capsule, however small substrates and products can go freely across the semi permeable membrane. Numerous materials have been utilized to form microcapsules are in range of 10-100 μm in diameter; such as, nylon and cellulose nitrate. Rupture of the membrane is a problem associated with diffusion may be result if products from a reaction accumulate rapidly.

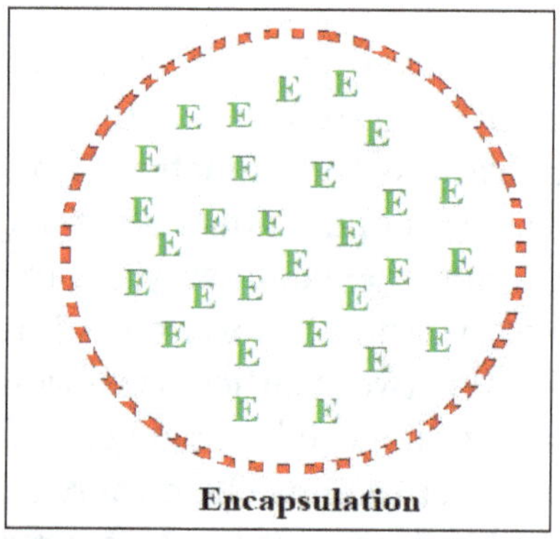

Figure: Immobilization by Encapsulation.

Biological cells also may be used as capsules, and a famous example of this is the use of erythrocytes (red blood cells). The membrane of the erythrocytes is normally just permeable to small molecules. However, when erythrocytes are placed in hypotonic solution, they swell, expanding the cell membrane and substantially expanding the penetrability. In this condition, erythrocytes proteins go out of the cell and enzymes can inter into the cell. Returning these erythrocytes, swollen, to the isotonic solution enables the cell membrane to return to the normal state, and the enzymes inside the cell can't leak out. A distinct advantage of this technique is co immobilization. Cells and/or enzymes may be immobilized in any type of combination to be suitable for particular application.

Cross Linking

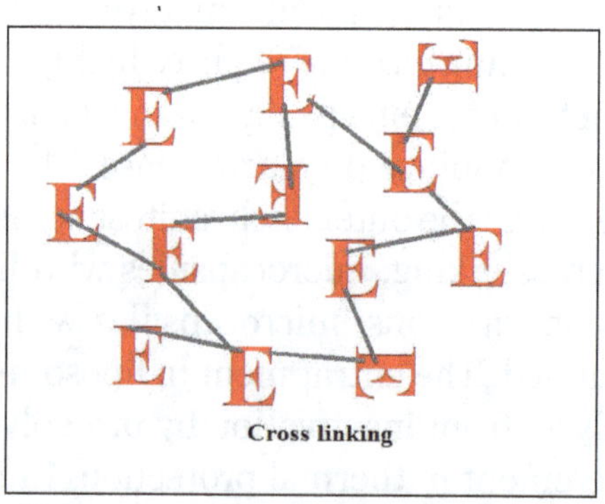

Figure: Immobilization by crosslinking.

This method of immobilization depend only on enzyme and it is support- free as it done by joining the enzyme (or the cells) to each other to prepare a large, three dimensional complex structure, and it can be done chemically or physically. Chemical type of crosslinking normally includes formation of covalent linkage between the cells by means of a bi- or multifunctional reagent, for example glutaraldehyde and toluene diisocyanate. However, limiting factors can be used in this method for living cells and many enzymes because of harmful materials. To minimize the close problems that can be found because of crosslinking of single enzyme, both albumin and gelatin have been used.

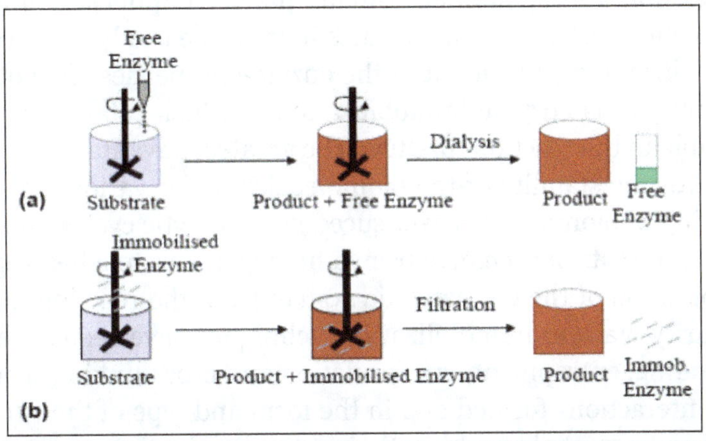

Figure: a & b Schematic diagram of free and immobilized enzyme reactions.

(a), Reaction of free enzyme with substrate and formation of product, which has to be separated via dialysis; (b), Reaction of immobilized enzyme with substrate and formation of product, which can be separated via filtration.

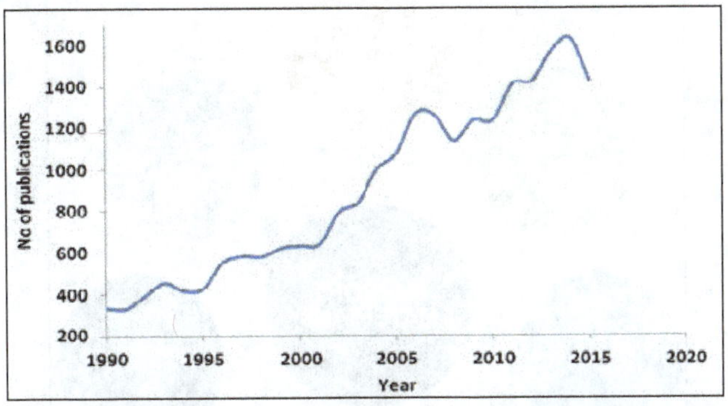

Figure: Annual number of articles on enzyme immobilization indexed in Scopus over the 1990–2015 period.

This technique uses a bi- or multifunctional compounds, which serve as the reagent for intermolecular cross-linking of the biocatalyst.

Covalent binding or crosslinking methods are done under relatively severe conditions in comparison with those of physical adsorption or encapsulation.

Hence, in the previous cases, conformational change of the enzyme structure and partial destruction of the active site may occur. Accordingly, unless covalent binding method is done under well controlled conditions, immobilized enzyme having high activity cannot be obtained. Also, the enzyme cannot easily be lost from carriers because of the strong binding forces between the enzyme and carrier.

Support Materials for Enzyme Immobilization

Enzymes are well-known as highly effective and efficient catalysts of a wide variety of processes characterized by high selectivity and activity. Additionally, enzymes may reduce the number of reaction steps and quantities of hazardous solvents needed and thus make a process more inexpensive and environmentally friendly. For these reasons enzymes have become extremely important catalysts which exhibit great potential in many practical applications in industries ranging from food to pharmaceuticals. The use of enzymes in multiple catalytic processes has resulted in studies leading to significant improvement of the enzyme properties. One of the most important and widely used techniques is enzyme immobilization in which catalysts are attached to a solid support that is insoluble in the reaction mixture. The greatest advantage of immobilization is that it significantly improves the stability of the biomolecules under various reaction conditions and enhances the reusability of biomolecules over successive catalytic cycles. Moreover, after binding the enzyme molecules, the catalysts change from a homogeneous to a heterogeneous form, which facilitates simple separation of the bio-catalytic system from the reaction mixture and results in products of higher purity. Various immobilization techniques have been developed, including adsorption, covalent binding, entrapment, encapsulation and cross-linking. These differ in the type and character of the interactions formed and in the form and type of the support materials used. Selection of the most appropriate immobilization method and support material depends strongly on the type and conditions of the catalytic process as well as the type of the enzyme. However, it should be emphasized that the selection of the support materials is the most crucial challenge due to the major impact the support material may have on the properties of the bio-catalytic system.

Figure: Main features of support materials used for enzyme immobilization.

A very broad variety of materials of various origins can be used as supports for enzyme immobilization. These materials may, in general, be divided into organic, inorganic and hybrid or composite. The support should protect the enzyme structure against harsh reaction conditions and thus help the immobilized enzyme to retain high catalytic activity. Moreover, use of a suitable material, for example hydrophobic carriers in lipase immobilization, may additionally increase the activity of the biocatalyst. However, there are some limitations in this area, because the matrix must not have a negative effect on the structure of the enzyme and should not disturb the enzyme more than is required to create stable enzyme–matrix interactions. Additionally, there should be affinity between the functional groups of the two materials to allow the formation of these enzyme–matrix interactions and effective binding of the enzyme to the support. This is particularly important in the case of covalent immobilization. The carrier should expose the active sites of the catalyst for easy attachment of substrate molecules and to reduce diffusional limitations of the substrates and products. The main required features of support materials for effective enzyme immobilization are summarized in Figure above.

Nevertheless, it should be remembered that the appropriate selection of a matrix is directly related to the type of enzyme and to the process in which the biocatalytic system will be used. The materials used as supports for enzyme immobilization have been classified as Classic materials, which are the most commonly used and New materials, which offer especially desirable properties. The latter type is also currently used in immobilization and not only allow effective enzyme binding but also increase the applicability of the resulting biocatalytic systems.

Classic Support Materials for Enzyme Immobilization

Since the beginning of the work to develop the immobilization techniques, there has been a need to define a group of materials to which enzymes may be attached. In general, materials have been sought which offer high stability, availability, relatively low price and high affinity to the bound enzymes. A wide variety of materials of both inorganic and organic origin have been evaluated as effective supports for biocatalysts as Classic materials. Although Classic materials have been less frequently applied in recent years, they remain an important group of materials used for the immobilization of enzymes.

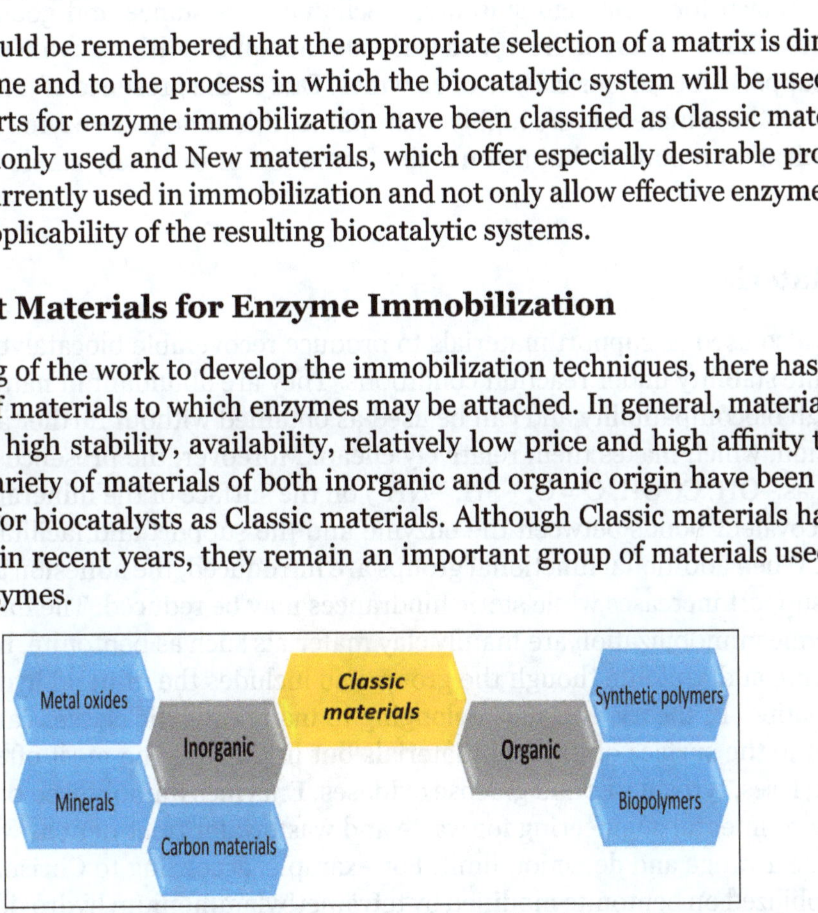

Figure: Selected examples of Classic materials of inorganic and organic origin used for enzyme immobilization.

Inorganic Materials

Silica and Inorganic Oxides

Silica is one of the most frequently used inorganic support materials for enzyme immobilization. Its high thermal and chemical resistance and good mechanical properties make it a suitable material for many practical applications. Silica offers good sorption properties due to its high surface area and porous structure. These properties allow effective enzyme attachment and reduce diffusional limitations. Moreover, the presence of many hydroxyl groups on the surface of silica facilitates enzyme attachment and favours its functionalization with surface modifying agents such as glutaraldehyde or 3-aminopropyltriethoxysilane (APTES). Another advantage of this material is that it can be used in many different forms. Enzymes belonging to many catalytic classes, for example oxidoreductases, transferases, hydrolases and isomerases, have been immobilized with the use of sol-gel silica, fumed silica, colloidal silica nanoparticles and silica gel as supports. The biocatalytic systems obtained demonstrate high catalytic activity retention and good thermal and pH resistance. For example, lipases immobilized on a silica gel matrix and on mesoporous silica retained respectively 91% and 96% of the activity of the free enzyme.

Among other inorganic oxides, titanium, aluminium and zirconium oxides have also been used for the immobilization of many enzymes, for example lipase, cysteine, urease and α-amylase. These supports are known for their high stability, mechanical resistance and good sorption capacity. Moreover, these materials are inert under various reaction conditions, which facilitates their application as supports for various classes of enzymes. Due to the presence of many hydroxyl groups on their surface, these materials are highly hydrophilic; this enhances enzyme immobilization and surface modification modified that favours the formation of relatively stable enzyme–matrix interactions.

Mineral Materials

Minerals are also used as support materials to produce recoverable biocatalytic systems with enhanced enzyme stability under reaction conditions. They are abundant in nature, are easily available, offer high biocompatibility and can be used as obtained without further advanced treatment and purification, which makes them relatively cheap. Moreover, the presence of many functional groups (such as $-OH$, $COOH$, $C = O$, $-SH$, $-NH_2$) on the surface of the minerals allows the formation even of covalent bonds between the enzyme and the support and facilitates modification of the minerals. When additional functional groups are introduced, the adhesion area and hydrophobicity of the support increases while steric hindrances may be reduced. The minerals used as supports for enzyme immobilization are mainly clay materials such as bentonite, halloysite, kaolinite, montmorillonite and sepiolite though the group also includes the mineral hydroxyapatite known as calcium apatite. In theory, enzymes belonging to many catalytic classes can be attached without limitation to the surface of mineral materials but in practice the most often immobilized are lipases, α-amylases, tyrosinases and glucose oxidases. Enzymes immobilized on minerals are used mainly in environmental engineering for waste and wastewater treatment as well as in biosensors to improve linear range and detection limit. For example, according to Chrisnasari et al., glucose oxidase immobilized on bentonite modified by tetramethylammonium hydroxide retains over 50% of its initial activity after five repeated catalytic cycles.

Carbon-Based Materials

Carbon-based materials such as activated carbons and unmodified and modified charcoals have been used as effective and valuable support materials in enzyme immobilization, especially during the last two decades. The well-developed porous structure of these materials, with pores of various sizes and volumes and the high surface area (up to 1000 m^2/g) mean that these materials contain numerous contact sites on their surface for enzyme immobilization. High adsorption capacity, the abundance of many functional groups and minimal release of fine particulate matter make carbon-based materials suitable carriers for the adsorption immobilization of various enzymes. For example, unmodified charcoal support was used for the immobilization of amyloglucosidase. The immobilized enzyme when used for starch hydrolysis without any additional treatment retained over 90% of the free enzyme catalytic activity. According to Silva et al. use of activated carbon for adsorption attachment of pancreatin allows a total immobilization yield that results in the creation of biocatalytic systems with good catalytic properties.

Organic Materials

It is well known that there is no universal support material suitable for all enzymes for all of their applications. Inorganic carriers have certain limitations, such as limited biocompatibility, lower affinity to biomolecules and reduced possibilities to create various geometrical shapes. Moreover, a cross-linking agent such as glutaraldehyde is usually required to create covalent bond between the enzyme and an inorganic support. Due to these reasons, some materials of organic origin are also used for the immobilization of various enzymes under different immobilization protocols. In general, organic support materials can be divided into two groups: (i) synthetic materials (mainly polymers) and (ii) renewable materials obtained from natural sources (biopolymers). Both groups have been widely used since the beginning of enzyme immobilization for the attachment of different types of biocatalysts.

Synthetic Polymers

The greatest advantage of synthetic polymers as support materials is that the monomers that build the polymeric chain can be selected according to the requirements of the enzyme and process in which the product of immobilization will be used. The type and quantity of the monomers determine the chemical structure and properties of the polymer. The composition of the monomers strongly affects the solubility, porosity, stability and mechanical properties of the polymer. A chemical feature that is directly related to the monomer structure is the presence of reactive chemical moieties in the polymeric chain. A very wide range of verified chemical functional groups may be observed in the structure of polymers. They include, for example, carbonyl, carboxyl, hydroxyl, epoxy, amine and diol groups, as well as strongly hydrophobic alkyl groups and trialkyl ammine moieties. These groups facilitate effective enzyme binding and also functionalization of the polymer surface. The type of functional groups determines whether the enzyme is anchored to the matrix via for example adsorption or by the formation of covalent bonds, since it is mainly these two types of immobilization that take place when synthetic polymeric supports are used. Additionally, the type and quantity of functional groups determine the hydrophobic/hydrophilic character of the matrix and therefore its ability to form polar or hydrophobic interactions with the enzyme. Moreover, by using polymeric supports, control of the length of the matrix–enzyme spacers has been achieved. Longer spacers allow the enzyme to retain higher conformational flexibility, while shorter spacers can protect the biomolecules against thermal inactivation and reduce leaching of the enzyme.

Various polymer materials can be used as effective supports and improve properties of the immobilized enzyme such as thermal stability and reusability. The polymer layers play a very important role in protecting the active sites of the enzyme from negative effects of the ingredients of the reaction mixture and the process conditions. However, it should be noted that synthesis of a polymer with the desired properties and functional groups is usually a time-consuming and costly process. Different polymers containing various functional groups have been used for enzyme immobilization. For example, α-amylase was covalently immobilized on polyaniline via –NH groups, while tyrosinase was immobilized via –NH and C = O groups on polyamide 66 (Nylon 66) without any linkers. In another study, commercial lipase was immobilized by covalent binding on strongly hydrophobic polystyrene microspheres activated by epoxy groups. In a hydrophobic environment, lipase exhibits extremely high catalytic properties which are related to the phenomenon called interfacial activation. Furthermore, polyurethane foam has been used for covalent immobilization of inulinase. Bai et al. used polyvinyl alcohol modified by glutaraldehyde as a support for the immobilization of laccase via –OH groups. After immobilization, as a result of the strong interactions, the product was characterized by good storage stability and reusability which make it suitable for use in biosensors to detect bisphenol A. Glucose oxidase, an antimicrobial enzyme, was immobilized on amino- and carboxyl-plasma-activated polypropylene film. The introduction of these groups enhanced the affinity of the polymer to the enzyme. Commercially available ion exchange resins—for example Amberlite and Sepabeads—have also been used, respectively, for the immobilization of enzymes such as α-amylase and alcohol dehydrogenase.

Biopolymers

An alternative to the use of synthetic polymers as matrices for enzymes is the use of biopolymers—polymers of natural origin. Biopolymers include carbohydrates but also proteins such as albumin and gelatin. Materials such as collagen, cellulose, keratins and carrageenan as well as chitin, chitosan and alginate are examples of biopolymers used for immobilization. Biopolymers possess a unique set of properties, from biodegradability to harmless products, biocompatibility and non-toxicity, to an outstanding affinity to proteins, which make them suitable supports for enzymes. Their natural origin and biocompatibility minimizes their negative impact on the structure and properties of enzymes and thus the immobilized proteins retain high catalytic activities. Furthermore, the availability of reactive functional groups in their structure—mainly hydroxyl but also amine and carbonyl moieties—enables direct reaction between the enzyme and matrix and facilitates modification of their surface. Above all, however, these materials are renewable and easy to obtain; in many cases they are by-products of various industries, which makes them inexpensive and reduces the costs associated with the immobilization process. Biopolymers are used for immobilization by adsorption and covalent binding; however, their ability to create various geometrical configurations and propensity for gel formation mean that they are also used for immobilization by encapsulation and entrapment.

Chitosan can be applied in various forms and shapes. For example, Shi et al. used chitosan microspheres cross-linked by glutaraldehyde for the immobilization of nuclease, which is an important enzyme in genetic engineering. In another study, glucose isomerase was adsorbed in macroporous chitosan beads prepared by chelation with various metal ions. As reported by Kim et al. cellulose nanocrystals obtained from cotton linter cellulose can be used for immobilization by non-specific adsorption interactions of Candida rugosa lipase with high loading efficiency, whilst lipase was immobilized by entrapment by Tumturk et al. using κ-carrageenan hydrogels. Vegetable and marine sponges characterized by an open fibrous network that reduces diffusional limitations have also been used as matrices for the immobilization of lipases, mainly via hydrogen bonds. It may be concluded that biopolymers can be used to immobilize enzymes belonging to various catalytic classes with the retention of good catalytic properties. Moreover, the produced biocatalytic systems offer improved thermal stability and in general are noted for their good reusability.

Special attention should also be paid to alginates. Their remarkable abilities for gelation, mainly using sodium or calcium ions and for the creation of capsules, in which single or multiple enzymes can be immobilized, mean that these materials are used principally for encapsulation and entrapment. However, due to the relatively low mechanical stability of alginate gels and diffusional limitations in the transport of the substrates and products, their utilization in immobilization is restricted to a few applications only. For example, Betigeri and Neau immobilized lipase in calcium alginate beads by entrapment, while Kocaturk and Yagar encapsulated polyphenol oxidase in copper alginate beads and obtained high immobilization yields. The products exhibited good catalytic activity retention but their reusability was poor due to leaching of the enzyme from the matrix.

Agarose is a popular choice among biopolymers for use in enzyme immobilization. This linear heteropolysaccharide biopolymer consists of β-D-galactose and 3,6-anhydro-α-L-galactose units, linked by β-1-4 glycosidic and α-1-3 glycosidic bonds. Like alginates, agarose also exhibits great ability for gelation which can occur at temperatures of agarose solution below 35 °C, without addition of any ions and results in formation of highly ordered stable and rigid structures. It is worth mentioning that the 3-D architecture of this hydrophilic material remains almost unaltered in the presence of various organic solvents and does not shrink or swell under such conditions. The ability of agarose gel to form various forms, such as beads, capsules or fibres, has led to this organic support material being of the great interest for industrial applications. For example, Prakash and Jaiswal used agarose beads for simple physical entrapment of thermostable α-amylase. The practical application of the resulting biocatalytic system was tested for removal of starch stains from clothes and the reusability of agarose immobilized enzyme was found to be up to five cycles. In another study, a cross-linked agarose bead support highly activated with aldehyde groups was applied for multipoint covalent attachment of the commercial enzyme Depol™ 333MDP (β-1, 4-endoxylanase). A high immobilization efficiency of over 85% was observed and the immobilized enzyme showed significant improvement of thermal stability compared to free protein.

Classic Materials

Inorganic support materials, such as inorganic oxides, minerals or carbon-based materials are characterized mainly by good thermal and chemical stability as well as by excellent mechanical resistance. These materials are also known for their good sorption properties which are a result of their well-developed porous structure and usually high surface area that ensures numerous contact sites for effective enzyme immobilization. In contrast, synthetic polymers and biopolymers, also grouped under Classic materials, offer numerous functional groups that facilitate even covalent binding of enzymes without cross-linking agents. Additionally, biopolymers are usually characterized by high protein affinity as well as biocompatibility that limits negative effects of the support on the structure of enzymes. Moreover, irrespective of their origin, Classic materials for enzymes immobilization are usually abundant in nature (mineral, biopolymers) or are easy to synthesize (inorganic oxides, synthetic polymers) which makes them relatively cheap. These facts have meant that these support materials still play an important role as carriers for use for immobilization of enzymes.

The Classic materials and types of enzymes that may be immobilized using of these supports are summarized in Table1together with information about immobilization type, cross-linking agents and binding group.

Table: Summary and selected examples of Classic materials of both inorganic and organic origin applied for enzymes immobilization.

Support Material	Binding Groups	Cross-Linking Agent	Immobilization Type	Immobilized Enzyme
		Inorganic Materials		
Sol-gel silica	–OH	–	adsorption	lipase from Aspergillus niger
Silica gel	–OH, C = O	glutaraldehyde	covalent binding	commercial lipase
γAl_2O_3	–OH	–	adsorption	cysteine proteinases from Solanum granuloso-leprosum
ZrO_2	–OH	–	adsorption	α-amylase from Bacillus subtilis
Montmorillonite	–OH	3-aminopropyltriethoxysilane	covalent binding	glucoamylase from Aspergillus niger
Hydroxyapatite	–OH	–	adsorption	glucose oxidase from Aspergillus niger
Bentonite	–OH, $-NH_2$	tetramethyl ammonium hydroxide	covalent binding	glucose oxidase from Aspergillus niger
Commercial activated carbon	–OH, C = O	–	adsorption	cellulose from Aspergillus niger
Activated charcoal	–OH, C = O, COOH	–	adsorption	papain
Activated charcoal	–OH, C = O, COOH	–	adsorption	amyloglucosidase
		Organic materials		
polyaniline	–N–H, C = O	glutaraldehyde	covalent binding	α-amylase
polystyrene	C = O, epoxy groups	poly(glycidyl methacrylate)	covalent binding	lipase
poly(vinyl alcohol)	–OH, C = O	glutaraldehyde	covalent binding	laccase from Trametes versicolor
polypropylene	–OH	plasma activated	covalent binding	Glucose oxidase
Cellulose nanocrystals	–OH	–	adsorption	lipase from Candida rugosa
Luffa cylindrica sponges	–OH, C=O, COOH	–	adsorption	lipase from Aspergillus niger
chitosan	–OH, – NH_2	–	entrapment	lipase from Candida rugosa
agarose	–OH	–	entrapment	α-amylase

New Support Materials for Enzyme Immobilization

Possibilities for practical applications of immobilized enzymes are continuing to grow. For this reason, discovery and use of New materials with desired properties, tailored to particular enzymes, has recently become extremely important. These materials, of both organic and inorganic origin, are characterized by exceptional thermal and chemical stability and very good mechanical properties. Moreover, these support materials are produced in various morphological shapes with controllable particle sizes, usually at nanoscale, which make them suitable for use with enzymes. Furthermore, these materials possess remarkable quantities of various functional groups, corresponding to the chemical groups of the proteins, which enhance enzyme binding and surface modification. However, particularly during the last decade, scientific attention has been directed towards hybrid and composite materials, which combine properties of both composite precursor types and thus maximize their advantages. Hence, with the use of New materials as enzyme supports, control of the technological process is improved, the immobilized enzymes exhibit enhanced catalytic efficiency and the purity and quality of the reaction products increase compared to the processes catalysed by Catalysts enzymes immobilized using the Classic materials.

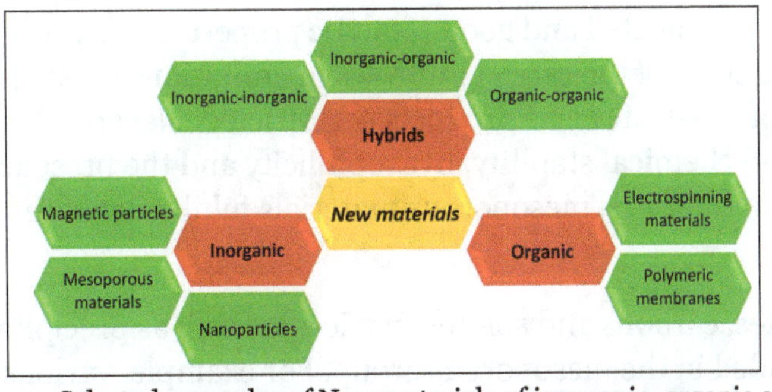

Figure: Selected examples of New materials of inorganic, organic and hybrid origin, applied for enzyme immobilization.

Inorganic Materials

Magnetic Particles

The separation of biocatalysts from the reaction mixture after the catalytic process is one of the crucial problems that must be solved when immobilized enzymes are used. One possible solution is attachment of the enzyme molecules to magnetic iron oxide nanoparticles (MNPs) and simple separation of the biocatalytic system with the use of an external magnetic field. MNPs are also known for their large surface area and the abundance of hydroxyl groups on their surface which enables their easy modification and strong (covalent) binding of the enzyme. These are very important features. High mechanical stability and low porosity, however, which minimize steric hindrances, are also relevant for the creation of a stable enzyme–matrix biocatalytic system. According to Netto et al. many enzymes grouped within the oxidoreductases, hydrolases or transferases can be immobilized on the surface of magnetic nanoparticles to create generally stable systems offering high reusability and easy separation from the reaction mixture. There are several examples that demonstrate these advantages. Mehrasbi et al. immobilized lipase on magnetic nanoparticles functionalized with 3-glycidoxypropyltrimethoxysilane. The immobilized enzyme when used to catalyse the production of biodiesel from waste cooking oil, maintained 100% of its initial activity even after six reaction cycles. In another study, Aber et al. immobilized glucose oxidase by adsorption on unmodified MNPs and used the resulting system for decolourization of Acid Yellow 12. After 15 catalytic cycles the immobilized glucose oxidase retained more than 90% of its initial properties. Atacan et al. used magnetic nanoparticles, pre-treated with gallic acid, for the covalent immobilization of trypsin. Their results showed that the MNPs-trypsin biocatalyst can degrade bovine serum albumin with high efficiency.

Mesoporous Materials

A feature that distinguishes mesoporous supports from other materials used for enzyme immobilization is the possibility of tailoring the properties of the support to the biomolecules by adjusting the synthesis conditions and obtaining matrices with a desired pore structure.

These kinds of carriers contain mesopores with diameters of usually 2 to 50 nm, with a narrow and regular pore arrangement, surface areas as high as 1500 m^2/g and pore volumes of around 1.5 $cm^3/$ g, which make them suitable supports for various biomolecules. Enzymes can be immobilized on mesoporous materials by covalent binding or by encapsulation. In both techniques of immobilization, however, the enzyme is placed in the pores of the support, which means that the structure of the protein is protected and good catalytic properties are generally retained. When the enzyme is located in the pores of the carrier, diffusional limitations must also be taken into account because the transport of substrates and products is restricted. Nevertheless, in view of their water insolubility, thermal and chemical stability, hydrophilicity and the presence of sufficient chemical groups for the binding of catalysts, mesoporous materials fulfil most of the requirements for effective enzyme matrices.

Materials such as zeolites, carbons and sol-gel matrices, as well as precipitated and ordered mesoporous oxides are included in the mesoporous group. For example, various mesoporous silica materials are frequently used. Lipases from Candida antarctica and from Candida rugosa have been immobilized via non-specific interactions without any intermediate agents on SBA 15 and MCM 41 mesoporous silica, respectively and used as catalysts in organic synthesis. The reusability of the immobilized enzymes was found to be significantly improved, because they could be reused for at least five reaction cycles without significant loss of their activity. SBA 15 and MCM 41 silicas were also used for adsorption immobilization of alkaline protease. The immobilized enzyme attained its maximum loading (589.43 mg/g) when SBA 15 silica was used. The products of immobilization also had good pH and temperature stability. In another study, Mangrulkar et al. immobilized tyrosinase on the mesoporous silica MCM-41 and used the resulting biocatalytic system for the detection of phenol. The lowest concentration of phenol detectable by the immobilized tyrosinase was found to be 1 mg/L. As mentioned above, mesoporous materials can also be used for enzyme encapsulation. Wang and Caruso used mesoporous silica spheres for the immobilization of catalase, protease and peroxidase and showed that after immobilization the lifetime of all tested enzymes was improved compared to their free forms.

Nanoparticles

Nanoparticles of both inorganic and organic origin with diameters of up to 30 nm have been extensively studied in recent years as potential supports for enzyme immobilization. Attention is focused on nanoparticles of inorganic origin as these materials are attracting growing interest due to the fact that they generally significantly improve the immobilization yield and the efficiency of the biocatalytic system obtained. Nanoparticles provide a large surface area for enzyme binding that leads to higher loading of the enzyme on the matrix surface and increased immobilization yield. The greatest advantage, however, of nanoparticles over other inorganic materials is their ability to minimize diffusional limitations. Enzyme molecules are attached to the surface of the nonporous particles and their active sites are exposed for wide contact with substrates. This means that biocatalytic systems based on nanoparticles usually provide high catalytic activity retention.

Various inorganic nanomaterials such as nanogold and graphene can be used as matrices for enzyme immobilization. Most frequently, however, inorganic oxide nanoparticles are used. For example, Hou et al. used titania nanoparticles for the immobilization of carbonic anhydrase by glutaraldehyde and immobilized over 160 mg of the enzyme per gram of the matrix. The product was used for biomimetic conversion of CO_2. In another study, lipase from Rhizomucor miehei was covalently immobilized on silica nanoparticles modified by octyltriethoxysilane and glycidoxypropyltrimethoxysilane. The immobilized lipase proved to be a very thermostable biocatalyst. Notably, both biocatalytic systems achieved over 90% catalytic activity retention.

Ceramic Materials

Ceramic materials are known for their extremely high resistance to temperature, pressure and chemicals (organic solvents, bases, acids). These features make them very promising materials for use as supports for industrial applications of immobilized enzymes. Ceramic supports also offer good mechanical stability. Hence, when the enzymes become catalytically inactive, they can be relatively easily regenerated and used for the immobilization of a new biocatalyst. Hydroxyl groups are mainly present on the surface of these materials. This favours mainly adsorption immobilization of enzymes, based on non-specific interactions. For covalent attachment of biomolecules, additional surface modification is necessary. Ceramic materials such as alumina, zirconia, titania, silica, iron oxide and calcium phosphate have been used as biomolecule carriers. It should be added that these materials can also be used in the form of ceramic foam or composite (TiO_2/Al_2O_3) ceramic membranes.

Ebrahimi et al. physically immobilized β-galactosidase using a ceramic material in the form of a membrane as a support. The system was used to catalyse a transgalactosylation reaction of lactose to produce galactosyl-oligosaccharides. The immobilized enzyme was used for continuous production of oligosaccharides and its efficiency reached 40% under optimal process parameters. Wang et al. immobilized horseradish peroxidase on a ceramic material (cordierite) which had been modified by N-β-amino-ethyl-γ-aminopropyl-trimethoxysilane for covalent attachment of the enzyme. The biocatalytic system was used to remove oil from wastewater. The highest recorded removal efficiency was close to 92% and the system exhibited good reusability.

Carbon Nanotubes

Carbon nanotubes are a new type of support material that has been more and more widely used in recent years. Both single-walled and multi-walled carbon nanotubes are characterized by an ordered, nonporous structure, large surface area and biocompatibility. Moreover, they exhibit outstanding thermal, chemical and mechanical resistance. Carbon nanotubes are also relatively amenable to functionalization to further increase their affinity to enzymes and favour the formation of strong enzyme–matrix interactions. Unlike other materials, carbon nanotubes enhance the transfer of electrons between the substrate and the immobilized enzyme. They are thus most frequently used for the immobilization of oxidoreductases and are applied in biosensors for the detection of various compounds such as phenol and its mono- and multi-substituent derivatives, bisphenols or pharmaceuticals, i.e., diclofenac or tetracycline. However, other groups of enzymes (transferases and hydrolases) have also been immobilized with the use of carbon nanotubes.

Glucose oxidase was immobilized on carbon nanotubes and further cross-linked with chitosan. The results confirmed that the transfer rate of electrons was strongly enhanced. In another study α-glucosidase was covalently immobilized on multi-walled carbon nanotubes functionalized by amine groups. The system so obtained was used in a biosensor for measuring the antidiabetic potential of medicinal plants. The immobilized enzymes used in both biosensors exhibited greatly improved sensitivity and time of response and thus better detecting properties. The biosensors also offered relatively good storage stability as indicated by their activity which remained almost unaltered over 30 days.

Graphene and Graphene Oxide

Graphene and graphene oxide (GO) among carbon-based materials have also attracted great attention as support materials for enzymes. This interest reflects their unique features such as biodegradability, two-dimensional structure, high surface area and pore volume as well as good thermal and chemical stability. Additionally, the presence of many various functional groups, such as carboxylic (COOH), hydroxyl (–OH) or epoxide groups, facilitates creation of strong enzyme-matrix interactions without linking agents or modification of the graphene surface. Due to these features, enzymes like lipases or peroxidases can be immobilized on GO surfaces mainly by adsorption, covalent binding or entrapment. It is also worth mentioning that graphene-based materials may even enhance enzyme biocatalytic activity. Moreover, graphene-based supports are characterized by antioxidant properties and may enhance removal of free radicals (i.e., hydroxyl or dithiocyanate) from reaction mixtures. This results in improved protection of enzyme molecules from inactivation.

For example, horseradish peroxidase (HRP) was covalently immobilized on reduced graphene oxide nanoparticles functionalized with glutaraldehyde. Kinetic parameters (turnover number (k_{cat}) and catalytic efficiency (k_{cat}/K_M)) of the immobilized enzyme increased after its attachment, thus demonstrating enhancement of the catalytic properties of the HRP. Moreover, reusability was also significantly improved, as indicated by immobilized enzyme which maintained over 70% of its initial activity after 10 catalytic cycles. In another study, D-psicose 3-epimerase (DPEase) was immobilized on non-modified graphene oxide and applied for production of the rare sugar D-psicose, an epimer of D-fructose. After immobilization, the efficiency of biocatalytic conversion of D-fructose to D-psicose was improved. Thermal stability of immobilized DPEase was also significantly enhanced, as the graphene-bounded enzyme exhibiting a half-life of 720 min that was 180 times higher than the half-life of free D-psicose 3-epimerase (4 min).

Organic Materials

As already mentioned, the properties of the immobilized enzyme are strongly dependent on the structure and characteristics of the matrix. In general, the organic materials classified here as New materials are mainly the same materials and containing very similar chemical groups. However, here these organic materials appear in completely different forms and as a result have very different properties. These materials may for example be used in the form of single particles, fibres of nanometre size, or membranes. These supports increase the performance, efficiency and stability of the immobilized biomolecules and render the latter more reusable. Moreover, control of the process is improved and processes can be carried out in a continuous manner, which makes them cost-effective. Hence, biocatalytic systems based on these materials can become attractive for applications in industrial-scale processes. Selected cases of the use of new materials of organic origin for enzyme immobilization are discussed below:

Electrospun Materials

The great potential of electrospun materials as supports for enzyme immobilization results from their many functional and structural advantages. These materials are known for their length (electrospun nanofibers), uniformity of diameter and diversity of composition. They also have high porosity and surface areas, which lead to high enzyme loading. The nanometre sizes of these materials provide additional benefits mainly related to low hindrance of mass transfer and reduced diffusional limitations. As a result, the efficiency of the immobilized enzyme can be increased. Electrospun support materials are also known for other useful properties, for example their biocompatibility, nontoxicity, biodegradability, high mechanical strength and hydrophilicity, which make them suitable matrices for various types of biocatalysts. It should also be noted that due to the presence of various functional moieties on their surface, electrospun nanomaterials can be easily modified to favour enzyme attachment and increase the activity of the enzyme. A very wide range of synthetic polymers such as poly(vinyl alcohol), polystyrene, polyacrylamide and polyurethane, as well as biopolymers such as chitin, chitosan, alginate and cellulose, may be used to produce electrospun carriers. Their great advantage over other matrices is the fact that because of the variety of materials used to produce electrospun supports, they can be obtained with properties tailored to the enzyme and the process. Enzymes can be immobilized on these materials by adsorption and covalent binding (surface attachment) as well as by encapsulation carried out at the same time as the support is formed, which additionally reduces the costs of the process.

For example, Canbolat et al. used poly(ε-caprolactone) to immobilize catalase by encapsulation using layering methods. Moreover, to enhance the catalytic activity of the enzyme, its molecules were combined with cyclodextrin before immobilization. The results indicated that the addition of cyclic oligosaccharides had a positive effect on the catalytic properties of the biocatalyst. It should also be noted that immobilization by encapsulation results in higher stability of the catalase because the enzyme is protected from negative effects of the reaction conditions. Weiser et al. used polyvinyl alcohol nanofibers for the immobilization by entrapment of five different types of lipase for the kinetic resolution of racemic secondary alcohols by acylation in inorganic media. The study proved that the activity of all tested lipases was enhanced by stabilization of the active conformation of the enzyme. Furthermore, after immobilization of the biomolecules, the turnover frequency of the reaction increased due to a reduction in mass transfer limitations, which was directly related to the structural properties of the polymeric nanofibers. In another study, poly(vinyl alcohol) or polylactic acid nanofibers in the form of a membrane were used for adsorption immobilization of Candida antarctica lipase B. The biocatalytic system was used for the kinetic resolution of 1-phenylethanol and 1-phenyl acetate. Polyvinyl alcohol membrane with immobilized lipase may be used in both organic and water media; however, the lipase immobilized on polylactic acid fibres preserved higher activity and exhibited higher enantiomer selectivity. Nevertheless, due to the protection afforded by the nanofibers, both immobilized systems demonstrated excellent stability even over 10 reaction cycles.

Polymeric Membranes

As has been described above, polymeric materials can be used in enzyme immobilization in the form of beads, powders, fibres or foams. However, in recent years there has been growing interest in the use of commercially available polymeric membranes. This is mainly because they have easily tuneable properties which make them suitable for many groups of enzymes. These membranes have a large surface area that allows efficient enzyme attachment. The membranes also offer good porosity as well as well-defined pore sizes and structure, which facilitate the immobilization of biomolecules not only on the surface of the support but also in its pores. The reaction mixture passing through the membrane therefore has relatively easy access to the active sites of the enzyme, which reduces diffusional limitations. Moreover, membranes with tailored properties can easily be prepared in different shapes and various geometrical configurations. The membranes are also generally insoluble and are known for their mechanical stability; hence they fulfil the requirements for performing as suitable carriers for enzyme immobilization.

Many types of synthetic polymers can be used for the preparation of membranes, for example poly(vinyl alcohol), polyurethane and poly(vinylidene fluoride). However, membranes made from biopolymers, for example from chitosan or cellulose, can also be used as enzyme support materials. Depending on the type of material used, various functional groups are present in the membrane that not only favour effective enzyme immobilization but also enable modification of the membrane for covalent binding of biomolecules. It should also be noted that in addition to membranes obtained on a laboratory scale, commercially available ultra- and nanofiltration membranes can also be used for enzyme immobilization and examples such as GR51PP, NF270 and NTR7450 may be mentioned. However, the greatest advantage of polymeric membranes as biomolecule support matrices, in comparison with other materials, is the fact that no additional separation and purification of the reaction mixture is required; the catalytic process can take place and products can be isolated from the reaction mixture in the same step. It is also worth noting that polymeric membranes, as supports for enzymes, play a crucial role particularly in the case of enzymatic membrane reactors (EMR).

Hybrid and Composite Materials

In the light of the above-mentioned unique features and properties of inorganic and organic matrices, many studies have been carried out with the aim of combining them to maximize their benefits. The possibility of selecting and combining precursors to meet the requirements of a given enzyme and the process in which it is to be used facilitates more precise control of the enzyme immobilization process. Additionally, the biocatalytic systems thus produced can be used in a wider range of practical applications. Combined and reinforced materials usually exhibit properties not observed for their individual components. Hybrids and composites usually make it possible to stabilize the interactions between an enzyme and a support and make biocatalysts more mechanically resistant and stable under reaction conditions. It should be added that hybrid supports in general provide a suitable environment for biomolecules that favours the retention of high catalytic properties by the immobilized enzyme, makes the biocatalytic system reusable and protects it against conformational changes during storage. An additional benefit of the use of composite supports is the fact that these materials are suitable for enzymes belonging to all catalytic classes. Hence, in the quest for support materials for enzyme immobilization, particular attention should be paid to hybrid or composite materials obtained by the conjugation of: (i) organic-organic; (ii) inorganic-inorganic, or (iii) organic-inorganic precursors.

Organic-Organic Hybrids

As described above, many types of polymers of synthetic and natural origin can be used as supports for enzymes. However, to increase their usability, they can be combined to obtain products with enhanced properties which are better suited to the enzyme and the technological process. Organic-organic hybrids may be synthesized by connecting together the following: (i) two synthetic materials, for example polyaniline and polyacrylonitrile, polyethyleneimine, with epoxy-activated acrylate copolymer or poly(acrylic acid) and polyvinyl alcohol; (ii) a synthetic polymer with a biopolymer, such as poly(acrylic acid) and cellulose, polyvinyl alcohol and chitosan, or polyaniline and chitosan; (iii) two biopolymers, such as chitosan and alginate, chitin and lignin or cellulose and dextran. For instance, combining a synthetic polymer having good pH and thermal resistance and mechanical stability with a biopolymer known for its biocompatibility and high affinity to biocatalysts and using the resulting hybrid as an enzyme support, could result in a stable, reusable biocatalytic system with the retention of good catalytic properties.

Additionally, through appropriate selection of precursors, the hydrophobic/hydrophilic character of the matrix can be controlled to increase the strength of the interactions formed between the enzyme and support and to enhance the catalytic activity of the biomolecule. The greatest advantage of these materials is the possibility of forming them into various shapes and sizes. The materials can be used as enzyme supports in the form of particles, fibres, tubes, beads, membranes or sheets. The wide variety of available organic-organic hybrids makes these materials suited for immobilization of enzymes belonging to all catalytic classes by adsorption, covalent binding and also by entrapment or encapsulation. For example, two monomers, polylactic acid and polyethylene glycol, were used to produce micro- or nanofibers by an electrospinning technique. The material thus produced had high porosity, a large surface area, the ability to incorporate functional additives, which are all excellent properties as a matrix for enzyme attachment.

In the study, alkaline phosphatase was immobilized via biotin-streptavidin interactions on surface of the matrix. The bound enzyme exhibited good stability and reusability over a long storage time so this system may be a promising platform for the development of biosensors. Polyvinyl alcohol–hypromellose is an example of an organic hybrid produced by merging a synthetic polymer with a semisynthetic biopolymer which is an inert derivative of methylcellulose. This material was used for the immobilization of lipase from Burkholderia cepacia without any cross-linkers. The resulting biocatalytic system was used for the synthesis of phenethyl butyrate in nonpolar medium and proved to be a successful catalyst, achieving 99% conversion of the phenethyl alcohol and vinyl butyrate used as substrates. The protective effect of the polymeric hybrid support resulted in the retention of high catalytic activity by the immobilized lipase and the activation energy of the reaction was found to be lower when the immobilized enzyme was used. Matto and Husain employed a calcium alginate–starch hybrid gel as a carrier for adsorption immobilization and entrapment of peroxidase. The presence of many hydroxyl groups in the starch structure enhanced the surface attachment of the enzyme, while the capacity of alginate for gelation favoured the entrapment of peroxidase. The entrapped enzyme was found to be significantly more stable against pH, temperature, solvents and inhibitors like urea or heavy metals, compared to the surface-immobilized enzyme.

Moreover, this form of immobilized peroxidase retained over 70% of its original activity even after seven repeated reaction cycles. In a study by Abdulla and Ravindra, equal proportions of alginate and κ-carrageenan were used to produce a novel biopolymeric hybrid for lipase immobilization by entrapment. The resulting biocatalyst was employed in biodiesel production from Jatropha oil and ethanol. Under optimal process conditions, total transesterification of triglycerides to fatty acid ethyl esters was achieved. This biocatalytic system also demonstrated good reusability as indicated by retention of over 75% of its initial activity after six cycles. The results suggest that lipase immobilized on an alginate–κ-carrageenan hybrid could be an efficient and environmentally friendly biocatalyst for biodiesel production. Another application of a composite built from two biopolymers was reported by Nupur et al. who combined chitosan and calcium alginate and used the resulting hybrid in the form of beads for entrapment immobilization of penicillin G amidase (PGA). The immobilized PGA demonstrated high thermal and storage stability and good reusability. The enzyme entrapped in alginate–chitosan hybrid beads was found to have several advantages and could be used in the organic synthesis of 6-aminopenicillanic acid with high efficiency.

Organic-Inorganic Hybrids

A very wide range of materials of both organic and inorganic origin can be combined to create hybrid or composite supports for the immobilization of enzymes. The most frequently used inorganic precursors include silica, inorganic oxides such as zinc and titanium oxides, as well as minerals, carbon materials and magnetic nanoparticles. They can be combined with polymers of synthetic origin, for example polyacrylonitrile, polyethyleneimine and polyvinyl alcohol, as well as with biopolymers such as chitosan, lignin and alginate. These materials are mainly used for the adsorption or covalent immobilization of hydrolases, oxidoreductases and transferases but the encapsulation of these enzymes in carriers of this type has also been reported.

Organic-inorganic hybrids display great potential as support materials for enzymes; such hybrids provide good stability and mechanical resistance and very high affinity to biological molecules. The high stability and often also chemical inertness are related to features of the inorganic precursor. The good ability to bind enzymes is due to the organic components since synthetic polymers and biopolymers have many functional moieties in their structures that are able to interact with the chemical groups of biocatalysts. As a result of their stability and functionality, hybrids in this group can be used for many practical applications, which is their greatest advantage. Many different combinations of organic and inorganic substances have been used to produce functional organic-inorganic composites and hybrids for enzyme immobilization. For example, Zhao et al. combined the stability and mechanical resistance of silica with the biocompatibility and gelation properties of chitosan and achieved immobilization of glucose isomerase on silica–chitosan hybrid microspheres via simple in situ encapsulation. The immobilized enzyme was further used as a catalyst for the conversion of glucose to fructose.

The relative activity of the enzyme was found to be above 90% over a wide pH range of 6–8, a temperature range of 40–80 °C, a storage time of 3 months and after 15 repeated catalytic cycles. In another study, a silica–dialdehyde starch (SiO_2-DAS) mixed hybrid, offering high enzyme binding capacity due to the presence of polysaccharide in the structure, was used for the immobilization of cellulose. The product maintained higher activity over broader pH and temperature ranges than the free enzyme. Moreover, the immobilized cellulase exhibited higher affinity to the substrates as well as better reusability and storage stability. This product might therefore be used as an effective biocatalyst for cellulose bioconversion. Miranda et al. reported the synthesis of an eco-friendly poly-L-leucine-rehydrated hydrotalcite nanohybrid material and its effective use for tyrosinase immobilization without bifunctional linkers. They used the resulting biocatalytic system in the asymmetric epoxidation reaction of chalcone. The nanohybrid-based biocatalyst exhibited excellent activity and enantioselectivity.

The product also demonstrated good reusability, with unaltered activity after five consecutive runs. Thus, this biocatalytic system could find potential applications in protein engineering, biomedicine and catalysis. Chang et al. combined natural clay composed of montmorillonite and layer silicates with chitosan, which exhibits good gelation abilities. The inorganic-organic hybrid was further cross-linked with glutaraldehyde and finally applied in the form of wet and dry beads for the immobilization of β-glucosidase. The covalently bound enzyme exhibited high stability over wide pH and temperature ranges. Additionally, the properties of wet and dry beads were compared and it was found that use of the dried materials led to higher catalytic efficiency. In another study, the conductive properties of carbon nanospheres were combined with the gelation ability of sodium alginate for entrapment immobilization of glucose oxidase and its use for biosensing of glucose. Under optimal measurement conditions, the biosensor achieved very good performance for glucose over a wide linear concentration range, with a detection limit of 0.5 μM. The biosensor also exhibited satisfactory reproducibility and good long-term stability.

Inorganic-Inorganic Hybrids

Besides the types discussed above, inorganic-inorganic hybrid and composite materials also possess many features which make them interesting potential supports for enzyme immobilization. In general, inorganic hybrids exhibit good pH and thermal stability, mechanical resistance and chemical inertness. Moreover, their precursors are easily available and in many cases their synthesis is simple and hence relatively cheap. A variety of functional groups such as carbonyl (C = O), carboxyl (COOH), amine ($-NH_2$) and epoxy are present on the surface of inorganic-inorganic materials, although the most frequently observed are hydroxyl (–OH) groups. These groups determine the hydrophilic character of inorganic composite supports and increase their affinity to biomolecules. Additionally, the presence of many functional groups enhances immobilization efficiency and allows easy functionalization of the surface. Besides simple modification, the presence of various chemical moieties allows the production of combined materials with desired technological features and high affinity for enzymes, which makes these materials promising for practical applications. For example, sol-gel derived silica was combined with multi-walled carbon nanotubes (MWCNTs) and used for the non-specific immobilization of lipase from Candida rugosa. The biocatalytic system thus obtained was applied in esterification reactions in organic media, with high efficiencies. The immobilized lipase also exhibited good reusability as indicated by almost unaltered initial activity after five catalytic cycles due to the protective effect of the MWCNTs. In another study, Zhu et al. used carboxyl-functionalized silica-coated magnetic nanoparticles (SCMNPs) as a carrier for covalent immobilization of porcine pancreatic lipase. The addition of the magnetic nanoparticles to the composite enabled easy separation of the biocatalytic system from the reaction mixture using a magnetic field. The immobilized lipase exhibited enhanced activity compared to the free enzyme and good thermal resistance, with high catalytic efficiency at 70 °C.

The presence of many functional moieties enables the attachment of enzymes belonging to many catalytic classes, including hydrolases and oxidoreductases and the potential use of all known immobilization techniques. For example, glucose oxidase was entrapped in nanozeolites combined with magnetic nanoparticles and multi-walled carbon nanotubes. In another case, β-glucosidase, α-chymotrypsin and lipase B from Candida antarctica were successfully covalently immobilized on the surface of magnetic nanoparticles activated by N,N-disuccinimidyl carbonate. In general, silica is one of the most commonly used precursors for inorganic hybrids. However, other inorganic components such as inorganic oxides, minerals, clays, noble metal nanoparticles and carbon-based materials may also act as precursors for inorganic-inorganic enzyme supports.

An interesting example of an inorganic-inorganic composite used for enzyme immobilization is a combination of calcium carbonate and gold nanoparticles ($CaCO_3$-AuNPs). This support was used by Li et al. for adsorption immobilization of horseradish peroxidase and further to produce a mediator-free hydrogen peroxide biosensor. Due to the good electrical conductivity of the $CaCO_3$-AuNPs inorganic hybrid, as well as the favourable orientation of the enzyme molecules on the surface of the matrix, the biosensor exhibited strong activity toward H_2O_2 reduction and achieved a good linear response over a wide range of hydrogen peroxide concentrations and a relatively low limit of detection (1.0×10^{-7} M). In another study, the ability of carbon nanotubes to direct electron transfer and the high surface area of titanium dioxide were exploited to create TiO_2–carbon nanotube microparticles for adsorption immobilization of glucose oxidase by non-specific interactions. The immobilized enzyme was further used to build a biosensor for glucose detection. Electrochemical analysis showed the biosensor to have high efficiency, sensitivity and reproducibility and the capacity to detect glucose up to concentrations of 3 mM. In another such system, multi-walled carbon nanotubes provided a highly porous conductive network that enhanced electrochemical transduction, while $CaCO_3$ acted as a host carrier for immobilization of tyrosinase by entrapment. This inorganic hybrid with immobilized enzyme was used as a highly effective and sensitive dopamine or catechol biosensor. The biocatalytic system was shown to be resistant to the effects of inhibitors and interferents and its parameters remained unaltered in the presence of uric and ascorbic acid.

New Materials

Table summarizes the most important types of New materials and enzymes that may be immobilized by using these carriers. Moreover, information about binding groups, immobilization type or cross-linking agents are also presented.

Table: Summary and selected examples of New materials of inorganic, organic and hybrid origin applied for enzymes immobilization.

Support Material	Binding Groups	Cross-Linking Agent	Immobilization Type	Immobilized Enzyme
		Inorganic Materials		
magnetic nanoparticles	epoxy groups	3-glycidoxypropyltrimethoxylsilane	covalent binding	lipase from Candida antarctica
	–OH	–	adsorption	glucose oxidase from Aspergillus niger
silica SBA-15	–OH	–	adsorption	alkaline protease
mesoporous silica	–OH	–	encapsulation	catalase
silica mesoporous nanoparticles	epoxy groups	3-glycidoxypropyltrimethoxylsilane	covalent binding	lipase from Rhizomucor miehei
TiO_2 nanoparticles	–OH	–	adsorption	carbonic anhydrase
cordierite	$–NH_2$	N-β-aminoethyl-γamino-propyltrimethoxysilane	covalent binding	horseradish peroxidase
multi-walled carbon nanotubes	$–NH_2$	3-aminopropyltriethoxysilane	covalent binding	α-glucosidase
reduced graphene oxide	C = O	glutaraldehyde	covalent binding	horseradish peroxidase

Organic materials				
electrospinning fibres of polycaprolactone	C=O	–	adsorption	catalase
electrospinning nanofibers of polyvinyl alcohol	–OH	–	encapsulation	lipase from Burkholderia cepacia
polyethersulphone membrane	–	–	adsorption	Phosphotriesterase lactonase from Sulfolobus solfataricus
NTR7450 membrane	–	–	adsorption	casein glycomacropeptide
Hybrid/composite materials				
polyaniline-polyacrylonitrile composite	–N–H	–	encapsulation	glucose oxidase
cellulose-poly(acrylic acid) fibres	–OH, COOH	–	covalent binding	horseradish peroxidase
chitosan-alginate beads	$-NH^2$, –OH	–	entrapment	amyloglucosidase
graphene oxide-Fe3O_4	OH, C = O	cyanuric chloride	covalent binding	glucoamylase
silica-lignin	–OH, C = O	–	adsorption	glucose oxidase form Aspergillus niger
polyacrylonitrile-multi-walled carbon nanotubes	–N–H, C=O, –OH	N-Hydroxysuccinimide	covalent binding	catalase
silica-graphene oxide particles	–OH, C = O	N-Hydroxysuccinimide	covalent binding	cholesterol oxidase
$ZnO-SiO_2$ nanowires	–OH	–	cross-linking	horseradish peroxidase
$CaCO_3$-gold nanoparticles	–OH, C = O	–	adsorption	horseradish peroxidase

New materials with tailored properties are increasingly frequently used as supports for enzymes both due to limitations in application of Classic materials and also to improve the properties of the immobilized enzymes. Materials belonging to the New materials group can facilitate easy separation of bio-catalytic systems from reaction mixtures (magnetic nanoparticles) or enable avoidance of enzyme particles overloading on the surface of the carrier (nanoparticles and mesoporous materials). Moreover, materials such as graphene or graphene oxide enhance transfer of electrons between immobilized enzyme and substrate and result in increased catalytic activity of the biomolecules. The greatest advantage of the materials of organic origin classified as New materials is that they can be formed in various geometrical shapes such as fibres or membranes and reduce diffusional limitations and improve the efficiency of the bio-catalytic processes. Within New materials, there is continuing and growing interest in hybrid materials. Hybrid supports may be synthesized through the combination of precursors of different origin and their properties can be tailored to the requirements of the biocatalysts as well as to the technological process in which the product will be used after immobilization. Hybrid supports are characterized by good thermal and chemical stability and mechanical resistance and usually ensure stable, covalent binding of the enzyme.

Properties of Immobilized Enzymes

The properties of immobilized enzymes are determined by the characteristics of carrier material as well as by the nature and number of interactions between the enzyme and the support. As a consequence of enzyme immobilization, the stability and kinetic properties of enzymes are usually changed, mostly due to the microenvironment and modifications imposed by the supporting matrix.

This modification in the properties may be caused either by changes in the intrinsic activity of the immobilized enzyme or by the fact that the interaction between the immobilized enzyme and the substrate takes place in a micro-environment that is different from the bulk solution. So, one of the main problems associated with the use of immobilized enzymes is the loss of catalytic activity, especially when the enzymes are acting on macromolecular substrates. Because of the limited access of the substrate to the active site of the enzyme, the activity may be reduced to accessible surface groups of the substrate only. This steric restriction may in turn, change the characteristic pattern of products derived from the macromolecular substrate. There are several strategies to avoid these steric problems such as: the selection of supports composed by networks of isolated macromolecular chains, the careful choice of the enzyme residues involved in the immobilization, and the use of hydrophilic and inert spacer arms.

The observed changes in the catalytic properties upon immobilization may also be due to changes in the three-dimensional conformation of the protein provoked by the binding of the enzyme to the matrix. These effects have been demonstrated and, to a lesser extent exploited for a limited number of enzyme systems. Quite often, when an enzyme is immobilized, its operational stability at higher temperature and in the presence of organic solvents is highly improved. The concept of stabilization has thus been an important driving force for immobilizing enzymes. True stabilization at the molecular level has been demonstrated, such as the case of proteins immobilized through multipoint covalent binding. Studies carried out by several authors using different methods have demonstrated that there is a correlation between stabilization and the number of covalent bonds to the matrix.

Advantages and Disadvantages of Immobilized Enzymes

The use of immobilized enzymes normally offers several advantages over free enzymes, such as increased stability, localization, and retention of the molecules at the material surface, which enables easier handling, repeated use, and decreased cost. Other important advantages of using therapeutic immobilized enzymes are the prolonged blood circulation lifetime without the loss of specific activity and the lower immunogenicity. This advantage is particularly important for delivering enzymes or other biomolecules and may constitute an alternative and suitable method for the enzyme replacement therapy. However, some limitations have been attributed to the use of immobilized enzymes in biomedical applications, such as mass transfer resistances (substrate in and product out), adverse biological responses of enzyme support surfaces (in vivo or ex vivo), fouling by other biomolecules, greater potential for product inhibition, and sterilization difficulties. Although the preparation of sterile immobilized enzyme systems may be complex, sterilization may be achieved by filtrating all the reagents and protein solutions through 0.2-mm filters and working under aseptic conditions.

A very important issue regarding the use of enzymes or other products derived from biological or biotechnological processes in medical applications is to ensure that these therapeutic products do not contain any pyrogenic material, toxins, or infectious agents able to cause harmful effects. For that, it is necessary to perform a complete examination of the products to test their safety in terms of local tolerance, toxicity, carcinogenicity, and immunogenicity, among other pharmacological safety tests. Taking into account the diversity in the range of products and the uncertainty about the regulatory status of some of them, it is necessary to design safety evaluation programs to provide useful information to the responsible of clinical trials and to ensure patient safety.

Applications of Enzyme Immobilization for Textile Processing

Enzymes as Industrial Catalysts

Most of the recent scientific developments in the textile industry have focused on biotechnological tools – with particular emphasis on enzymology. Enzymes have a long history in textile processing and fibre modification, and one of the most important criteria for specific application at industrial scale is resistance to high temperature, alkaline pH and the presence of denaturing agents normally employed during bleaching, washing or dyeing. Utilization of highly specific enzymes for various textile-processing applications is becoming increasingly accepted because of their ability to replace harsh chemicals currently used in the industry, and consequent reduction of water, chemicals and energy usage. Enzymes can be applied to virtually all manufacturing steps in textile processing, from fibre and fabric processing, to laundry detergents and effluent treatment. Enzymatic treatment of cotton fibres is a non-toxic, environmentally benign process used for a large variety of cotton processing conditions. Such applications includes: desizing or removal of starch size with amylases; bioscouring or dissolution of waxes, proteins, pectins and natural fats from the surface of cotton fibres with pectinase, lipase and cellulase solutions; bleaching cleanup or effluent treatment to remove residual H_2O_2 with catalases; bio-finishing which improves the appearance of cotton fabrics and garments by removal of fibre fuzz and pills from the substrate surface with cellulases; biostoning or 'stone- washing' of denim fabrics to produce the fashionable aged appearance with cellulases; laundering of garments or removal of various types of stain with mixtures of enzymes; and bio-bleaching or in situ generation of H_2O_2 for cotton bleaching with glucose oxidase. In the case of wool fibre, there are efforts to substitute the conventional chlorine treatment by an enzymatic process with proteases capable of providing the fabric with the same characteristics, like anti-shrinking and anti-pilling behaviour.

Furthermore, textile and several other industries also generate a number of visual pollutants, which they discharge to the surrounding environment without any further treatment. These pollutants not only add colour to water but they also cause extensive toxicity to the aquatic and other forms of life. Therefore, the treatment of industrial effluents containing aromatic compounds becomes necessary prior to their final discharge to the environment. Several peroxidases and/or phenoloxidases (laccases and tyrosinases) have been used for the treatment of wastewater containing dyes.

Enzymes from Extremophiles

The industrial application of enzymes that can with- stand the harsh environments prevailing in textile processing has greatly increased over the past decades. This is mainly a result of the discovery of novel enzymes from extremophilic microorganisms which have found the most practical commercial use because of their overall intrinsic stability. Extremophilic microorganisms are adapted to survive in ecological niches such as at high temperatures, extremes of pH, high salt concentrations and high pressure. Therefore, technical applications of enzymes in the industry are feasible only if such biocatalysts are stabilized against temperature (e.g. washing from 60 to 70°C, textile desizing from 80 to 90°C, etc.), pH (e.g. bleaching at pH 10 or even more) and in the presence of salts, alkalis and surfactants usually employed in the textile processing conditions. Thus, the increase in the number of newly isolated extremophiles and related enzymes document the enormous potential within this scientific field. Furthermore, modern techniques like site-directed mutagenesis will lead to in vitro tailored enzymes that are highly specific for numerous industrial applications.

Immobilization of Enzymes

Besides the benefits of selecting enzymes from extremophiles, it seems rational to consider enzyme immobilization to improve stabilization. The industrial application of enzymes requires, in many instances, its recovery and reuse to make an economically feasible process. Moreover, the use of an immobilized enzyme permits simplification of the reactor design and easy control of the reaction. Thus, immobilization is typically a requirement for the use of an enzyme as a realistic industrial biocatalyst. Thus, several of the enzyme-mediated process limitations may be dramatically alleviated by enzyme immobilization, essentially due to tailor-made improvements in enzyme performance. Enzyme immobilization can be defined as the attachment of soluble enzymes to different types of support resulting in reduction or loss of mobility of the enzyme. Support materials play an important role in the usefulness of an immobilized enzyme as it should be low-cost and provide adequate large surface area together with the least diffusion limitation in the transport of substrate and product.

Enzyme orientation on the support surface may be critical if the active centre needs to be in appropriate association with the support. Thus, in many examples, the combined use of site-directed mutagenesis and immobilization has been used to control the orientation of the protein on the support surface. There are several reports showing the advantages of site-directed mutagenesis to improve enzyme properties directly, and how enzyme immobilization can be a powerful tool to improve the properties of biocatalysts.

Another complex problem is the stabilization of multimeric proteins (e.g. catalases), where dissociation of the subunits produces enzyme inactivation and even product contamination. Moreover, in the case of multimeric enzymes, not all the subunits may be covalently bound to the support, and thus, they are free to dissociate under certain circumstances promoting enzyme inactivation. These unbound subunits may be prevented from dissociating using defined experimental conditions that assist in the handling of enzymes (ions, polymers, etc.) or by genetic tools that might be used to crosslink (via disulfide bonds) or reinforce the subunit–subunit interactions.

Additionally, rational application of immobilized enzymes specifically to textile processing conditions should focus on interactions between the enzyme protein, textile chemicals and additives – which often destabilize the enzyme, and hence reduce its activity.

Advantages of Soluble Bio-Conjugates over Insoluble forms

In several textile processing conditions, enzymes need to act on macromolecular (insoluble) substrates particularly in fibre processing viz. starch, cellulose, protein and pectins. Thus, poor contact between the insoluble substrate and the immobilized enzyme, as well as incomplete separation of the immobilized enzyme from unreacted solid substrates is a common problem in heterogeneous reaction systems. Therefore, when enzymes are to be used with macromolecular substrates, the use of smart polymers as carriers has been suggested as a good solution to these problems. Enzymes linked to smart polymers combine the benefits of using soluble carriers (reduced mass transfer constraints) with the advantage of reusability. Smart materials respond to chemical or physical changes in their environment in a predictable fashion and can be used to design reversible soluble-insoluble bio- catalysts. Stimuli that are used to recover smart polymer-enzyme conjugates for reuse include changes in pH, temperature, ionic strength and addition of chemical species. End-group conjugation and site-specific conjugation are further strategies to to obtain catalysts with improved design. Such polymers occur naturally (e.g. alginate and chitosan), but can also be synthesized chemically (e.g. methyl methacrylate polymers, hydroxypropyl methylcellulose phthalate and hydroxypropyl methylcellulose acetate succinate). These smart biocatalysts have already been used in the bioconversion of starch, cellulose, chitin/chitosan, hydrolysis of xylan and for protein hydrolysates. Hence, the potential of smart polymers for employment in textile fibre processing establishes them as very attractive carriers.

Effect of Enzyme Immobilization on Interaction with the Surrounding Medium

In wet processing conditions many interactions occur between immobilized enzymes, dyes, chemical additives and processing aids, which may be of a great importance. Therefore, a balance between stabilizing and destabilizing interactions should be attained. Stability of laccases in immobilized form depends both on the composition of the dyeing liquor and the chemical structure of the dye itself. Anionic sulphonic dyes are known to protect enzymes from inactivation by forming ionic pairs between negatively charged sulphonic groups and positively charged amino acid residues.

The extremely heterogeneous character of textile fibres requires surfactants as auxiliary agents in textile processing – in which they serve as wetting, anti- foaming, dispersing, washing or level-dyeing agents. Ionic surfactants undergo a strong hydrophobic interaction with globular proteins in aqueous solution. Vasconcelos et al. and Zhang et al. demonstrated that the biological activity of immobilized proteases was not affected by exposure to anionic or non-ionic surfactants. Conversely, Costa et al. reported a notable reduction in activity of catalase immobilized on alumina during treatment of textile bleaching effluents with anionic and non-ionic surfactants; such a reduction reached 96%, and the type and size of the carbon chain were two important factors affecting enzyme stability.

The solution ionic strength is also an important factor for biocatalyst behaviour. Salt addition affects the prevailing electrostatic interactions in solution, and operations in textile mills are usually performed at salt concentrations above 0.5 M. Zille et al. observed a higher stability of laccase in the presence of relatively high levels of salt in the dyeing bath, probably due to enhancement of electro- static coupling of anionic dyes, and consequent formation of stable aggregates between positively charged enzyme and dyes. Nevertheless, the degree of stabilization/destabilization by salts depends to a great extent on the protein: salt ratio.

Another important issue in textile baths is the influence of bleaching stabilizers. Costa et al. concluded that such stabilizers exert some degree of inhibition upon immobilized catalase. Finally, the issue of interaction with organic solvents has been gaining importance, since wet treatment of textile materials is carried out using several of these solvents. Increasing the concentration of organic solvents leads to faster inactivation of enzymes due to reversible changes in their protein structure, while longer incubation promotes irreversible inactivation.

Current and Potential Applications of Enzyme Immobilization for Textile Processing

Several techniques may be applied to immobilize enzymes on solid supports. The basic aspects of enzyme immobilization in respect to their potential applications for the textile processing conditions are summarized in Table below.

Table: Enzyme immobilization for textile industry applications.

Enzyme	Origin	Support	Immobilization
Cellulase	Aspergillus niger Commercial acid cellulase (Rotta)	Polyvinyl alcohol coated chitosan beads Chemically modified pumices particles	Epichlorohydrin- Adsorption $ZrOCl_2$ - Adsorption
Pectinase	Aspergillus niger	Sodium bis-2-ethylhexysulpho-succinate	Encapsulation
Amylase	Aspergillus niger (Diastase)	Alkylamine glass beads coated with zirconia	Adsorption followed by GLUTAL
	Non specified	Fe_3O_4/Poly(styrene-co-maleic anhydride) Magnetic composite microspheres	Covalent binding
Protease	Subtilisin (Novozymes)	Eudragit S-100	Covalent linked by carbodiimide coupling
	Subtilisin (Novozymes)	Eudragit S-100 and L-100, Kollicoat MAE 100P, HPMCP-55, cellulose acetate phthalate and AQOAT AS-LG	Covalent linked by carbodiimide coupling
	Subtilisin (Sigma)	PEG	Covalent binding
	Subtilisin (Sigma)	Eudragit S-100	Covalent linked by carbodiimide coupling
	Subtilisin (Sigma)	Eudragit S-100	Covalent linked by carbodiimide coupling
Glucose oxidase	Aspergillus niger	Alumina and glass	Covalent binding
	Commercial (Serva GmbH)	Cotton	Covalent binding
Catalase	Bacillus SF	Alumina pellets	Covalent-GLUTAL
	Bacillus SF	Alumina pellets	Covalent-GLUTAL
Catalase	Bacillus SF	α- and γ-Alumina balls, Novalox saddles and Raschigrings	Covalent-GLUTAL
	Bovine liver	Cotton fabric or Nylon 6	Adsorption and covalent-GLUTAL

	Bovine liver	Poly(ethylene terephthalate) or polyamide 6.6	Covalent -Photo chemical
	Bovine liver	Poly(ethylene terephthalate)	Chemical and Covalent- Photo-chemical
Peroxidase	Horseradish	Ion exchange resins	Covalent binding
	Horseradish	Eupergit → C	Covalent binding
	Momordica charantia	Con A-Sephadex	Adsorption
	Momordica charantia	Con A-layered calcium agi-nate-starch beads	Adsortion followed by GLUTAL
	Saccharum spon-taneum	Polyethylene	Covalent-GLUTAL
Laccase	Myceliophthora ther-mophila	Sepabeads EC-EP3 and Dilbeads NK	Covalent binding
	Rhus Vernificera	PEG	Covalent binding
	Sclerotium rolfsii	Alumina	Covalent-APTES- GLUTAL
	Sclerotium rolfsii and	Alumina	Covalent-APTES- GLUTAL
	Trametes hirsuta		
Laccase	Trametes hirsuta	Alumina	Covalent-APTES- GLUTAL
	Trametes hirsuta	Alumina	Covalent-GLUTAL followed by coating with polyelectrolytes
	Trametes hirsuta	poly amide 6, 6	Cross linking- GLUTAL and spacer
Trametes pubescens	Alumina		Covalent-APTES- GLUTAL
	Trametes versicolor	Silica, amberlite IRA-400, glass ceramic and montmorillonite	Covalent-imidazol, covalentcar-bodiimideGLUTAL and cova-lent-APTESGLUTAL
	Trametes versicolor	Poly(4-VP) grafted and/or Cu(II) ions chelated magnet-ic beads	Adsorption
	Trametes villosa	Alumina	Covalent-APTESGLUTAL
Tyrosinase	Solatium tuberosum	Celite 545	Adsorption

Textile Fibre Processing

Cellulases. Cellulases constitute a group of enzymes that are able to catalyze cellulose hydrolysis via de- grading β-(1-4) linkages of the biopolymer, and concomitantly release reducing sugars. In nature, cellulolytic systems belong to three major classes, which act in a synergistic fashion: endo-glucanases, exoglucanases and β-glucosidases. These enzymes achieved worldwide success in textile processing because of their ability to modify cellulosic fibres in a controlled manner, producing better quality fabrics without excessive inner structural damage. The multicomponent nature of the cellulase system finds major application in biopolishing (removal of fluffy and protruding fibres from cotton fabrics resulting in clear fabric surface), as well as in denim garment treatment to achieve the desired stonewashed final look.

Various methods and supports have been reported to date for immobilization of cellulases for different purposes. Pazarlioglu et al. used a commercial cellulase immobilized onto $ZrOCl_2$-activated (pumice) particles for denim washing. They concluded that such immobilized enzymes could efficiently abrade indigo-dyed denim fabrics, and because strongly bound enzymes remained on the pumice, staining was detected at lower levels after the second usage with immobilized cellulase than its free counterpart. Dinçer & Telefoncu improved the stability of an acidic cellulase in the neutral pH range by immobilizing it on chitosan beads coated with maleic anhydride-modified polyvinyl alcohol. The major reason behind replacing acid cellulases with neutral ones is that the former are more aggressive and produce more abrasion that reduces the tensile strength of the fabric.

The applications of cellulases in immobilized form for textile fibre processing are still scarce, probably because of substantial constraints associated with limited substrate diffusion, as discussed above. Eudragit L-100 a soluble–insoluble reversible polymer was successfully used for cellulase immobilization and could be an alternative to apply in bio- polishing and/or bioscouring.

Pectinases

Pectinases are enzymes able to breakdown complex pectins, which occur as structural polysaccharides in plant tissues, into simpler molecules, viz. galacturonic acids. They include polygalacturonases, pectin esterases, pectin lyases and pectate lyases, depending on their mode of action.

Many non-cellulosic impurities are found in the cellulose matrix of the primary wall, and to a lesser extent the secondary wall of a cotton fibre. This barrier limits the dyeing and finishing processes. Hence, scouring, which remove the cuticle and the primary wall compounds from the cotton fibre is essential to achieve satisfactory wettability. Bioscouring has become a feasible alternative to classic chemical scouring, because of the far better eco-friendliness of the former. Besides pectinases, several other types of enzyme can be used for bio- scouring – viz. lipases, proteases, cellulases, xylanases and cutinases.

Amylases

Amylases are enzymes that cleave starch molecules and related compounds, by hydrolysing α- 1, 4- and/or α-1, 6-glucosidic linkages, in either endo or exo-locations. Modern production processes for textiles introduce a considerable strain on the warp during weaving. The yarn must, therefore, be prevented from breaking. For this purpose a removable protective layer (size) is applied to the threads. Starch is a very attractive size, because it is cheap, easily available in most regions of the world, and it can be removed quite easily. Good desizing of starch sized textiles is achieved by the application of amylases, which selectively remove the size and do not attack the fibres.

Despite the various forms, and many studies on immobilized amylases, the general enzymatic actitivity is usually reduced, so practical applications of immobilized amylases in textile desizing are rare, since they act upon macromolecular substrates. Nevertheless, Qiu et al. reported the immobilization of α-amylase on magnetic composite microspheres. Owing to easy separation under a magnetic field for repeated use, a potential for a wide range of industrial applications was claimed – including textiles. Further- more, an amylase from Aspergillus niger was immobilized via glutaraldehyde-mediated coupling onto zirconia-coated alkylamine glass beads. The usefulness in removal of starch stains from cotton was tested using various detergents. The immobilized enzyme was used in removal of starch stain from cotton clothes along with detergents. All the detergents gave better washing in presence of immobilized amylase than that with detergent alone.

Proteases

Proteases are a class of enzymes which bring about protein degradation via hydrolysis of the peptide bonds in the polypeptide backbone. The conventional wool processing techniques to achieve shrink-resistant wool uses chlorination steps which pose ecological problems. Replacing these harsh chemicals with pro- teases for processes such as shrink-resistance, while simultaneously improving whiteness, dyeability and handle are well recognised. However, despite some beneficial effects, such enzymatic treatments may cause excessive damage to the fibre cuticle, with a consequent high degree of weight and strength losses of the wool fibres.

Immobilization of pro- teases (or chemical modification thereof) typically increases their molecular size, constraining proteolytic attack to the cuticle and thus reducing the problem. To elucidate the pattern of diffusion of serine proteases into wool fabrics and yarns, free subtilisin and subtilisin covalently linked to polyethylene glycol (PEG) were studied. The modified protease remained on the surface of the cuticle layer region (hydrolyzing just the cuticle layer of wool), hence producing a higher tensile strength and a lower felting of the fibres. Silva et al. used a commercial protease (Esperase) covalently linked to Eudragit S-100. This novel approach is a promising alternative for wool shrink-resist finishing, replacing the conventional chlorine treatments. Under optimised conditions (Eudragit, 2.5% w/v, carbodiimide, 0.2% w/v, coupling time, 1 h and blocking agent concentration, 0.05%), the conjugate activity yield was about 45% and its operational stability at 60°C was increased by 1.7 times. These data agreed with those by Shen et al., and the dyeing properties of the wool fabric appeared to be unaffected by treatment with immobilized Esperase. Zhang et al. also used Eudragit S-100 via carbodiimide coupling; their results made it apparent that the ethoxylated alkyl phosphate (i.e. an anionic surfactant) acts synergistically with the immobilized enzyme, in the bioscouring processes. Different enteric polymers can also be successfully coupled with Esperase using carbodiimide coupling. Treatment on wool showed that such enzymes modified with Kollicoat, hypromellose acetate succinate, hydroxylpropyl methylcellulose phthalate or cellulose acetate phthalate produced effects similar to those of Esperase modified with Eudragit S-100 and L-100, in terms of improved shrink-resistance, but with less fibre damage in terms of weight loss and tensile strength.

Textile Bleaching and Treatment of the Bleaching Liquor

Hydrogen peroxide is a substrate (catalase and per- oxidase) or product (glucose oxidase) of several interesting industrial textile enzymes. These biocatalysts may have to act in the presence of high H_2O_2 concentrations during textile bleaching or in the treatment of the bleaching liquor. Under these conditions, such enzymes may be sensitive to H_2O_2, because it can oxidize some residues in the proteins and denature them. Hence, an increase in the resistance to this chemical reagent has major practical relevance. This might be addressed from different perspectives: (i) multipoint and multisubunit covalent immobilization, which reduces any conformational change involved in enzyme inactivation and greatly increase the enzyme stability. Supports like agarose beads, zeolites, porous glass, epoxy resins, offer large areas for enzyme-support interactions; (ii) immobilization on porous supports with large internal surface areas stabilizes the enzyme against interaction with molecules from the surrounding medium; and (iii) chemical modification by crosslinking with polyfunctional-polymers, which could reduce the effective concentration of H_2O_2 in the proximity of the enzyme.

Glucose Oxidase

Glucose oxidase is a dimeric glycosylated flavoprotein – which is able to catalyze oxidation of glucose to glucolactone, which in turn spontaneously yields gluconic acid, as well as H_2O_2 as a side-product. In the last decade, due its biodegradability, H_2O_2 has almost entirely replaced conventional chlorine treatments in the textile bleaching processes. Therefore, glucose oxidase has been considered to be a possible method for producing H_2O_2 for biobleaching, aimed at improving fibre performance before dyeing, and destroying pigments initially present in the fibres that account for cotton greyness. Implementation of this enzymatic system in cotton pretreatment is beneficial for reusing the desizing waste baths as an additional source for glucose, and to decrease waste water pollution and water consumption. Higher quantities of enzymatically-produced peroxide were needed to accomplish the same bleaching effect as in the standard procedure. Probably, the added glucose protects the cotton from being bleached, acting as a substrate for bleaching and oxidation, thereby consuming the peroxide, or as a stabilizer. On the other hand, increasing the bleaching temperature causes the protein to denaturate and deposit on the fabric surface, due to hydrophobic fabric/enzyme interactions. These problems could be overcome by using immobilized instead of free enzyme. The stability of glucose oxidase has been improved via several immobilization procedures on various supports .The first textile application was reported by Opwis et al., with cyanuric chloride-mediated attachment to cotton – with subsequent release of H_2O_2 at levels high enough to support adequate bleaching. This biotechnological approach allowed important economic and ecological savings – in terms of water and chemicals, as desizing liquors include glucose from cotton pre-treatment with amyloglycosidase. Glucose oxidase from Asper- gillus niger was also covalent immobilized on two in- expensive and commercially available supports, viz. alumina and glass. High protein immobilization yields, as well as sufficient bleaching levels were attained. Glass permitted a higher rate of generation of H_2O_2 (0.35 g/L) due to its more porous morphology and thus bound more protein, however, alumina produced a higher operational stability and permitted reuse for at least three consecutive cycles.

Catalases

Catalases are ubiquitous enzymes in aerobic organisms, in which they catalyze breakdown of H_2O_2 into water and oxygen.

As a strong oxidizing agent, H_2O_2 oxidizes reactive dyes – which constitute a major reason for dye loss and non-uniformity in dyeing. Consequently, residual H_2O_2 in bleaching baths must be removed from the fabric prior to dyeing. A critical problem is that bleaching is carried out at temperatures above 60°C and under alkaline conditions (pH 9 and above), so most commercial catalases cannot withstand these harsh conditions. An alternative is to use catalases from alkalothermophilic and thermophilic microorganisms. However, direct application of free catalase promotes interaction between the protein and the dye, which in turn reduces dye uptake by the fabric.

Immobilization of catalase overcomes this constraint, and concomitantly enables reuse of the enzyme. Recycling bleaching baths, pre-treated with immobilized catalase, for subsequent dyeing leads to savings up to 50% in overall water consumption. Some natural polymers – e.g. gelatin and chitosan, as well as some synthetic polymers – e.g. polyacrylamide, were found not to be appropriate for bleaching treatments.

The treated liquid was reused for dyeing fabrics with various dyes, resulting in accept- able colour differences for all the dyes used.

However, Costa et al. confirmed that immobilized catalase could be inhibited by anionic stabilizers or surfactants added to the H_2O_2 substrate solution. Opwis et al. used low-cost synthetic textile fabrics as alternative carrier materials for covalent catalase immobilization by a photochemical process (UV light) in the presence of different cross-linking agents. The great advantage of using fabrics as carriers is that they can be removed very quickly from the reactor without a filtration step and without protein residues after the enzymatic reaction. Even after 20 or 30 applications the immobilized enzyme still showed a distinct activity, and the integral activity was higher by a factor of 3.5 and 1.5, respectively.

Decolorization of Dyes and Effluent Treatment

Peroxidases: Peroxidases are oxidoreductases that consume H_2O_2 to catalyze oxidation of a wide variety of organic and inorganic chemicals, including a large variety of aromatic compounds.

Huge volumes of potentially hazardous compounds are released at various stages of textile manufacturing including dyes, alkalis, chromium, phenol, oils and waxes. Therefore, transformation of such potential toxic compounds is essential, prior to their disposal. Most of the physicochemical methods currently used have technical and economic restrictions, since they are expensive, produce large amounts of sludge and are inefficient for some soluble dyes. Therefore, white-rot fungal peroxidases have attracted particular interest due to the low yield and high cost of production of plant peroxidases and since they participate in the degradation of a broad range of substrates.

Peroxidase immobilization, for wastewater treatment, was successfully achieved on various natural and synthetic carriers. However, only a few reports have been linked with direct decontamination of textile industry effluents. The fundamental features of peroxidase immobilization with regard to potential applications for the textile industry are summarized in Table.

The majority of the matrices currently used for the immobilization of enzymes such silica, controlled pore glass, polyvinyl alcohol, polyacrylamide and chitosan beads were not suitable, because of dye adsorption onto the matrices, probably inactivating the enzyme. Therefore, immobilization of horseradish peroxidase (HRP) on various supports, and application in treatment of textile effluents, was assessed by Peralta-Zamora et al. Among them, Amberlite IRA-400 ion exchange resin produced the best results for enzyme immobilization, efficiency of colour removal and degradation of phenol species. The decolorization process was more efficient with combined photo- enzymatic procedures which appear to be an efficient decontamination method for industrial effluent treatment. Moreover, the peroxidase from the plant Saccharum spontaneum can be used as an alternative to the commercial HRP and soybean peroxidase for the decolorization of textile dyes. Four textile dyes at initial concentration of 50 mg/L were completely degraded within 6–8 h at pH 3 by the enzyme immobilized on a modified polyethylene matrix. The immobilized enzyme was also used in a batch reactor and the reusabilility was studied for 15 cycles, demonstating a half-life of around 60 h (equivalent to 10 cycles).

Bio-affinity-based immobilization can be exploited for the direct immobilization of enzymes from partially purified preparation or even from crude homogenates. Simplicity of immobilization, lack of chemical modification and an enhancement in stability is usually achieved. In this way, Matto

& Husain immobilized bitter gourd peroxidase on concanavalin A layered calcium alginate-starch beads for the effective decolorization of textile industrial effluents. They evaluated its efficiency at large scale by using a continuous two-reactor system, one reactor containing immobilized enzyme and the other, activated silica. Immobilized enzyme could remove more than 90% of the colored compounds over 10 days of operation. The reactor was capable of removing 40% of the color from the textile effluent even after 60 days of operation. An enhancement in decolorization was probably due to increased adsorption on adsorbent activated silica.

Nevertheless, amongst the large number of supports available for enzyme immobilization, epoxy-activated carriers are being widely used and have received great attention for large scale applications such as textile industry effluents. One of the most interesting epoxy-activated supports is Eupergit C, because it is commercially available, resistant to mechanical and chemical stress and adaptable to a variety of configurations and specific processes carried out in reactors. Pramparo et al. reported direct HRP binding on Eupergit C as an effective method for the preparation of stable biocatalysts in removing phenol compounds from wastewater. In this case, a maximum immobilised enzyme activity of 127 U/g support was found using an enzyme loading on the support of 35.2 mg/g support.

Laccases: Laccases are copper-containing oxidoreductases, which belong to the group of small blue oxidases. They are widely distributed in higher plants, fungi and bacteria. Such enzymes are able to oxidize various aromatic compounds (particularly phenols) and inorganic compounds, with concomitant reduction of oxygen to water. Laccases have been extensively studied, and fundamental aspects have been duly reviewed. They have found industrial and biotechnological applications in the food, pulp and paper industries, as as well as the textile industry. The use of laccase in this sector is growing fast, since besides decolorization of textile effluents, laccase is used to bleach textiles and even to synthetise dyes. Applied aspects of immobilized laccases, using a wide range of methods and substrates, have been described. The basic aspects of laccase immobilization with respect to potential applications for textile processing conditions are summarized in Table.

Most laccases that are efficient in decolorization of dyeing effluents are produced by white-rot fungi. Therefore, laccases from Trametes hirsuta, Sclerotium rolfsii, T. villosa, T. modesta and T. pubescens have been successfully immobilized on alumina pellets, and could efficiently decolorize several synthetic dyes. Research has shown that the subsequent coating of the alumina-immobilised lac- case with polyelectrolyte layers considerably increased laccase stability. However, although studies have been performed with immobilized lac-case at laboratory-scale using reactors with small volumes, these have not addressed the large wastewater volumes generated by the textile industry.

Peralta-Zamora et al immobilized laccase from T. versicolor on various supports viz. silica, amberlite IRA-400, glass-ceramic and montmoril- lonite. Imidazole modified silica (SiIm) allowed almost total laccase immobilization. In the initial stages, decolorization was mainly due to adsorption of the dyes onto the support surface, but later the support became saturated and enzymatic decolorization was apparent. The use of a brief photochemical pre-treatment permitted a significant increase in the efficiency of the enzymatic decolorization process.

For the treatment of wool dyeing effluents at low pH, an acid stable laccase would be desirable. Ryan et al. (2003) confirmed that an acidic laccase could be used for continuous decolorization of wool dyeing effluents which allowed water recycling. Eighty-three percent of the protein was bound to the carrier (0.13 mg/g alumina) and 72% of the laccase activity was retained on that carrier material.

Silva et al. demonstrated the potential of woven nylon (polyamide 6, 6) as a carrier for laccase immobilization and application in a membrane reactor. Woven nylon offers several advantages such as low cost, chemical inertness, non-toxicity, and mechanical stability, insolubility in water, ready availability and availability in a number of forms. Thus, it seems to be a promising system about future application.

Recently, Kunamneni et al. investigated decolorization of synthetic dyes by laccase from Myceliophthora thermophila covalently immobilized on polymethacrylate-based polymers activated with epoxy-groups. The enzyme immobilized in Sepabeads EC-EP3 exhibited notable activity (203 U/g) and retained 41% of its activity in the decolorization of methyl green in a fixed-bed reactor after five cycles. High mechanical stability and the absence of swelling in water make them very interesting alternative carriers for dye degradation in both batch and continuous fixed-bed bioreactors. Magnetic particles have gained significant attention for large-scale applications due to their potential for rapid separation in a magnetic field. The immobilized laccase was more stable during operation (degradation of three individual dyes in batch systems) and storage compared to the free enzyme.

Besides their potential for detoxification of textile effluents, laccases are also effective in preventing dye redeposition on fabrics and garments – during washing and finishing processes, by degradation of the solubilized dye. However, the action of those enzymes must be restricted to solubilized dyes to avoid decolorization of fabrics. In order to prevent dye redeposition on already dyed fabrics, laccases have been covalently attached to PEG; this led to enhanced stability and limited adsorption onto fibres and garments, thus preventing their enzyme-mediated decolorization.

Tyrosinases: Tyrosinases also known as polyphenol oxidases or catecholases, catalyze two sequential oxygen-dependent reactions: the o-hydroxylation of monophenols to o-diphenols (i.e. a cresolase activity), and the subsequent oxidation to o-quinones (i.e. a catecholase activity) which undergo spontaneous (non-enzymatic) polymerization to yield water-insoluble substances.

At present, several peroxidases and laccases are used specifically for the treatment of textile dyeing effluents – but decolorization promoted by the former adds considerably to the costs of industrial processing, because of the (expensive) H_2O_2 required as co- substrate. Furthermore, the limited availability and high purification cost of lac-cases constrains their use for this purpose. Tyrosinases, mainly extracted from plants, are simpler to use, and less expensive when decolorization of aromatic-rich effluents is considered, as only molecular oxygen is required as oxidant.

Particularly for phenol removal from synthetic and industrial effluents, using a wide range of methods and with various supports, e.g. chitosan, chitosan with alginate-filled pore space sodium and calcium aluminosilicates, silica gel, polysulfone capillary membrane and polyclar. Khan & Husain took advantage of a tyrosinase preparation from Solanum tuberosum, immobilized on Celite in view of its high stability and cost of preparation, for treatment of wastewater/ dye effluents contaminated with reactive textile and non-textile dyes; this was more effective in terms of decolorization efficiency, pH tolerance and total organic carbon removal than the corresponding free form, for treatment of either individual or complex mixtures of dyes. The maximum decolorization was found between pH 3.0 and 4.0. These observations suggested that industrial effluents could be successfully treated with immobilized tyrosinase.

Enzyme Immobilization in Biomedical Applications

Biomaterials can be combined with biomolecules, such as enzymes and growth factors, to yield biologically functional systems. There is a wide and diverse range of materials and methods available for enzyme immobilization on or within the biomaterial. Furthermore, the criteria for selecting the immobilization methods should also take into account that the immobilized enzyme should retain an acceptable level of activity over a certain period of time in terms of economic or clinical aspects.

The methods used for the administration of immobilized enzymes may be divided into two principal groups: immobilized enzymes that are intended for prolonged circulation and enzymes that must be necessarily present in different tissues and organs of the body. In the second case, the immobilized enzyme is intended for local deposition during the treatment of discrete lesions (e.g., thrombi, tumors, atherosclerotic injuries) or of the individual organs.

Table below describes some examples of immobilized enzymes in various biomaterial supports for different biomedical applications.

TABLE: Examples of Immobilized Enzymes in Various Biomaterials for Different Biomedical Applications.

Enzyme	Biomaterial (Carrier)	Immobilization Method	Application
Alkaline phosphatase	pHEMAa Glassy carbon	Entrapment Protein fusion to affinity tag	Induction of bone and cartilage mineralization Biosensor/enzyme-linked immunoassays
α-amylase	pHEMA[a] Starch polymeric blend	Covalent binding Encapsulation	Fixed-bed reactor Tailor the degradation rate of starch based biomaterials
β-galactosidase	Gelatin	Encapsulation	Not specified
Glucose oxidase	Proteins	Enzymatic conjugation	Biosensor
α-chymotrypsin	Cellophane	Plasma modification/ covalent binding	Not specified
Acetylcholinesterase	pHEMA[a]	Entrapment Covalent binding	Biosensor
Choline oxidase		Covalent binding Ionic interactions	
Papain	Polyethylene Glass surfaces	Plasma modification Covalent binding	Not specified
Lysozyme	Gelatin and succinylated gelatin Poly(ethylene glycol)–poly(aspartic acid) copolymer	Adsorption Ionic interactions Entrapment/ionic binding	Reduction of prosthetic valve endocarditis Delivery of lysozyme as a lytic enzyme

Heparinase	Sepharose	Covalent binding	Bioreactor for extracorporeal elimination of heparin from blood
Phospholipase A_2	Agarose	Not mentioned	Bioreactor for treatment of hypercholesterolemia (reduce plasma cholesterol)
Bilirubin oxidase	Agarose	Covalent binding	Bioreactor for removal of bilirubin from blood
Prolidase	PLGA[b] microspheres	Encapsulation	Enzyme replacement therapy

[a] Poly(2-hydroxy ethyl methacrylate).
[b] Poly(D,L-lactide-co-glycolide).

Biologically Functional Surfaces

Biomaterials, especially when used in tissue engineering applications, must have the capacity to induce tissue regeneration/repair in order to achieve a more rapid recovery of the defect. At present, the existing scaffolds are not satisfactory in achieving rapid and full recovery of the defect. The attachment of cells to biomaterials, and their subsequent spreading, is mediated by extracellular matrix (ECM) glycoproteins such as fibronectin, collagen, etc. ECM glycoproteins contain short sequences with cell attachment properties, which interact with the integrin family of cell surface receptors. The peptide sequence Arg-Gly-Asp (RGD) has been identified as being capable of interacting with cell surface receptors. Thus, the immobilization of biologically active molecules on the surface of biomaterials for presenting effectors to target cells or to induce a particular effect is of great interest, since the immobilization of active agents presents the advantage of providing a continuous and localized stimulus for cell proliferation. Unlike nonimmobilized active agents, which are often consumed by cells, immobilized biomolecules remain bound to a substrate that is not consumed by cells and thus remain available to stimulate growth of additional cells. This is particularly useful in perfusion cultures in which growth medium is continuously added and removed to allow long-term cell proliferation.

Biodegradable polymers have been used as scaffolding materials for various tissue engineering applications because they can provide the support on which cells and tissues can adhere, but they can also guide and regulate the proliferation and activities of the adhered cells. However, the intrinsic hydrophobic property of some of these polymers restricts their applications as cell colonizing materials. Many methods have been used to modify the properties of polymer surface, such as plasma treatment-induced grafting polymerization, ozone oxidation, and immobilization of enzymes, and special biologically active agents have been used to introduce reactive groups onto polymeric surfaces.

Various strategies have been developed to incorporate bioactive agents on the surface of biomaterials for controlling cell and tissue responses. The immobilization process can be involved, enriching surfaces for enhancing the cellular adhesion. Biomolecules such as enzymes, antibodies, antigens, peptides, or drugs have been immobilized on or within polymeric systems. An example of an adhesive protein is fibronectin, which is able to promote cellular adhesion through binding to integrin receptors, and this interaction has also been shown to play a role in cell growth, differentiation, and overall regulation of cell function. The incorporation of the adhesion peptide RGD into a non-adhesive hydrogel proved to be useful for tissue resurfacing. There are numerous other adhesion peptides for targeting particularly desirable cell types and to modulate biological responses.

Urokinase has been widely used for the clinical treatment of thrombogenetic disease and hemorrhoidal disease. Artificial organ materials, on which urokinase was immobilized for its fibrinolytic activity, have been developed for blood-compatible materials. For example, Liu et al. immobilized urokinase by encapsulation in poly(2-hydroxyethyl methacrylate) and König et al. introduced urokinase on the surface of the polytetrafluoroethylene using plasma modification technique by covalent bond. Another example of immobilized urokinase application was reported by Kato and co-workers, who had used urokinase immobilized in a Teflon catheter for treatment of thrombosis.

The surface functionalization of biomaterials with biological molecules, include the incorporation of adhesion and differentiation factors. The same approach may be used to incorporate specific enzymes able to regulate a number of cell functions. For instance, it is known that mitogen activated protein (MAP) kinase, upon activation by dual phosphorylation at threonine and tyrosine residues, is able to activate downstream targets that have been implicated in controlling gene expression, cell differentiation, and proliferation. This enzyme may be immobilized on the surface of biomaterials to control cell response, but other possibilities using different enzymes remain unexplored.

Enzyme Delivery

Although purified enzymes are now available for some enzyme deficiency diseases, there are many problems in delivering the enzyme to the required site under such conditions that it will remain stable and active for a reasonable time. Normally, quite large amounts of enzyme are necessary with high level of purity and in a non-immunogenic form. In addition, many enzymes when administered are inactivated or degraded fairly rapidly. The delivery of therapeutic molecules requires, therefore, efficient strategies to have a precise control on their release profile according to specific locations. It might be possible to control the release of such molecules by creating delivery systems sensitive to changes in pH, temperature, or salt concentration or to the feedback provided by cells. The concept of enzyme-activated drugs in therapy is scientifically, as well as clinically, attractive, as it allows the chemist and enzymologist full intellectual rein in designing interlinked systems. As therapeutic drugs, enzymes possess several attributes such as high specificity toward substrate, high solubility for preparing liquid formulations, and optimum activity under physiological conditions. The administration of enzymes, in cases of enzyme deficiency and inborn errors of metabolism and in the treatment of certain types of cancer, appears to offer a successful form of therapy. Cancer therapy based on the delivery of enzymes to tumor sites has advanced in several directions since antibody-directed enzyme/prodrug therapy was first described. Nanospheres, Nano-capsules, liposomes, micelles, and other Nano-particulates are frequently referred to as carriers for delivery of therapeutic and diagnostics agents.

Asparagine is an essential amino acid for certain types of leukemias that lack asparagines synthetase activity. The activity of L-asparaginase is to degrade asparagine into aspartate and ammonia. Therefore, asparaginase has been of interest to biochemists and clinicians as a possible cancer therapeutic agent. Some success has been achieved in administering asparaginase in capsules made of nylon and polyurea to mice and rats. Although asparaginase has been found to be effective in the treatment of some patients, it may have several serious side effects. Relatively high concentrations of the enzyme are needed for it to be clinically effective. These levels cause a wide range of toxic effects on several organs including the liver, pancreas, kidneys, and brain. The enzyme may also be recognized as foreign by the body and potentially severe immunogenic responses will be stimulated, resulting in hypersensitivity reactions. To overcome these problems, immobilized enzyme derivatives have been prepared on various supports for extracorporeal treatment.

Enzymes may be also used in cancer therapy as prodrug activators. This therapy consists basically in using a drug that has been chemically modified so that it remains inactive until specifically activated by an enzyme at the target site.

Another example of an anticancer enzyme is hyaluronidase.50 Hyaluronidase is a globular enzyme of endoglycosidase action, which can depolymerize hyaluronic acid in the organism, decreasing its viscosity and increasing tissue permeability. Hyaluronidase has been utilized extensively as an adjunct in anticancer chemotherapy regimens, suggesting that hyaluronidase has intrinsic anti-cancer properties against tumor growth. In vitro studies in tissue culture with tumor spheroids and in vivo tests using animal models demonstrated the beneficial effect of hyaluronidase for the penetration of drugs into tumor tissue. Later, in a prospective clinical trial, hyaluronidase significantly improved the outcomes of patients with bladder carcinoma if the enzyme was administrated topically together with mitomycin C. Hyaluronidase is also used for local application (subcutaneous injections) during treatment of joint disease, in dermatology, and in ophthalmology. Hyaluronidase had been used in ophthalmology with the aim of formation of a thinner scar and to prevent necrosis after paravataes with zytostatics.

Trypsin has also been used to remove dead tissue from wounds, burns, and ulcers to speed the growth of new tissue and skin grafts, as well as to inhibit the growth of some contaminant organisms. The inappropriate activation of trypsinogen within the pancreas leads to development of pancreatitis. Once trypsin is activated, it is capable of activating many other digestive pro-enzymes. These activated pancreatic enzymes further enhance the auto-digestion of the pancreas. Many materials, such as nylon, polysulfone, glycidyl methacrylate, chitosan, cellulose, and cellulose derivatives, have been used for trypsin immobilization. Enzyme therapy has also been tested for pancreatic insufficiency and cystic fibrosis. Pancreatic insufficiency can be alleviated by administrating orally enteric-coated microspheres containing lipase, amylase, and proteases. A special polymer coating protects the enzymes at low pH, such as in the stomach, and then releases them in the intestine at physiological pH.

Lysozyme is a good example of an enzyme that catalyzes chemical reactions in the cell. Lysozyme acts to kill bacteria by cleaving the covalent bond between the alternating polysaccharides that compose peptidoglycan in bacterial cell walls. The human salivary defense proteins and lysozyme are known to exert a wide antimicrobial activity against a number of bacterial, viral, and fungal pathogens in vitro. Therefore, these proteins, alone or in combinations, have been incorporated as preservatives in foods and pharmaceuticals as well as in oral health care products to restore saliva's own antimicrobial capacity in patients with dry mouth. These antimicrobials used in oral health care products, such as dentifrices, mouth rinses, moisturizing gels, and chewing gums, have been purified from bovine colostrum. Other studies had been reported with lysozyme bound to chitosan, silica gel by means of physical adsorption, crosslinking to a polystyrene divinylbenzene matrix by the formation of ionic bindings, and by covalent attachment to nonporous glass beads. Harada and Kataoka28 described lysozyme immobilized into poly(ethylene glycol)–poly(aspartic acid) micelle. Lysozyme was selected as a model protein to incorporate into the micelle because it has a high isoelectric point (pI = 11), is positively charged over a wide range of pH, and has practical usage in drug delivery application as a lytic enzyme. Chen and Chen118 prepared immobilized lysozyme by carbodiimide method to form amide bonds with an enteric coating polymer (hydroxypropyl methyl-cellulose acetate succinate [AS-L]) as the carrier, which shows reversibly soluble-insoluble characteristics with pH changes.

The glucose-6-phosphate dehydrogenase enzyme catalyzes the oxidation of glucose-6-phosphate to 6-phosphogluconate while concomitantly reducing the oxidized form of nicotinamide adenine dinucleotide phosphate ($NADP^+$) to nicotinamide adenine dinucleotide phosphate (NADPH). NADPH, a required cofactor in many biosynthetic reactions, maintains glutathione in its reduced form. Reduced glutathione acts as a scavenger for dangerous oxidative metabolites in the cell. With the help of the enzyme glutathione peroxidase, reduced glutathione also converts harmful hydrogen peroxide to water. Red blood cells rely heavily on glucose-6-phosphate dehydrogenase activity because it is the only source of NADPH that protects the cells against oxidative stresses. People deficient in glucose-6-phosphate dehydrogenase are not prescribed, therefore, with oxidative drugs because their red blood cells undergo rapid hemolysis under this stress. Glucose-6-phosphate dehydrogenase deficiency is the main cause of severe neonatal jaundice. The deficiency of this enzyme affects all races; the highest prevalence is among persons of African, Asian, or Mediterranean descent. Study of immobilized glucose-6-phosphate dehydrogenase has been reported. Kotorman et al. immobilized glucose-6-phosphate dehydrogenase from yeast on polyacrylamide beads possessing carboxylic functional groups activated by a watersoluble carbodiimide. They verified highest operational stability of immobilized glucose-6-phosphate dehydrogenase.

In relation to the use of immobilized enzymes with therapeutic purposes, catalase is one of the most interesting because it is employed to accelerate healing as well as to correct hereditary deficiencies and, in combination with hydrogen peroxide, as an antiseptic against anaerobes. The catalase enzyme has the ability to decompose hydrogen peroxide into oxygen and water, playing a central role in controlling the hydrogen peroxide concentration in human cells. More than 98% of blood catalase is localized in erythtocytes. These cells, with their high catalase level, provide a general protection against the toxic concentration of this small hydrogen peroxide molecule. The deficiency of catalase could cause acatalasemia. Several methods have been developed for the immobilization of catalase. Immobilization is often accompanied by changes in the enzymatic activity, optimum pH, affinity to the substrate, and stability. The extent of these changes depends on the enzyme, carrier support, and the immobilization conditions. The shift in the optimum pH, from acidic or alkaline to neutral pHs, may be useful for biomedical applications since it will allow the use of some enzymes (more active at low or high pHs) under more physiological conditions.

The approach developed by Sakiyama-Elbert et al., consisting in a cell-triggered growth factor delivery system, may also be used for the release of other important therapeutic molecules.

Bioreactors for Extracorporeal Enzyme Therapy

Extracorporeal shunts have been proposed for the treatment of several clinical conditions. The most likely applications for enzymatic treatment are the removal of urea during kidney failure, removal of toxins (e.g., paracetamol) during liver failure, or the reduction of key metabolites from the circulation to treat cancer.

Urease is one of the most important enzymes in biomedical applications. Urease is an enzyme that catalyzes the hydrolysis of urea to form ammonia and carbon dioxide. Urea is one of the main metabolic end products, and the removal of its excess has been a major problem for patients suffering from renal failure. Hence, its immobilization by entrapment has been investigated by many workers for applications in biosensors and as artificial kidneys. The most attention has been given to the development of enzyme reactors, where the urea would be removed and the dialysis fluid prepared for further use. The use of this enzyme is often limited due to its high cost, availability in small amounts, instability, and the limited possibility of feasible recovery of these biocatalysts from a reaction mixture. Numerous synthetic and natural polymeric supports have been used for urease immobilization, and their uses in medical and technical fields are well reported. The covalent bond of urease in different supports has been reported in many studies. Some commonly used supports are chitosan-poly(glycidil methacrylate), carboxymethylcellulose, polyurethane, sepharose-2B, polyacrylamide, ion exchange resins, copolymers of polyglycidylmethacrylate, calcium alginate beads, poly(vinyl alcohol) (PVA), hydroxyapatite, 2-dimethylaminoethylmethacrylate, poly(ethylene glycol dimethacrylate/2-hydroxy ethylene methacrylate) microbeads, poly(caprolactone)/starch, and poly(orthoesters).

L-asparaginase has been used for treating leukaemia and disseminating cancers that require asparagines for growth, but this treatment presents several serious side effects. To overcome these problems, immobilized enzyme derivatives have been prepared on various supports for extracorporeal treatment. Blood can be passed over the immobilized enzyme, thus depleting the asparagine supply needed by the cancer cells. The enzyme does not come into direct contact with the organs to which it is toxic, and hypersensitivity reactions do not occur. With this type of treatment, however, the blood plasma must be first separated from the cells to minimize cell damage and then passed through a separate column containing the immobilized enzyme. This process requires that the blood remain outside the body for relatively long periods of time, resulting in the denaturation and depletion of many plasma proteins. Some techniques have been developed to minimize this problem, such as the use of a porous hollow-fiber plasmapheresis device. With this system, the plasma can be separated from the whole blood and contact with immobilized enzyme in one passage, thus minimizing its time outside the body and then reducing the damage to the plasma proteins. Adsorption techniques have been used to immobilize asparaginase onto hollow fibers after first coating the fibers with albumin and then crosslinking the enzyme with glutaraldehyde. The use of L-asparaginase covalently attached to nylon tubing may constitute a useful system to be used in clinical applications.

Bilirubin oxidase is also an example of enzyme used in extracorporeal applications. All human newborns accumulate bilirubin to levels greater that those in adults, and 20% accumulate enough to stain their skin, resulting in jaundice. Bilirubin binds to cellular and mitochondrial membranes, causing cell death in a variety of tissues. Clinically, bilirubin toxicity may lead to mental retardation, cerebral palsy, deafness, seizures, or death. The most common treatments for jaundiced infants are phototherapy and exchange transfusion. This technique presents serious problems such as hypoglycemia, hypocalcemia, acidosis, transmission of infectious, etc. Lavin et al. reported the use of a highly specific enzyme to remove bilirubin from the bloodstream using a small reactor (extracorporeal circuit) containing bilirubin oxidase covalently immobilized in agarose beads. These researchers obtained good results for the removal of the bilirubin in humans and in genetically jaundiced rats.

Heparinase, an enzyme that degrades heparin into small polysaccharides, has also been immobilized into an extracorporeal device (artificial kidney bioreactor) to eliminate the anticoagulant properties of heparin (used to prevent clotting in the device) before the blood returns to the patient.

Diagnostic Assays and Biosensors

Isolated or combined enzymes are being used in medicine as useful tools for clinical analysis. About 50 different enzymes are used in different aspects of clinical diagnoses, and for most of these, much higher levels of purity are required than for most industrial enzymes. Two of the major enzymes used are peroxidase from horseradish and alkaline phosphatase from beef intestinal mucosa, both being required for immunoassays. The enzymes may be used in test strips, ELISA, biosensors, and auto-analyzers. The serum uric acid concentration is an important index for clinical diagnosis of gout, leukemia, toxemia of pregnancy, and severe renal impairment. A number of enzymes are assayed in serum and urine for diagnostic purposes; the more frequently used ones are discussed below.

Alkaline phosphatases (ALP) are a group of enzymes found primarily in the liver (isoenzyme ALP-1) and bone (isoenzyme ALP-2). There are also small amounts produced by cells lining the intestines (isoenzyme ALP-3), the placenta, and the kidney (in the proximal convoluted tubules). What is measured in the blood is the total amount of alkaline phosphatase released from these tissues into the blood. As the name implies, this enzyme works best at an alkaline pH (pH 10), and thus the enzyme itself is inactive in the blood. Alkaline phosphatase acts by splitting off phosphorus (an acidic mineral), creating an alkaline pH. The primary importance of measuring alkaline phosphatase is to check the possibility of bone or liver diseases.

Another application of immobilized enzymes is the development of improved sensing devices. Because of their high specificity for given substances, enzymes and monoclonal antibodies are particularly suitable for use as sensors. The membrane-covered electrode described by Clark in 1959 is the dominating sensor for the measurement of dissolved oxygen. Numerous modifications of the original concept have been developed. For instance, biosensors using enzymes have been used to detect the presence of various organic compounds, and recent developments have proven to be both rapid and highly selective. They have been used in important applications such as in clinical laboratories, fermentation processes, and pollution monitoring. Most of them have used a free or immobilized enzyme and an ion-sensitive electrode that measures indirectly (e.g., by temperature or color changes produced by an enzymatic reaction) the presence of a product whose formation is catalyzed by the enzyme. The biosensors usually have immobilized biological molecules attached to the surface of a transducer that allows an electronic or optical signal to be converted into an appropriate signal. This type of biosensor could be used to measure glucose, sucrose, lactose, L-lactate, galactose, L-glutamate, L-glutamine, choline, ethanol, methanol, hydrogen peroxide, starch, uric acid, etc., by using specific enzymes.

Glucose oxidase is normally used to assay glucose concentration. Glucose sensors are the biosensors that have attracted much interest in both research and applications fields. One particularly important medical application of improved biosensors could be in the treatment of diabetic patients for whom proper levels of insulin and glucose must be maintained. For instance, small implantable devices for sampling blood to determine the levels of glucose and regulate the delivery of insulin could be developed using this enzyme. A great variety of immobilization methods (e.g., encapsulation, entrapment) and transducers have been developed to construct the glucose sensors with better performance and practicability since the work of Clark and Lyons in 1962. Several materials have been used, such as polyethylene terephthalate (PET), polyacrylamide, N- isopolyacrylamide, sol-gel, poly(2-hydroxyethyl methacrylate), alginate, artificial resins, glass, etc.

For example, Zhang and Cass have also immobilized alkaline phosphatase on a nanoporous nickel-titanium film for sensor applications.

A number of other enzymes have been described with great potential for medical applications, including carboxypeptidases, collagenase, fibrinolysin, pepsin, streptokinase, subtilisin, thrombin, tissue plasminogen activator, α-amylase, α-galactosidase, glucoamylase, lactase (-galactosidase), pectinase, pancreatin, phospholipases, cholesterol esterase and other DNases, RNases, phosphatases, esterases, sulfatases, isomerases, glucose isomerase, superoxide dismutase, cholesterol esterease, creatine kinase, and penicillin acylase.

Use of Immobilized Enzymes in the Food Industry

Various applications of immobilized enzymes can be found in the food industry, medicine and research. Other industrial applications of immobilized enzymes include laboratory-scale organic synthesis and analytical and medical applications.

Table below lists a selection of currently used immobilized enzymes in the food industry. In any commercial application, the choice of a specific immobilized enzyme or mode of immobilization must be based on a compromise of all advantages and disadvantages between free and immobilized enzymes. This suggests that the development and application of immobilized enzyme products requires a good understanding of both technical and economic factors, as well as a good sense of the larger market forces at play.

Applications of Immobilized Protease

The broad class of hydrolases includes proteases of various types which are of economic consequence to segments of the food industry. In general, proteases are very expensive enzymes and many of them are not listed as generally recognized as safe. The practical applicability of immobilized proteases in food processing is restricted to fluid processing, such as the chill-proofing of beer, continuous coagulation of milk and the oxidative rancidity in milk.

Table: Some important applications of immobilized enzymes in the food industry.

Enzyme	EC number	Substrate	Product
β-Galactosidase	3.2.1.23	Lactose	Lactose-free milk
Lipase	3.2.1.23	Triglycerides	Cocoa butter substitutes
Nitrile hydratas	4.2.1.84	Acrylonitrile	Nicotinamide
Aminoacylase	3.5.1.14	D, L-aminoacids	L-aminoacids
Raffinase	3.2.1.22	Raffinose	Galactose and sucrose
Invertase	3.2.1.26	Sucrose	Glucose and fructose mixture
Thermolysin	3.4.24.27	Peptides	Aspartam
Glucoamylase	3.2.1.3	Starch	D-Glucose
Papain	3.4.22.2	Proteins	Removal of 'chill haze' in beers
Tyrosinase	4.1.99.2	Pyrocathecol	L-DOPA

Proteases immobilized to an ethylenemaleic anhydride copolymer have been used to chill-proof beer on a pilotplant scale. In the manufacture of beer, soluble proteins from the malt produce a clouding of the beer when the product is chilled. This defect is called 'chillhazing' and can be overcome by hydrolysing the proteins. The immobilized-enzyme treatment was carried out during the fermentation step. Glutaraldehyde cross-linked papain has also been used in attempts to chill-proof beer. Researcher also prepared enzymically active coimmobilized papain and its activator (Cysteine) using cyanogen-bromide-activated sepharose as the matrix. Papain was also immobilized on gold nanorods as a heterogenous biocatalyst. As well as used immobilized pepsin to study the feasibility of this system for the continuous coagulation of milk, another area where proteases can be effectively utilized is in the production of protein hydrolysates. Amino peptidases hydrolyse peptides splitting off N-terminal residues, whereas carboxy peptidases hydrolyse peptide, splitting off C-terminal residues. The amino peptidases and carboxy peptidases have been immobilized and have been used in conduction with general proteases to produce protein hydrolysates.

The use of immobilized enzymes finds particular advantages using proteolytic enzymes. Papain, a thiol protease present in the latex of Carica papaya, exhibits a broad proteolytic activity and is an enzyme of industrial use and of great research interest. Papain has a variety of industrial applications, especially in the food industry. It is used to tenderize meat and meat products, in the manufacture of protein hydrolysate, in the brewing industry for clarifying juice and beer, in the dairy industry for cheese, in the baking industry and in the extraction of flavour and colour compounds from plants. Papain can also be used in forage industry to increase the utilization and transformation rate of proteins and in resolving plant and animal protein to make high-quality health products. The potential uses of papain include amino acid esters and peptide synthesis, treatment of acute destructive lactation mastitis, treatment of red blood cells prior to use in antibody-dependent cell-mediated cytotoxicity assays with lymphocytes and enzyme inhibition-based biosensors for food safety and environmental monitoring.

Inorganic immobilization supports such as particulate alumina have also been examined for immobilization of the proteolytic enzyme, papain. Organic phosphate linkers have been used for the creation of free carboxyl groups in a two-step process. Papain binding to these alumina derivatives was performed using the watersoluble carbodiimide 1-ethyl-3-(dimethylaminopropyl) carbodiimide. It has been shown that immobilized papain had similar kinetic constants to papain in solution. By fluorescence measurements, it was concluded that the hydrophobic environment of the active site remained unchanged, while the structure of the rest of the protein was perturbed by its association with the negatively charged surface.

Identification of proteins by liquid chromatography followed by mass spectrometry (LC-MS) is normally used as a suitable analytical method. In practice, proteins are subjected to proteolytic cleavage and a complex mixture of peptides is produced. The peptide mixture is then separated by high-resolution LC-MS. Protein digestion is normally performed in a homogeneous aqueous solution containing the proteolytic enzyme and the sample (enzymeto-protein ratio, 1:50). However, this method suffers from several problems that may interfere with identification of sample proteins. Some of these problems include long digestion times (up to 24 hours), autodigestion byproducts and limited enzyme-to-substrate ratio.

Recent experiments performed with papain immobilized covalently with sepharose revealed that the immobilization of papain did not significantly alter any of the kinetic parameters of the enzyme.

This study suggested that the immobilization of papain conferred exceptional conformational stability on the enzyme and that the immobilized enzyme was only reversibly denatured, whereas the soluble enzyme was irreversibly denatured. There therefore exists a great potential for the use of proteases in processing liquid substrates; exactly when this potential will be realized is not known.

Applications of Immobilized Amino Acylase

Developmental work on the first commercially successful process using an immobilized enzyme was initiated by Researcher in 1966 and consisted of an L-amino acid acylase. The reaction catalysed by the enzyme L-amino acid acylase can be represented as follows:

$$\text{DL-acyl amino acid} + H_2O \xrightarrow{\text{amino acylase}} \text{L-amino acid} + \text{D-acyl amino acid}$$

In the production of amino acids by microbial fermentation, a racemic mixture of acyl amino acids is produced. The desired product is the L-amino acid, which is used to supplement proteins in foods and in pharmaceutical preparations. The resolution of racemic mixtures was performed by a soluble enzyme in a batch operation. The most satisfactory method of immobilizing the amino acid acylase was by linking the enzyme to DEAE Sephadex →A-25 at pH 7.0. Approximately 333 units of acylase could be bound per millilitre of the carrier, and yield of enzymic activity was reported to be 47%. Immobilization of the enzyme enhanced the thermal stability of the catalyst, and the Sephadex→-aminoacylase column could be operated continuously at 50 °C for 32 days with less than 40% loss of initial activity. The lost activity was compensated for by the addition of an amount of fresh enzyme, corresponding to the deteriorated activity. The continuous resolution of racemic mixtures of amino acids is depicted in Figure. The acylated DL-amino acids are tempered to 50 °C prior to passage through immobilized enzyme reactor. The reaction mixture after enzymic treatment is concentrated, and the L-amino acids are crystallized and separated. The acylated D-amino acids are then allowed to undergo chemical racemization.

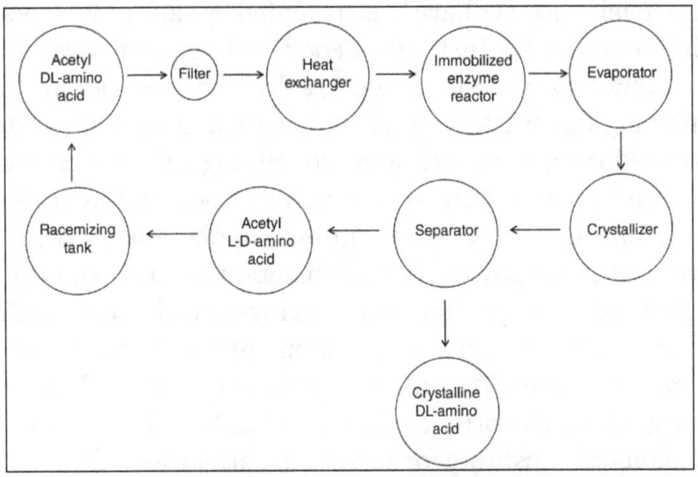

Figure: Schematic for the process of L-amino acid production using immobilized aminoacylase.

Researcher also observed conversions with racemic methionine mixtures to be 88% of the calculated theoretical yields. The use of an ion exchange rather than covalent linking provided an advantage in regenerating the system. Circulation of a fresh enzyme charge through the reactor was sufficient to regenerate the system. The substrate used was easier to work with as it was devoid of macromolecules, such as proteins and nucleic acids. The economies affected by the use of immobilized amino acylase show the cost of producing a given quantity of the product by the immobilized enzyme system was only 60% of that of a conventional batch process using soluble enzyme. In the batch process, purification steps have to be included in order to separate the product in a pure form from contaminating proteins and pigments, and this led to a decrease in the yield of the product. Automation of the process using immobilized acylase and the elimination of the purification step helped to lower labour costs and increased yields of the product. Researcher later covalently coupled amino acylase to halogenacetyl groups on cellulose and reported this system to be as effective as the DEAE Sephadex -acylase system.

Applications of Immobilized GlucoseIsomerase

Another immobilized enzyme that has been applied successfully on a commercial scale utilizes glucose isomerase. This enzyme isomerizes glucose to fructose and is used in the corn syrup industry for producing HFCS. Glucoamylases introduced in the 1950s were capable of hydrolysing corn starch to yield corn syrups with 95–97 dextrose equivalent (DE) sweeteners, but these high-DE syrups were still not as sweet as sucrose and the product tended to crystallize at room temperature. The isomerization of glucose to fructose not only eliminates crystallization problems encountered during shipping and storage, but also enhances the sweetening power of the syrup. Researcher reported that organisms of the Streptomyces species isolated from soils could produce high levels of glucose isomerase when grown in culture media containing sources of xylan, such as wheat bran, corn cob or corn hull. This led to the economical production of the enzyme which could be used on an industrial scale to produce HFCS. Both the application requirements and the biochemical properties of glucose isomerase have driven the development of immobilized forms of the enzyme.

Glucose isomerase could be fixed in the cells by heating the cells to 65 °C for 15 min. Other methods of fixing glucose isomerase in cells include using entrapping cells in polyacrylamide gels or covalently coupling the enzyme to porous glass beads. Glucose isomerase preparations used in a continuous manner enhanced stability of the immobilized enzyme, whereas batch processing was detrimental to enzyme stability. Commercial processes have been patented by various workers. Based on the success and economics of such processing, the Institute of Food Technologists awarded their 1975 Industrial Achievement Award to Clinton Corn Processing Co., a division of Standard Brands Inc. Clinton first produced a HFCS containing 15% fructose by a batch method using soluble glucose isomerase. This process was later improved to produce a corn syrup containing 42% fructose on a continuous process using immobilized glucose isomerase. Cornstarch was liquified and hydrolysed to dextrose which was refined, isomerized, refined again and concentrated to obtain the syrup. The process is shown schematically in Figure. The entire process is a combination of batch and continuous operations in which hydrolysis of starch and subsequent preparation for enzymic isomerization are batch processes, whereas the isomerization, refining and evaporation are continuous operations and the remainder of the process is semi-continuous. The entire operation is automated and controlled by extensive on-line process-control systems. The price of the HFCS has been lower than sucrose, making these syrups an attractive economical substitute for sucrose in various food formulations.

Applications of Immobilized Glucosidases Enzymes

Glycoside hydrolases or glycosidases are categorized under the Enzyme Commission (EC) number 3.2.1 and catalyse the cleavage of individual glucosyl residues from various glycoconjugates including α- or β-linked polymers of glucose. Glycoside hydrolases have a variety of uses in the food industry.

One of the glycosidases used in the wet corn-milling industry is glucoamylase. This enzyme hydrolyses α-1, 4 glucan links in non-reducing ends of the starch chains. Weetall and Havewala conducted preliminary studies with glucoamylase immobilized on porous silanized support. Although only two or three immobilized enzyme processes are being used commercially by the food industry, research is in progress to develop newer systems. Monsan and Durand immobilized invertase by a combination of adsorption and cross-linking using bentonite as the support.

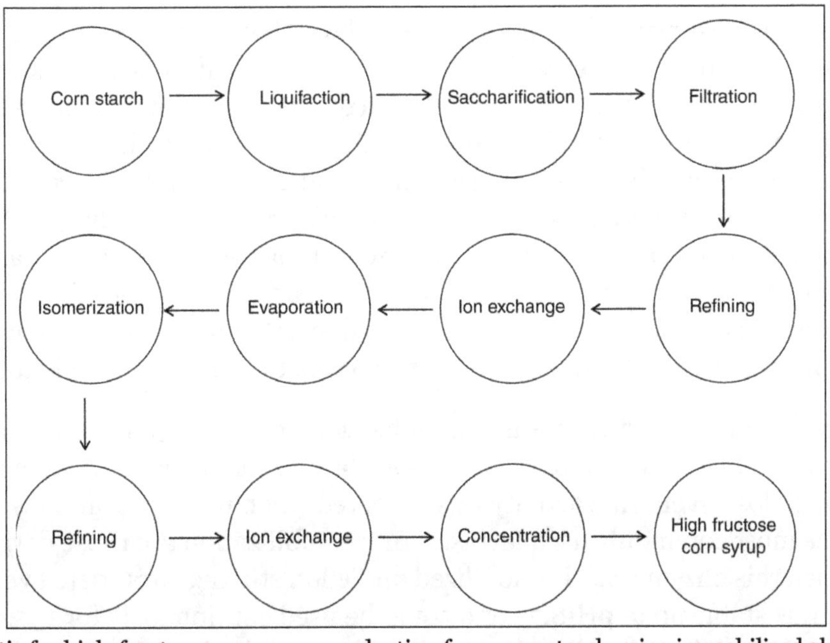

Figure: Schematic for high-fructose corn syrup production from cornstarch using immobilized glucose isomerize.

Invertase or β-fructofuranosidase, now called β-D-fructofuranoside fructohydrolase, hydrolyses sucrose to glucose and fructose. Researcher adsorbed invertase on DEAE cellulose, whereas others have covalently bound this enzyme to polystyrene tubes and porous glass. However, the use of immobilized invertase is still in the experimental stages, and the demand for an industrial process to invert sugar by immobilized invertase is not that great.

Amylases are important to the starch industry and attempts have been made to adsorb α-amylase on cellophane, prior to cross-linking with glutaraldehyde, onto diazotized polyamino styrene and on a variety of other supports including polyacrylamide entrapment. Two problems that have been encountered are loss of activity of the enzyme post-immobilization and the leaching of enzyme from the support. α-amylase attacks the α-1, 4 glucan linkages in polysaccharides (e.g. starch) to produce oligosaccharides, glucose, maltose, maltotriose, etc. Immobilization of this enzyme led to a change in specificity of the enzyme as evidenced by the enhanced rate of glucose production from starch. The use of α-amylase in ultrafiltration reactors has also been investigated extensively, and it has been suggested that coupling of α-amylase to form soluble-polymer derivatives is more desirable.

Another amylase, called β-amylase, degrades α-1, 4 glucan linkages starting from the non-reducing ends of starch to produce maltose. This enzyme has been immobilized on polyacrylamide derivatives. Marshall and Whelan suggested that the degradation of potato starch by β-amylase in the presence of pullulanase to yield maltose was a feasible proposition. Pullulanase cleaves β-1, 6 glycosidic linkages and has recently been purified and immobilized by cross-linking to a copolymer of acrylamide and acrylic acid. Further development of such a process would lead to the utilization of amylopectins to produce amylose. The overall developments in starch-degrading enzymes have resulted in many potential uses. The obvious benefits of utilizing these enzymes in the corn-milling, brewing, grain-milling and baking industries and in the production of distilled alcoholic beverages can be readily visualized. However, some other potential uses for these enzymes lie in the area of waste treatment of starchy effluents from these industries and others such as effluents from paper mills.

Lactase or β-galactosidase hydrolyses the glycosidic bond between glucose and galactose in the milk sugarlactose. This interest stems from reports that indicate individuals of certain ethnic groups suffer from a malady called lactose intolerance. This condition is believed to arise from a lack of β-galactosidase in the intestinal microvilli. Lactose also poses technological problems in the processing of dairy products, such as ice cream and concentrated milk. More recently, intensive efforts have been made to utilize lactase in treating cheese whey. Cheese whey is a byproduct of the cheese-making industry and consists of 63–69% lactose. It has been estimated that in the US alone nearly 30 billion pounds of whey are produced. Of this, nearly 40% is not utilized and is therefore wasted as an effluent. This wasted whey imposes a high biochemical oxygen demand on receiving waters. Apart from pollution potential by whey, it also constitutes a loss of valuable nutrients.

In an effort to reclaim these lost nutrients, immobilized lactases have been prepared and evaluated. Lactase has also been immobilized on porous glass beads, with a 75% retention of original activity. Further, no loss in enzyme activity was observed when the preparation was stored at 4 °C for 2 months. The most commonly used lactases for immobilization are those derived from E. coli and A. niger. When this enzyme was immobilized on Teflon stirring bars coated with a polyisocyanate polymer, it was stable up to pH 8.75 and could be used continuously for 137.6 hours without appreciable losses in activity. As well as immobilized lactase obtained from two different sources: one was from a fungus (called lactase M) and the other from a yeast. The lactase Y which had a near-neutral pH optimum in the soluble form showed a dramatic shift of more than 3 pH units towards the acidic side upon immobilization. Lactase has also been immobilized on collagen membranes, in which case an additional factor contributing to the inhibition of bound lactase has been suggested to be the presence of divalent cations. Additionally, Wierzbicki et al. immobilized lactase from various microbial sources onto porous glass and concluded that many lactases, including one from A. niger, possessed enough activity and stability to be of economic use in the hydrolysis of lactose.

Agarose has been used as a matrix to immobilize the neutral pH optimum lactase. Agarose was derivatized to the succinyl amino ethyl form, and the enzyme was linked to the spacer arm by using N.Ndicyclohexyl carbodi imide. These techniques resulted in an insoluble derivative of lactase in which there were no significant changes in either pH and temperature optima or the kinetic parameters. This enzyme preparation was stable when stored at pH 7.5 and 4 °C for periods of up to 1 year, with no apparent loss in activity. Further, this immobilized lactase was capable of hydrolysing lactose in whey and milk. The safety aspects of many immobilized-enzyme systems, including immobilized lactase, are not very clear according to the Food and Drug Administration (FDA). In many processes for immobilization, however, bound toxic factors are converted to non-toxic forms during chemical reactions, and products which are not converted to nontoxic compounds are subsequently washed out. It is possible that at some future date, if the processes are economical, the FDA may grant approval for the use of such systems.

Applications of Immobilized Enzymes in the Flavour Industry

Apart from proteases and carbohydrases, upases are also important to the food industry. Lipases are used extensively for the enzymic modification of flavour of foods. A search of the literature revealed very little work pertaining to the immobilization of lipase. Kosugi and Suzuki were successful in fixing cell-bound lipase of Pseudomonas mephitica var. lipolytica. The fixed lipase retained 83% of the soluble lipase activity, and the optimum temperature and heat stability of the fixed lipase were the same as those of the soluble lipase extracted from the microbial cells. The fixed lipase could be used continuously for the hydrolysis of tributyrin or triacetin. Porcine pancreatic lipase was immobilized by entrapment in polyacrylamide gel. When the number of coupling groups on the surface of the beads was increased, a concomitant decrease in the specific activity of the bound enzyme was recorded. Additionally, the diazotization of lipase prior to coupling was also detrimental to enzyme activity. In another study, a fungal lipase on succinylaminoethyl agarose was immobilized. The immobilization of this lipase did not significantly alter any kinetic properties of the enzyme. However, this particular enzyme preferentially hydrolysed olive oil rather than butter oil. These experiments, preliminary in nature, may lead to newer methods of flavour modification. Among the other uses for enzymes in the modification of flavour are the use of naringinase and adenosine deaminase. Naringen is a glycoside of the flavonoid family which imparts bitterness to fruits, such as grapefruits and sour oranges. Naringinase is the enzyme which hydrolyses naringin to yield naringin-7, β-glucoside (also called prunin), which is not bitter.

Naringinase immobilizes on different support, including: tannin aminohexyl cellulose; κ-carrageenan; packaging film; and silicate for fruit and vegetable juices debittering. The use of hollow fibre reactors for hydrolysis of naringin in unclarified grapefruit juice has also been evaluated. The other flavour-modifying enzyme studied is adenosine deaminase. Adenosine deaminase converts adenosine to inosine and has been immobilized using polyacrylamide gels. Inosine so produced could be chemically phosphorylated to form inosine 5′ monophosphate (IMP) which is a strong flavour potentiator. Alternatively, an immobilized ATP deaminase has been employed to convert adenosine triphosphate to inosine monophosphate.

References

- Methods-of-Enzyme-Immobilization-313366444: researchgate.net, Retrieved 14, February 2020
- Application-of-immobilized-enzyme-technologies-for-the-textile-industry-A-review-232083060: researchgate.net, Retrieved 23, May 2020
- Enzyme-Immobilization-and-its-Application-in-the-Food-Industry-279524145: researchgate.net, Retrieved 01, March 2020

Chapter 5

Enzyme Engineering and its Applications

The process through which the sequence of amino acids is changed by using various techniques to design valuable and useful enzymes is known as enzyme engineering. Enzymes are used in numerous industries such as pharmaceutical industry, food industry, light industry and biotechnology. The diverse applications of enzyme engineering have been thoroughly discussed in this chapter.

Enzyme engineering (or enzyme technology) is that the application of altering an enzyme's structure (and thus its function) or modifying the catalytic activity of isolated enzymes to produce new metabolites, to allow new (catalyzed) pathways for reactions to occur or to convert from some certain compounds into others (biotransformation).

The study of industrial enzymes and their uses is called enzyme technology. Enzyme technology broadly involves production, isolation, purification, and use of enzymes (in soluble or immobilized form) for the ultimate benefit of humankind. In addition, recombinant DNA technology and protein engineering involved in the production of more efficient and useful enzymes are also a part of enzyme technology. Enzyme technology today thus mostly refers to the application of altering an enzyme's structure (and thus its function) or modifying the catalytic activity of isolated enzymes to produce new metabolites, to allow new (catalyzed) pathways for reactions to occur or to convert from some certain compounds into others (biotransformation).

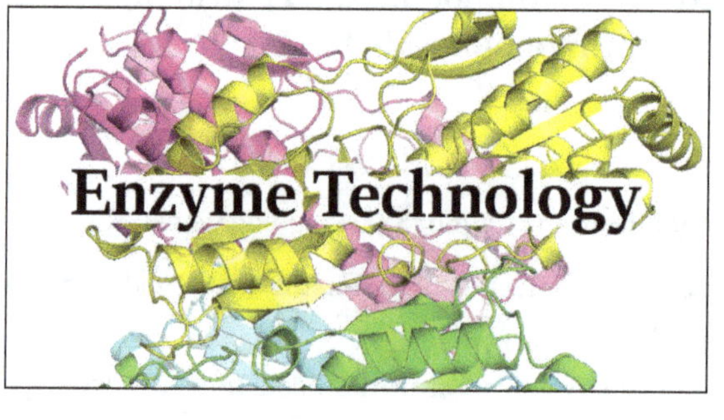

Goal

To develop new and more sustainable products, processes and services to meet the human needs or to improve processes to produce existing products from new raw materials and biomass.

Significance of Enzyme Technology

Enzymes have a wide range of applications. These include their use in food production, food processing and preservation, washing powders, textile manufacture, leather industry, paper industry, medical applications, and improvement of environment and in scientific research.

As per recent estimates, a great majority of industrially produced enzymes are useful in processes related to foods (45%), detergents (35%), textiles (10%) and leather (3%).

The products of enzyme technology will be useful for chemicals, pharmaceuticals, fuel, food or agricultural additives. A number of properties which may be improved or altered by genetic engineering as well as the yield and kinetics of the enzyme, the ease of downstream process and various safety aspects. Enzymes from dangerous or unapproved microorganisms and from slow-growing or limited plant or animal tissue may be cloned into safe high-production microorganisms. In the future, enzymes may be redesigned to fit more appropriately into industrial processes. Strategies that target mutagenesis to particular regions of a protein or use recombination to launch large sequence changes can complement full-gene random mutagenesis and pave the way to achieving ever more ambitious enzyme engineering goals.

Applications of Enzyme Technology

In enzyme technology – a subfield of biotechnology – new processes have been and are being developed to manufacture both bulk and high added-value products utilizing enzymes as biocatalysts, in order to meet needs such as food (e.g., bread, cheese, beer, vinegar), fine chemicals (e.g., amino acids, vitamins), and pharmaceuticals. Enzymes are also used to provide services, as in washing and environmental processes, or for analytical and diagnostic purposes

Advances in Enzyme Technology

New enzymes are being sought in the natural environment and by strain selection Established industrial enzymes are being used in as wide a variety of ways as can be conceived; Novel enzymes are being designed and produced by genetic engineering; New organic catalysts are being designed and synthesized using the 'knowhow' established from enzymology; and More complex enzyme systems are being utilized.

Future Applications

The exploitation of enzymes as electrocatalysts (specific biosensors), Enzymes as analytical tools to measure specific compounds, for the regeneration of specific metabolites, Enzyme utilization in the synthesis of bulk organic materials and the production of fragrances and cosmetics, Enzyme utilization in the formation of food flavors and aroma compounds, The use of enzymes as tools for the detoxification of pesticide residues, Enzymes as monitors of toxic chemical levels in food and water.

Biomedical applications of Enzyme Technology will include:

- The synthesis of new anti-microbial compounds.
- Enzyme replacement therapy.
- Enzymes in the treatment of cancer.
- Enzyme graft and dermatological applications.
- Enzymes as activators of precursor biomolecules.
- Enzyme technology in the prevention of dental cavities.

Advancements in Enzyme Engineering

Many worldwide corporations have recognized the bio-based technologies as one of the key drivers of sustainable growth. However, the biological process is often considered only when the chemical arsenal has failed to achieve synthesis of the target molecule. This is primarily because the unavailability of the desired enzyme to catalyze the reaction in an efficient manner. The exploitation of new types of enzymes, improvements of enzyme properties and of the production process are overall goals of innovation in the enzyme manufacturing industry. Accordingly, systematic methods in the field of enzyme and reaction engineering have allowed access to means to achieve the ends, i.e.1) screening for novel enzymes from natural samples with improved characteristics as a good starting point, 2) engineering the existing enzymes using genetic engineering approaches, 3) fining the enzyme processes in the enzyme manipulation to overcome catalyst limitation, e.g. downstream processing in enzyme manufacturing, formulation of enzyme preparations and enzyme immobilization, etc.

Metagenomic Analysis for Novel Biocatalysts Recovery

In order to obtain microorganisms with special characteristics, traditional enrichment culture technique and screening of a wide variety of microorganisms for the desired activity are widely employed. However, Numerous microbes inhabit the biosphere, more than 99% of which are uncharacterized or unculturable due to difficulties in enriching and isolating microorganisms in pure culture. Recent success of genome sequencing programmes has resulted in an explosion of information available from sequence databases, thus creating an opportunity to explore the possibility of finding new enzymes by database mining. Additionally, tools of metagenome analysis obviate the need of culturable microorganisms by directly extracting DNA from environmental samples and systematically screening for the open reading frames potentially encoding putative novel enzymes. Metagenomic libraries screening are mostly based on the function-driven approaches: i) direct phenotypical detection, ii) heterologous complementation and iii) induced gene expression. By accessing different locations such as volcanic vents, deep oceans beds, arctic tundra, etc, this method could help people to find more than a million previously unknown genes coding for novel enzymes. Numerous studies by metagenome screening have yielded enzymes with potential for biocatalytic applications, such as nitrilase, lipase, β-lactamase, protease and many others, which could serve as a good start for protein engineering. However, the metagenome analysis method also suffers from disadvantages, such as missing constitutively expressed genes, sensitive to orientation of genes and not for substrates that do not migrate to the cytoplasm. The rapid development of cloning-independent DNA sequencing technologies and the combination of metagenomics, metatranscriptomics, and metaproteomics will promote the new biocatalysts discovery.

Surface Display for High-Throughput Protein Screening

Enzymes obtained by screening approaches often force the process engineers to alter process parameters due to inadequacies of the enzyme, such as instability, inhibition, narrow substrate specturm, low yield or selectivity. Utilization of enzyme improvement tools therefore provides a feasible means to fine tuning for adaptability to industrial scale production. Many different enzyme engineering and process engineering methods are available today, such as directed evolution, de novo protein design, use of nonconventional media, using new substrates for old enzymes, immobilization, active-site imprinting, etc.

Phage display is a powerful technique for screening large libraries of proteins which relates phenotypes with their corresponding genotypes. This method is particularly used in “synthetic binding protein engineering”, where libraries of synthetic binding proteins were developed with antigen-binding sites constructed from man-made diversity. Jung et al generated a random library of lipase displayed on the surface of Escherichia coli using the ice-nucleation protein (INP) as an anchor. The top mutant showed 29-fold increase in whole-cell activities compared to wild-type enzyme, and the lipase displaying cells can serve as an alternative immobilized biocatalyst in aqueous-organic solvent reaction medium. Yeasts are another attractive host strains for cell-surface display systems due to their safety, ease of high-cell density cultivation and the capability of eukaryotic proteins folding and glycosylation. These properties make enzyme displayed yeast be of great value in bio-catalytic process, especially in the production of biodiesel, polyester and esters using lipase displaying yeast as whole-cell biocatalysts. the approaches that yeast surface display coupling with fluorescence activated cell sorting (FACS) would be novel platforms for rapidly discovering new protein catalysts in industrial settings, due to without requiring soluble expression and purification.

Cell-Free System Applications in Enzyme Engineering

Cell-free expression systems were also described as important tools for protein engineering and production. By removing the native genomic DNA and transport barriers (cell membrane/wall), processes occurring during the manipulation of cellular systems, like the cloning of a gene-construct into an expression vector system, cell transformation, cell culturing, cell lysis and the removal of cellular compartments are simply bypassed. The limitations of recombinant protein expression in living cells, such as protein degradation and aggregration will also be avoided. Nakano developed a novel strategy in an array format for generation and screening of protein library by affinity selection or enzymatic activity. Through ribosome display, another perspective cell-free technology for in vitro selection and evolution of proteins from large libraries, affinity of a human interleukin-13-neutralising antibody was improved by over 200-fold. Additionly, in vitro compartmentalization (IVC) techniques in cell-free systems showed a strong momentum for the directed evolution of enzymatic activity. One striking example is an extremely fast phosphotriesterase was obtained by IVC lead to a 63-fold increase in k_{cat}. A high-throughput IVC screening platform for oxygen-tolerant [FeFe] hydrogenases evolution was also developed and used to identify mutants. The mutant hydrogenase retains significantly higher methyl viologen reduction activity than the wildtype enzyme after oxygen exposure.

Enzymes in Pharmaceutical Industry

Majority of the industrial enzymes come from microorganisms as they are the most convenient sources, which gives the propensity of faster production, easy scale up, recovery and purification, strain manipulation for over-expression, enzyme activity, specificity modulations, and so on. At present, around 200 types of microbial enzymes from 4,000 known enzymes are used commercially. Nearly 75% of the total enzymes are produced by three top enzyme companies, that is, Denmark-based Novozymes, US-based DuPont, and Switzerland-based Roche.

Several drugs or pharmaceutical formulations are comprised of active pharmaceutical ingredients (APIs) that are synthesized using enzymes as important components of the manufacturing process. The usage of biocatalysts for pharmaceuticals' production as well as rising interest in the production of chiral intermediates and green synthetic processes has substantiated the interest in the applications of biocatalysts in these fields. These have been possible due to improvements in the technologies for biocatalyst selection, screening, improvement, production, and supply.

Production of Pharmaceutical Enzymes

Most of the industrial enzymes including those used in pharmaceutical industries are produced through fermentation of suitable microbial strains mainly belonging to bacteria and fungi due to their easy handling, fast growth rates, and convenient scale up in large vessels (fermenters). Bacteria like Escherichia coli, Bacillus subtilis, lactic acid bacteria, and the filamentous fungi such as Aspergillus oryzae, Aspergillus niger, Trichoderma atroviride, and Yeasts, for example, Saccharomyces cerevisiae, Pichia pastoris, and so on are the most exploited microorganisms for enzyme production by the biotechnology industries over the world. As the strains have been improved through genetic engineering, they produce enzymes in very high yields. Selection of suitable microbial strains for the production of various industrial enzymes is a very important aspect for their successful industrial applications. Ideally, the enzymes produced should be secreted out in the fermentation medium by the producing microbial strain as it makes the downstream processing more convenient and economically feasible; however, this may not be the case with most of the industrial strains. Pharmaceutical enzymes are produced by employing the fermentation technology, mainly using the microorganisms (bacteria and fungi), which comes under the Generally Recognized as Safe category. This production is mainly carried out employing two processes, that is, submerged fermentation (SmF) and solid-state fermentation (SSF). Both the processes have their own benefits and limitations. Most industries have adopted SmF process for enzyme production; but there has been a renewed interest in SSF for certain specific industries.

Enzyme Production Through SmF

Submerged fermentation, as the name suggests, is carried out in a liquid medium, wherein the nutrients and other medium components are either dissolved or remain suspended in an aqueous medium. The microorganisms also grow and proliferate in the liquid fermentation medium in the suspended state. Generally, aerobic microorganisms are grown in submerged cultures in a stirred-tank reactor for industrial production processes for extracellular enzymes and even for the intracellularly overexpressed enzymes as in the case of recombinant E. coli and other bacterial and yeast strains. The fermenters for SmF applications have been developed successfully and are being used for the production of enzymes and other biotechnological products. The main reason behind the success of SmF is feasibility of process scale up from pilot scale fermenters (~100 L) to production scale fermenters having capacities of millions of liters (vessel volumes) with well-developed offline and/or online control over several parameters such as, temperature, dissolved oxygen, pH, and foam formation. Further, there is no problem of mass transfer and heat dissipation; however, it is generally achieved at the cost of higher energy consumption.

Submerged fermentation for industrial production processes are manly carried out in four ways, namely batch culture, continuous culture, perfusion culture, and fed batch culture. In batch culture, the production strains are inoculated in a fixed volume of medium and thereafter no further addition of medium is carried out during the entire fermentation process (except the addition of acid, alkali, and antifoam). In continuous culture, the fermentation process is initiated in the batch mode. However, after certain time period fresh medium or nutrient concentrate is added into the fermenter at a rate which approximately matches the growth rate of the microorganism used and simultaneously the medium containing the product and biomass is continuously withdrawn from the overflow line of the vessel. A perfusion culture is somehow similar to a continuous culture process, in that it also involves the constant feeding of fresh media and removal of spent media and product. However, it differs from the previous in retention of high numbers of viable cells by using alternating tangential-flow and standard tangential-flow filtration or by binding the cells to a substrate (capillary fibers, membranes, microcarriers in fixed bed, and so on) in the fermenter. In the case of fed batch culture, the concentrated components of the nutrients are added to the batch culture in small lots at regular time points or based on the requirement of the process.

Enzyme Production Through Solid-State Fermentation

Solid-state fermentation has a great potential for enzyme production. This process offers several advantages over the SmF process, such as high product titer, lesser effluent generation, use of simple fermentation equipment, less trained labor, and so on. However, SSF is in general more suitable for those processes where the crude fermented products themselves are used as the final product rather than the isolated enzymes.

Agroindustrial residues are commonly used substrates for the SSF processes, including those used to produce enzymes. A variety of substrates have been used for the cultivation of enzyme producing microorganisms. Examples of the substrates used are wheat bran, rice bran, sugar cane bagasse, wheat straw, rice straw, saw dust, corncobs, banana waste, cassava waste, palm oil mill waste, oil cakes, and so on.

Applications of Enzymes in Pharmaceutical Manufacturing

Enzymes for the synthesis of antimicrobials:

Synthesis of 6-amino Penicillanic Acid by Penicillin Acylases

Penicillin acylases are a group of enzymes that cleave the acyl chain of penicillins to yield 6-amino penicillanic acid (6-APA) and the corresponding organic acid. Screening of penicillin acylase-producing microorganisms has shown that these enzymes are produced by bacteria, actinomycetes, yeasts, and fungi.

Based on their substrate specificity, penicillin acylases have been classified into three groups: penicillin G acylases, penicillin V acylases, and ampicillin acylases. Penicillin acylase-catalyzed synthesis of 6-APA has replaced the traditional chemical processes used for its synthesis. The conventional chemical method of 6-APA synthesis, started in 1970s, was dependent on a one-pot diacylation reaction of penicillin G (produced through fermentation), using hazardous reagents and solvents. This process of production continued for next 15–20 years and thereafter it was replaced substantially by penicillin G acylase-based method, as the enzymatic method allowed the production of 6-APA in good yields. Some researchers proposed an alternative process for the synthesis of 6-APA, based on penicillin V acylase and penicillin V. The advantage of this process relied on the fact that the penicillin V exhibited higher stability at lower pH during the extraction process from the fermented broth, which led to higher yield of 6-APA. In addition, penicillin V acylases-based processes afforded 6-APA production even at higher substrate concentrations compared with penicillin G acylases. Further, their wider optimal pH range alleviated the need of pH regulation during hydrolysis. Enzyme-based production of 6-APA is economically viable with immobilized forms of the enzymes. It results in great cost savings in terms of easy separation of the enzyme from the product, reuse of the enzyme and enhanced stability. Penicillin acylases from Bacillus megaterium, E. coli, and Alcaligenes faecalis have been immobilized for efficient application.

Synthesis of Semisynthetic Penicillins by Penicillin Acylases

Semisynthetic penicillins have been shown to have better properties than penicillin G and V, such as higher stability, easier absorption and lesser side effects and are also better candidates against the adaptive microbial resistance to antibiotics. Large-scale production of semisynthetic antibiotics that are derived from penicillin are based on the condensation of the β-lactam nucleus with appropriate D-amino acid catalyzed by penicillin acylases. Another application of penicillin G acylase is kinetic enantioselective acylation of the racemic azetidinone intermediate for the synthesis of the carbacephalosporin antibiotic Loracarbef, an analogue of Cefaclor and anti-platelet agent Xemilofiban. Penicillin G acylase has also been reported for enantioselective acylation of the l-enantiomers of methyl esters of phenylglycine and 4-hydroxyphenylglycine in organic solvents. The process resulted in easy isolation of the enantiomerically pure d-enantiomer, which is of practical use in the preparation of β-lactam antibiotics.

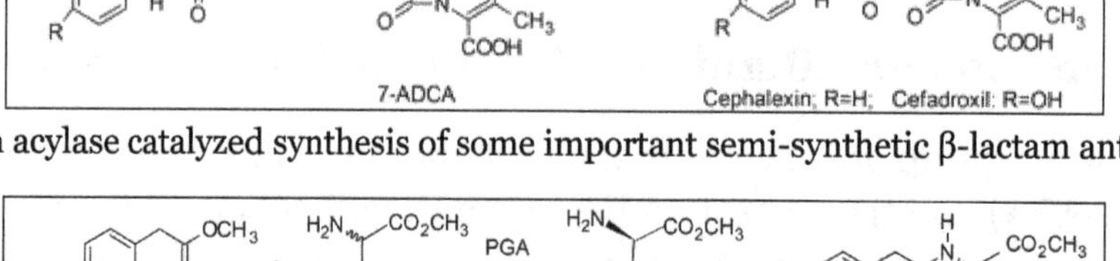

Penicillin acylase catalyzed synthesis of some important semi-synthetic β-lactam antibiotics.

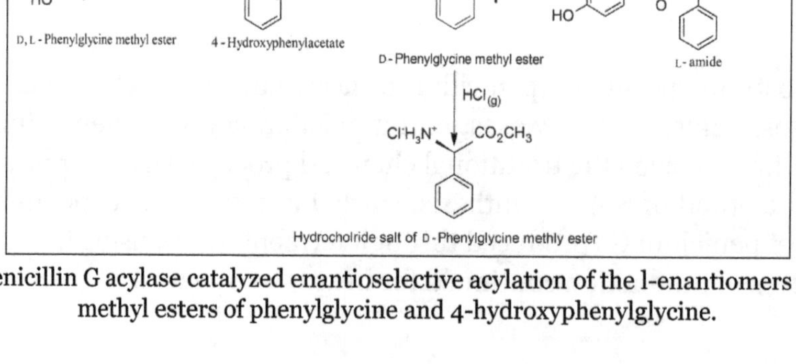

Penicillin G acylase catalyzed enantioselective acylation of the l-enantiomers of methyl esters of phenylglycine and 4-hydroxyphenylglycine.

Synthesis of β-lactam Antibiotic Key Intermediate, 7-aminocephalosporanic Acid

The cephalosporin-derived antibiotics are regarded as one of the most efficient and key drugs in the treatment of bacterial infections/diseases and presently a broad range of drugs have been derived from hetero-cyclic key moiety of the cephalosporins. Several cephalosporin-based semisynthetic β-lactam antibiotics have been commercialized such as cefotaxime, ceftriaxone, cefuroxime, and cefdinir. The lately developed ceftobiprol and ceftaroline fosamil are capable of treating the severe infections caused by MRSA (methicillin resistant Staphylococcus aureus). Cephalosporin C is the most feasible starting material for the synthesis of 7-aminocephalosporanic acid (7-ACA), as it is available in the desirable quantities at affordable cost, obtained through microbial fermentation. Additionally, cephalosporin C is a direct precursor leading to 7-ACA by amide bond cleavage. Enzymatic synthesis of 7-ACA involves a combined enzymatic and spontaneous reaction cascade, which consists of a d-amino acid oxidase-catalyzed oxidation of the amino acid side chain in to the corresponding α-keto acid. This is followed by a spontaneous decarboxylation of the α-keto acid. Finally, the resulting intermediate is hydrolyzed by glutaryl-7-ACA hydrolase to produce the 7-ACA.

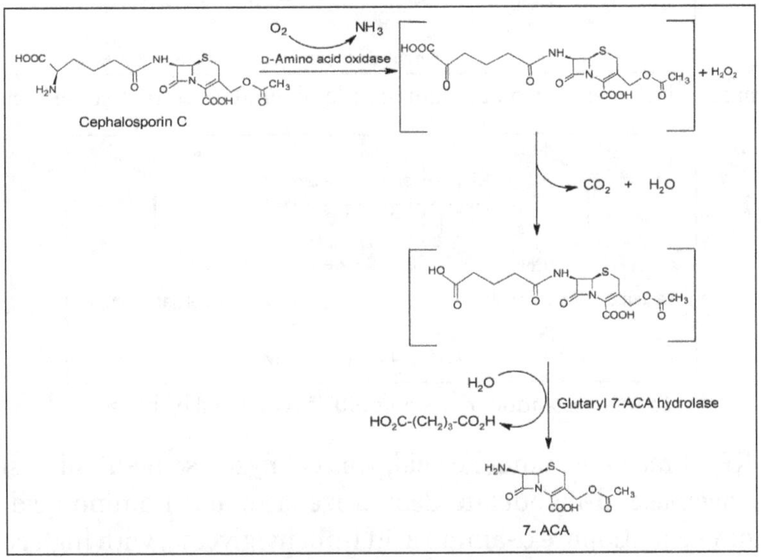

Enzymatic transformation of cephalosporin C into 7-ACA catalyzed by d-amino acid oxidase and glutaryl-7-ACA acylase.

Enzymes for the DKR of Drugs

Enzymes are exceptional in performing enantioselective reactions because of their high chemoselectivity. Enantioselective enzymes act one of the enantiomers (either R or S form) and selectively transform it into the corresponding product, whereas the other enantiomer remains untouched. At this point, the product can readily be separated from the other enantiomeric form of the substrate. However, the conventional method for resolution of racemates (i.e., the enzymatic resolution of a racemic mixture of a compound like amino acids or any synthetic or natural biomolecule) results in 50% yield of the required form of the isomers. To overcome this barrier, the dynamic kinetic resolution (DKR) principle has been used recently in a variety of enzymatic reactions. Researcher reported the production of optically pure α-amino acids via enzyme catalyzed DKR of α-aminonitriles. This process is performed in three sequential steps: (i) nonselective hydrolysis of racemic α-aminonitrile by a Nitrile Hydratase (NHase), (ii) breakdown of the resulting amide by a stereoselective amino acid amide hydrolase and (iii) simultaneous racemization of the α-amino acid amide by α-amino-ε-caprolactam racemase. Many compounds were hydrolyzed using the combinations of these enzymes; the results were promising wherein both (R)- and (S)-amino acids were obtained in excellent yields and ee's (>99%). An investigation with Burkholderiacepacia lipase to catalyze the DKR of racemic α-aminonitrile, resulted in enantioselective production of acetylated amine with substantial yield of 87%–90% and ee's of 85%–88%.

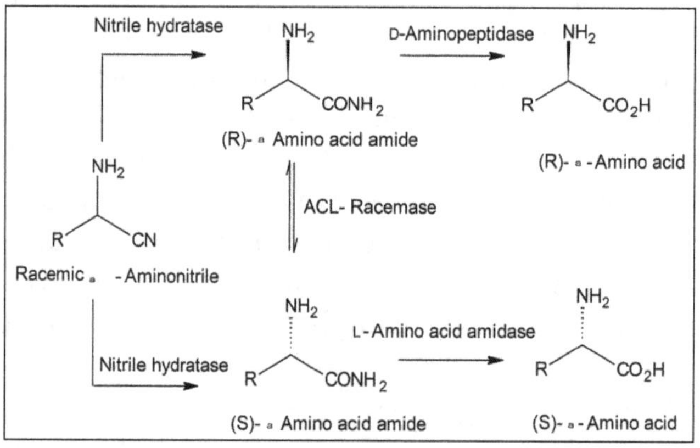

Dynamic kinetic resolution of α-aminonitriles to produce chiral α-amino acids.

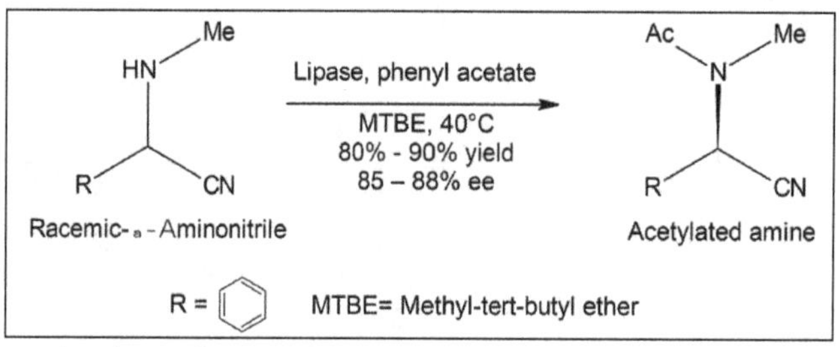

Kinetic resolution of α-aminonitriles catalyzed by lipase.

In another study DKR of racemic mandelic acid, was catalyzed sequentially using three enzymes, that is, mandelate racemase, d-mandelate dehydrogenase, and l-amino acid dehydrogenase to yield the corresponding enantiopure α-amino acid l-phenylglycine, with high conversion and ee in excess of 97%.

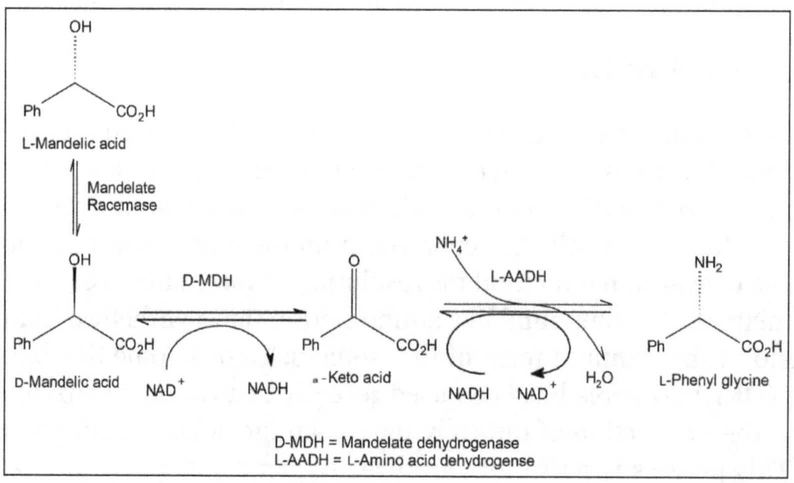

Enzymes for the Synthesis of Amino Acids

Amino acids are building blocks of life and thus are in dispensable for proper health and nutrition of both humans and animals. In addition, due to their chirality, they are biochemically very significant and are very useful in chemical synthesis. The nine essential amino acids, that is, l-leucine, l-isoleucine, l-valine, l-lysine, l-histidine, l-methionine, l-phenylalanine, l-threonine, and l-tryptophan are not synthesized in animals and humans and therefore must be ingested with food. Enzymes and whole cell biocatalysts have been important in the production of proteinogenic and nonproteinogenic d- and l-amino acids, and enantiomerically pure amino acid derivatives, which are used as important building blocks for active ingredients that are used in the manufacture of pharmaceuticals, cosmetics, and agrochemicals. The rapid expansion in the amino acid market from 1980s is mainly due to cost effective production and isolation of amino acid products. Out of different methods of amino acids' production the two methods, that is, the fermentation, and enzymatic catalysis are predominant, owing to their cost effective and eco-friendly nature.

Enzymatic Production of Proteinogenic Amino Acids

Enzymes have been exploited from the last 40 years for the production of l-amino acids in Japan. The production of l-methionine, which is used in special diets, has been carried out using enzymatic resolution with acylase of Aspergillus oryzae. Several hundred tons of l-methionine and l-valine are now produced each year using enzyme membrane reactor technology to minimize the enzyme loss.

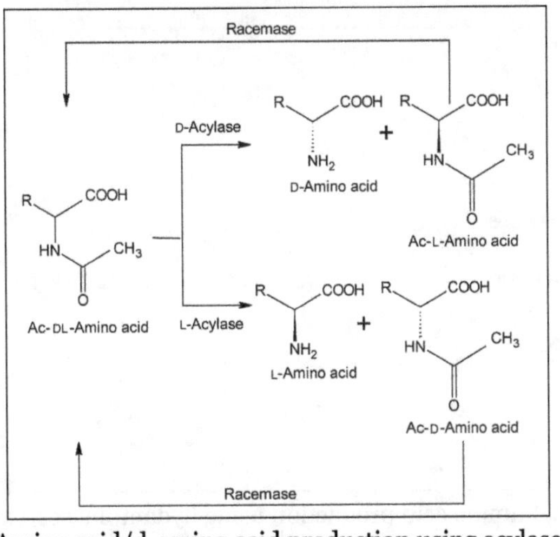

l-Amino acid/d-amino acid production using acylases.

Another amino acid that is preferably obtained through enzyme catalysis is l-aspartic acid. The enzyme aspartase catalyzes the addition of amino group from ammonia to fumaric acid to directly produce the l-aspartate, which is used on industrial scale to produce the artificial sweetener l-aspartame. l-Alanine is produced from l-aspartate using aspartate β-decarboxylase enzyme. l-Cysteine has been produced from l-cystine mainly through electrochemical reduction process. However, now its industrial production is feasible through enzymatic hydrolysis and recemization of the substrate dl-2-amino-2-thiazoline-4-carboxylic acid (ATC) carried out by three enzymes, that is, l-ATC hydrolase, S-carbamoyl-l-cysteine hydrolase, and ATC racemase. Recently, substantial production of l-cysteine has been achieved through fermentation of an E. coli strain with disrupted l-cysteine regulon, that is, the yciW gene.

Enzymatic Production of Nonproteinogenic Amino Acids

Enzymatic production of d-amino acids and nonproteinogenic l-amino acids are becoming a more acceptable and environmental-friendly method. d-Amino acids can be obtained as the byproducts through enzymatic resolution of racemic mixtures of dl-amino acids used to produce l-amino acids. However, it is also possible to produce d-amino acids directly, for example, using a d-specific acylase, from racemic acetyl amino acids. Still another method of industrial importance is the production of building blocks for the semisynthetic antibiotics ampicillin and amoxycillin, that is, the d-phenylglycine and p-hydroxy-d-phenylglycine, which are produced using the enzyme system hydantoinase/carbamoylase. The recent advances in modern molecular biological methods (directed evolution), has made it possible to switch the d-specificity of hydantoinases to l-specificity. Further, it is now possible to coexpress racemases with d- or l-selective hydantoinases and carbamoylases in highly potent recombinant whole-cell systems capable of producing a variety of d- and l-amino acids. Recently, a more cost-effective enzymatic method for d-amino acid production has been reported involving dynamic kinetic resolution of N-succinyl amino acids carried out by two enzymes namely the d-succinylase and N-succinyl amino acid racemase, which enantioselectively hydrolyzed the N-succinyl-d-amino acids to their corresponding d-amino acids.

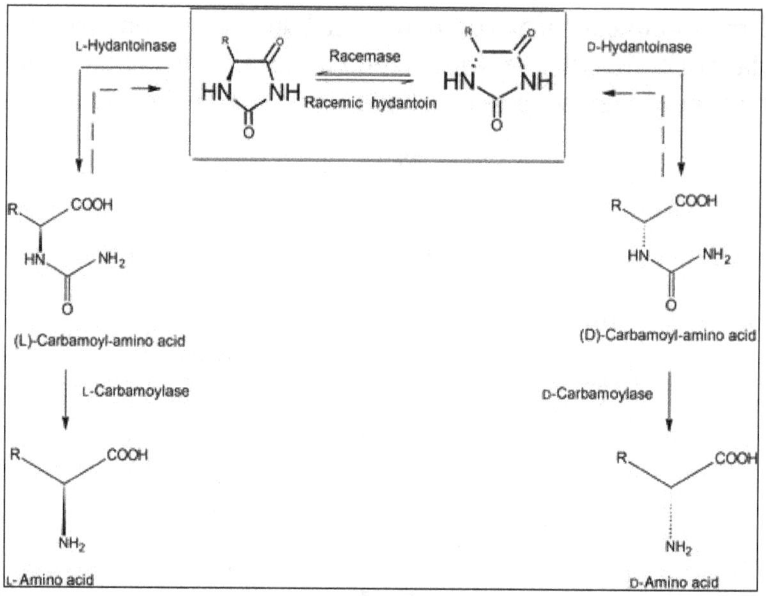

d-Amino acid/l-amino acid production using hydantoinases/carbamoylases.

Enzymatic Synthesis of Pragabalin Intermediate, (S)-3-Cyano-5-Methylhexanoic acid

Pregabalin is a powerful anticonvulsant, used to treat seizure disorders, neuropathic pain, and fibromyalgia. The pharmacological activity of pregabalin is confined to its (S)-enantiomer so its asymmetric synthesis has gained great importance in the pharmaceutical industry. Many chemocatalytic and biocatalytic routes have been used for the asymmetric synthesis of the key intermediate (S)-3-cyano-5-methylhexanoic acid of (S)-pregabalin. For instance, bisphosphine rhodium and nitrilase have been reported for the asymmetric synthesis of (S)-3-cyano-5-methylhexanoate in appreciable yields and good enantiomeric excess viz. 98% and 97% ee, respectively. However, both approaches were unsustainable in terms of environmental and cost effectiveness. In this context, an efficient biocatalytic route for the synthesis of (S)-3-cyano-5-methylhexanoic acid, has been reported wherein an esterase from Arthrobacter sp. ZJB-09277 (whole cell catalysis) was used to catalyze the kinetic resolution of racemic 3-cyano-5-methylhexanoic acid esters to yield (S)-3-cyano-5-methylhexanoic acid in 44.6 mM with 95.1% ee.

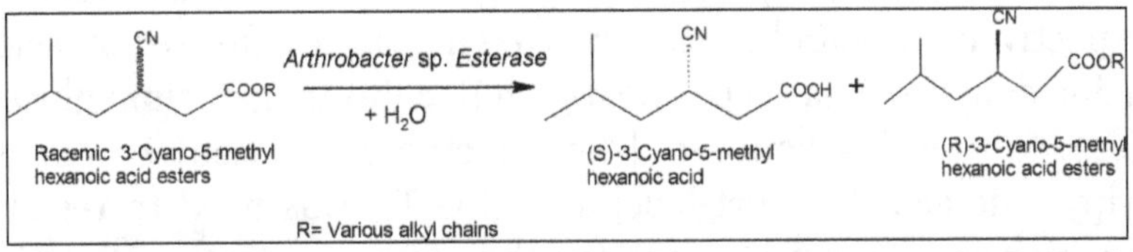

Enzymatic kinetic resolution of racemic 3-cyano-5-methylhexanoic acid esters using Arthrobacter sp. Esterase.

Enzymatic Synthesis of Sitagliptin Intermediate, (R)-3-Amino-4-(2,4,5-Triflurophenyl) Butanoic Acid

(R)-3-amino-4-(2,4,5-triflurophenyl) butanoic acid (3-ATfBA), a β-amino acid is a crucial intermediate for the synthesis of important antidiabetic drugs such as sitagliptin, retaglptin, and evogliptin. A two-step enzymatic process has been reported for the synthesis of 3-ATfBA, wherein the first step is Candida rugosa lipase mediated conversion of β-ketoester substrate to β-keto acid, which was subsequently aminated by the ω-Transaminase (ω-TA) of Ilumatobacter coccineus to its corresponding β-amino acid. The process resulted in about 92.3% conversion of 100 mM β-keto ester substrate to 3-ATfBA. Scaleup of the reaction showed excellent conversion (81.9%) and enantioselectivity (99% ee).

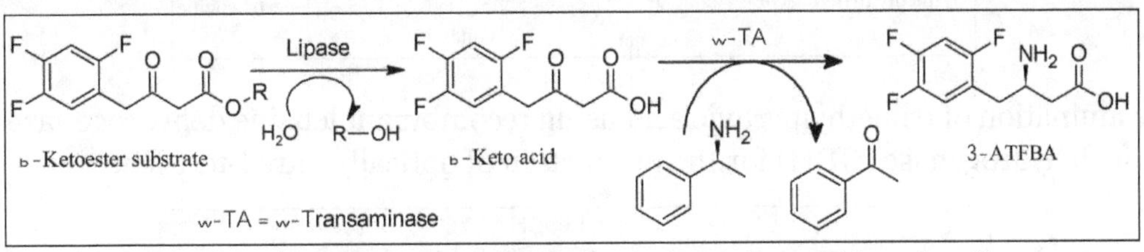

Enzymatic synthesis of 3-ATfBA [(R)-3-amino-4-(2,4,5-triflurophenyl) butanoic acid] from corresponding keto acid.

Enzymatic synthesis of another key intermediate of stigaliptin, Boc-(R)-3-amino-4-(2,4,5-trifluorophenyl) butyric acid, has been reported by Researcher. The reaction was performed at 100 mM scale, resulting in 82% conversion in 24 H. The amino ester product was further transformed to Boc-(R)-3-amino-4-(2,4,5-trifluorophenyl) butyric acid, the key intermediate of sitagliptin.

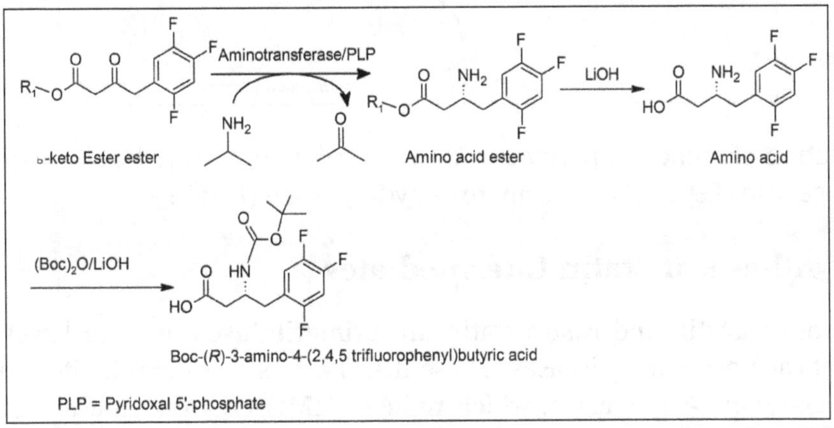

The chemo-enzymatic route for the synthesis of Boc-(R)-3-amino-4-(2,4,5-trifluorophenyl)-butyric acid.

Enzymatic Synthesis of l-Tert-Leucine

l-Tert-leucine is the basic chiral building block for many pharmaceutically active compounds such as atazanavir, boceprevir, and telaprevir, which are potent protease inhibitors for human immunodeficiency virus (HIV) protease, hepatitis C virus genotype 1 protease, and hepatitis C NS3-4A serine protease, respectively. Researcher has reported the successful coexpression of Exiguobacterium sibiricum leucine dehydrogenase (LeuDH) with Bacillus megaterium glucose dehydrogenase (GDH) in E. coli BL21 for the production of l-tert-leucine. Here, LeuDH was used to convert the substrate trimethylpyruvic acid to l-tert-leucine and GDH was used to regenerate the cofactor NADH from NAD^+, which was required to continue the synthesis reaction. By using the coexpressed whole cells, they were able to prepare decagram of l-tert-leucine at 1 L scale from 0.6 M (78.1 g L^{-1}) of substrate with 99% conversion after 5.5 H, resulting in 80.1% yield and >99% ee. In another report, recombinant LeuDH from Thermoactinomyces intermedius and recombinant format dehydrogenase (FDH) from Pichia pastoris were used to catalyze the synthesis of L-tert-leucine from trimethylpyruvic acid and NADH regeneration, respectively. Using this system, a reaction yield of >95% with an ee of >99.5% was obtained at 100 g/L substrate input.

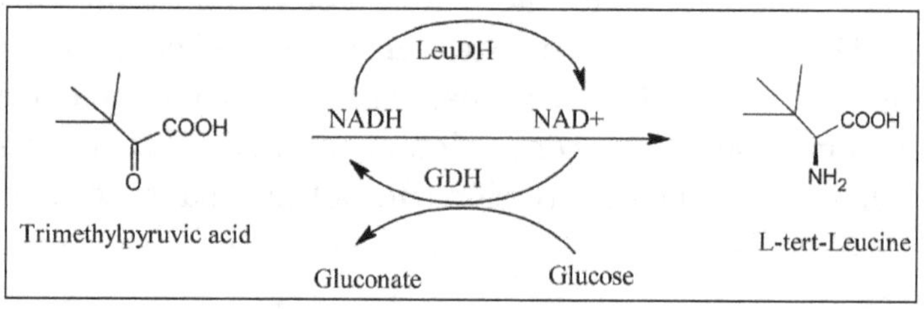

Reductive amination of trimethylpyruvic acid using recombinant leucine dehydrogenase (LeuDH) and glucose dehydrogenase (GDH) for the production of optically pure l-tert-leucine.

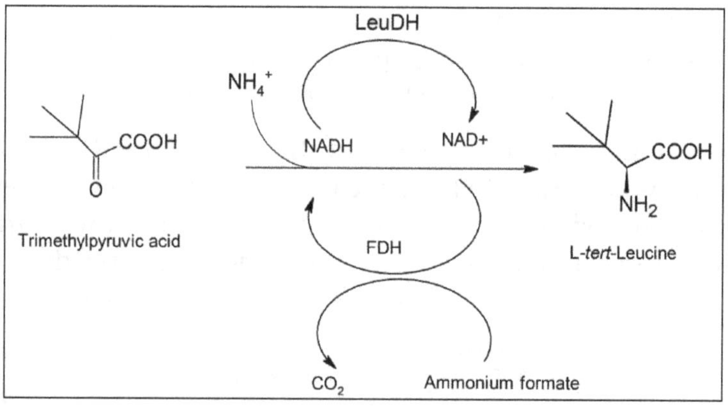

Reductive amination of trimethyl pyruvic acid to optically pure l-tert-leucine using recombinant leucine dehydrogenase (LeuDH) and formate dehydrogenase (FDH).

Enzymatic Synthesis of Statin Intermediates

Statins such as atorvastatin and rosuvastatin are primarily used as lipid lowering agents and for prevention of cardiovascular diseases. These drugs act as specific inhibitors of the hydroxylmethylglutaryl coenzyme A reductase, which reduces HMG-CoA into mevalonate, a rate-limiting step in the biosynthesis of cholesterol. Two intermediates that are converted to atorvastatin and rosuvastatin have been discussed here. First, is t-butyl 6-chloro-(3R,5S)-dihydroxyhexanoate and another is (3S,5R)-dihydroxy-6-(benzyloxy) hexanoic acid, ethyl ester. Synthesis of t-butyl 6-chloro-(3R,5S)-dihydroxyhexanoate has been carried out using recombinant E. coli cells harboring the alcohol dehydrogenase (KleADH) of Klebsiella oxytoca. KleADH catalyzed the conversion of t-butyl 6-chloro-(5S)-hydroxy-3-oxohexanoate to t-butyl 6-chloro-(3R,5S)-dihydroxyhexanoate with more than 99% diastereomeric excess (de) and 99% conversion in 24 H without adding any expensive cofactors. In another approach (3S,5R)-dihydroxy-6-(benzyloxy) hexanoic acid, ethyl ester or tert-butyl ester have been synthesized via enantioselective reduction of the corresponding diketoesters, 3,5-dioxo-6-(benzyloxy) hexanoic acid, ethyl or tert-butyl esters, catalyzed by recombinant E. coli cells harboring the ketoreductase of Acinetobacter calcoaceticus.

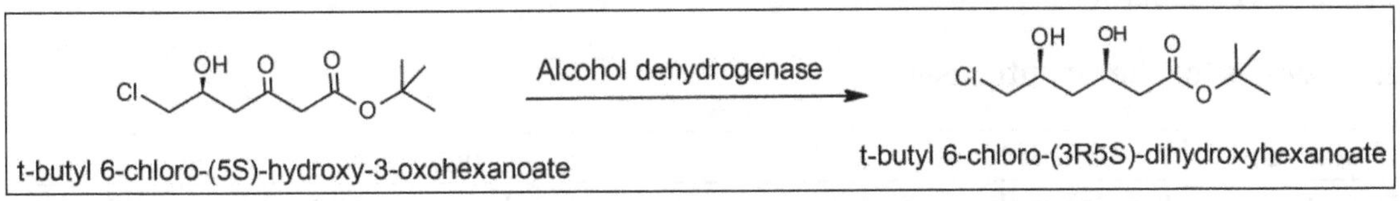

Enzymatic preparation of t-butyl 6-chloro-(3R,5S)-dihydroxyhexanoate by alcohol dehydrogenase of Klebsiella. oxytoca.

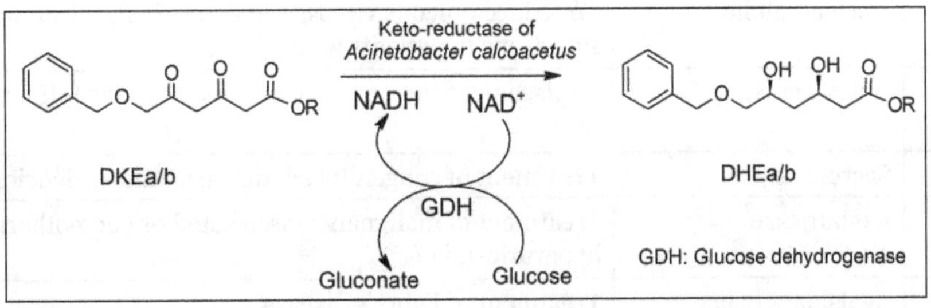

Enzymatic preparation of (3S,5R)-dihydroxy-6-(benzyloxy) hexanoic acid, ethyl (DHEa) or tert-butyl (DHEb) ester by keto reductases (Kred) of Acinetobacter calcoaceticus from the corresponding diketo ethyl ester (DKEa) or diketo tert-butyl ester (DKEb) (modified from Refs.

Enzymatic Synthesis of Antiplatelet Agent, Xemilofiban

A racemic mixture of ethyl 3-amino-5-(trimethylsilyl)-4-pentynoate was resolved through an enantioselective acylation catalyzed by penicillin G amidohydrolase (penicillin G acylase) from E. coli to yield the S-isomer, which can be used as a chiral synthon for the synthesis of the antiplatelet agent Xemilofiban. The first step is the synthesis of desilylated phenylacetamide racemic ethyl 3-amino-5-(trimethylsilyl)-4-pentynoate, which is a nonenzymatic step. In the second step, phenylacetamide is enzymatically hydrolyzed to produce (R)-amide in a quantitative yield and (S)-amine with yield of 90%.

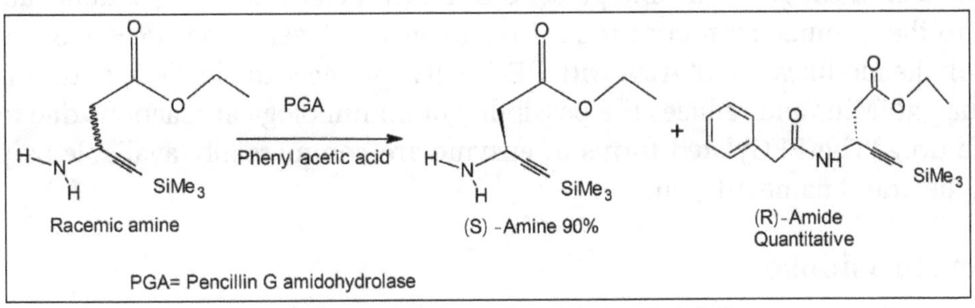

Acylation of racemic amine (ethyl 3-amino-5-(trimethylsilyl)-4-pentynoate) using penicillin G amidohydrolase (PGA).

Enzyme Therapy

Enzyme therapy refers to the applications of enzymes for treating enzyme deficiencies and other medical conditions in human beings. In humans, enzymes assist in food digestion, body detoxification, strengthening of the immune system, muscle contraction, reduction of stress on the vital organs like pancreas and others. In this regard, enzyme therapy has numerous possible medical applications, for example, treatment of pancreatic insufficiency and cystic fibrosis (CF), metabolic disorders, lactose intolerance, removal of dead tissues, cancers or tumors, and so on. The therapy may be systemic or non-systemic, and may be administered via multiple routes of administration, most often orally, topically or intravenously. A summary of enzymes used as therapeutic agents is presented in Table below.

Table: Enzymes for therapeutic uses.

Trade Name	Generic Name	Indication
ADAGEN	Pegademase bovine	For enzyme replacement therapy for ADA in patients with SCID
CEREDASE	Alglucerase injection	For replacement therapy in patients with Gaucher's disease type I
PULMOZYME	Dornase alpha	To reduce mucous viscosity and enable the clearance of airway secretions in patients with CF
CEREZYME	Imiglucerase	Replacement therapy in patients with types I, II, and III Gaucher's disease
SUCRAID	Sacrosidase	Treatment of congenital sucrase-isomaltase deficiency
ELITEK	Rasburicase	Treatment of malignancy-associated or chemotherapy-induced hyperuricemia
FEBRAZYME	Agalsidase beta	Treatment of Fabry's disease
NATTOKINASE NSK-SD	Nattokinase	Support healthy blood clotting, circulation, and platelet function

Adenosine Deaminase

Adenosine deaminase (adenosine aminohydrolase, or ADA) is an enzyme involved in metabolism of purine bases. It performs the breakdown of adenosine obtained through food and from the turnover of nucleic acids in tissues. Its primary function in humans is the development and maintenance of the immune system. This enzyme has emerged as therapeutic agent for the treatment of immunological disorders. Adenosinedeaminase (Adagen: pegadamase bovine) is used for the treatment of Severe Combined Immunodeficiency Disease (SCID). It is the first successful use of an enzyme for the treatment of an inherited immunological disorder. ADA plays an important role in the metabolism of excess adenosine present in the circulation of these patients and decreases the toxicity to the immune system of the raised adenosine levels. The success of the treatment depends upon the modification of ADA with PEG. PEG enhances the half-life of the enzyme (originally less than 30 Min) and reduces the possibility of immunological reactions due to the bovine origin of the drug. The PEGylated forms of enzyme are commercially available as pegadamase bovine with the brand name Adagen.

β-Glucocerebrosidase

β-glucocerebrosidase (d-glucosyl-N-acylsphingosinegluco hydrolase, EC 3.2.1.45) is an enzyme with glucosylceramidase activity that hydrolyze the β-glucosidic linkage of the chemical glucocerebroside, an intermediate in glycolipid metabolism. Mutations in the glucocerebrosidase gene cause Gaucher's disease, a lysosomal storage disease characterized by an accumulation of glucocerebrosides. Mutations in the glucocerebrosidase gene are also associated with Parkinson's disease. This enzyme was approved for use in enzyme replacement therapy of Gaucher disease in which an exogenous modified placental glucocerebrosidase was targeted to its correct compartment within the body. The commercial formulations of the enzyme were available as alglucerase injection with the brand name Ceredase.

Sacrosidase

Sacrosidase (β-fructofuranoside fructohydrolase) is another enzyme that is used for the treatment of congenital sucrase-isomaltase deficiency (CSID). CSID patients cannot digest the disaccharide sucrose. The sacrosidase-based drug hydrolyses sucrose, helping in its digestion allowing the consumption of regular food items having added sucrose or otherwise sucrose-containing fruits. A sacrosidase from S. cerevisiae that can be taken orally has been used for the treatment.

Enzymes used as Digestive Aids

A number of enzymes are employed to cure the digestive problems induced by some sugars. The α-galactosidase enzyme is taken as a digestive aid for people who develop the symptoms of bloating, gas, and diarrhea upon ingesting foods like beans and Brassica vegetables such as, cabbage, broccoli, Brussels, and so on. The hydrolase α-galactosidase assists in breakdown of terminal α-galactosidic residues of sugar substrates present in these foods, which when remain undigested cause the discomfort, probably due to the bacterial fermentation of the undigested sugars. Now-a-days α-galactosidase and lactase are available as a number of ready to use supplements.

Lactose intolerance is a condition in which the patient is unable to produce sufficient quantities of lactase enzyme that is responsible for the digestion of the milk sugar lactose. The lactose intolerant people cannot digest lactose-containing foods and therefore, suffer from stomach upset when they consume lactose containing food such as milk and milk products containing lactose. For such people, lactase supplements such as milk fortified with lactase, lactase powder, and so on are available. These supplements help to break down the lactose into its monomers, that is, glucose and galactose and alleviate the symptoms of lactose intolerance such as bloating, gas, and diarrhea.

Another disease (genetic disorder) is phenylketonuria (PKU), an inheritable disorder, requires strict compliance with a specialized diet. PKU occurs due to little or absence of phenylalanine hydroxylase enzyme. This enzyme helps in converting phenylalanine to tyrosine, thereby maintaining its normal levels in the body. An oral treatment for the disorder has been developed, which is composed of a plant phenylalanine ammonia lyase (PAL) overexpressed and obtained from a recombinant yeast, available in market with the trade name Phenylase. PAL has been demonstrated to hydrolyze phenylalanine in the gastrointestinal tract.

As a general digestive aid for the benefit of people with weak immunity a cocktail of pancreatic enzymes, including amylases proteases, and lipases, has been developed. The cocktail has been shown to be effective in the alleviating the problem of fat malabsorption in patients suffering from HIV. This enzyme cocktail is also beneficial in treating the pancreatic insufficiency, which is common in patients of CF. A cocktail of pancreatic enzymes with the trade name "TheraCLEC Total" is available commercially.

Proteolytic and Glycolytic Enzymes for Treating Damaged Tissue

Proteolytic enzymes of bacterial and plant origin have been investigated for debridement (removal of dead skin) of burns. However, due to inconsistent results and lower efficacy of these enzymes of natural origins, commercialization of the technology could not be done. However, with the intervention of recombinant DNA technology, efficient enzyme formulations like Debrase gel dressing, with consistent results is now available. Debrase gel dressing is composed of a mixture of enzymes extracted from pineapple, was approved by US FDA in 2002, and is now in phase IIb clinical trial, in patients with deep partial thickness or full thickness thermal burns, in Europe and the United States.

Enzymes for the Treatment of Infectious Diseases

Lysozyme a bactericidal enzyme produced naturally in human body is added in many food products. It has been discovered that this enzyme has activity against HIV, similar to RNase A and urinary RNase U, it selectively degrades viral RNA and therefore is a promising candidate for the treatment of HIV infection. Other examples of antimicrobial enzymes are chitinases. Chitin is a good target for antimicrobials because it is a major component of the cell wall of different pathogens like fungi, helminths, and protozoa. On the other hand, lytic bacteriophage-derived enzymes have been known for killing bacteria such as Streptococcus pneumonia, Clostridium perfringens, and Bacillus anthracis. Thus, the application of lytic bacteriophages as a treatment for bacterial infections is also under development and is assumed tobe effective against antibiotic resistant bacterial strains.

Enzymes for the Treatment of Cancer

Application of therapeutic enzymes in cancer treatment is an emerging field of research. Lately, it has been discovered that PEG immobilized arginine deaminase (an arginine-degrading enzyme) can inhibit human hepatocellular carcinomas and skin cancers which are deficient in arginine due to lack of arginosuccinate synthetase activity. A better enzyme therapy has been developed using PEGylated l-asparaginase with the name Oncaspar (pegaspargase). It has shown better results for the treatment of acute lymphoblastic leukemia, acute myeloid leukemia, and non-Hodgkin's lymphoma. In other words, the normal cells, that is, the noncancerous cells can synthesize asparagine, whereas the cancer cells cannot, and die in the presence of enzymes which degrade asparagine. Although the cost of the PEG–asparaginase formulations is higher than the native enzyme the overall cost of the treatment is very similar in both cases. In fact, asparaginase and PEG-asparaginase are better alternatives to standard chemotherapy.

Future Prospects of Enzymes in Pharmaceuticals

As enzymes have successfully replaced many conventional chemical catalysts-based pharmaceutical manufacturing processes, the future of enzymes in this sector is very promising. There are some enzymes that have been shown to have cellular detoxification function such as superoxide dismutase and catalase. Both these enzymes work in conjunction to detoxify the body. The superoxide dismutase transforms the highly toxic superoxide anion to moderately toxic hydrogen peroxide.

The catalase then converts hydrogen peroxide to harmless water and oxygen. However, both these enzymes could not be used for these desired activities even in their PEGylated forms. Nevertheless, these enzymes have been reported for prolonging the life of the nematode, Caenorhabditis elegans and it is assumed that they can be used for prolonging the human life as well. Improvement in the structure and functional properties of these enzymes shall be helpful in avoiding organ damage in hemorrhagic shock. A serum detoxifying enzyme as an example the human butyrylcholinesterase-has been shown to be effective in the treatment of cocaine overdose. Structural engineering of this enzyme has been shown to increase its detoxifying activity toward cocaine. In this respect, directed evolution is proposed to provide more efficient forms of the therapeutic enzymes and other enzymes that will catalyze the pharmaceutical biotransformations with great efficiencies.

Enzymes in Food Industry

The majority of currently used industrial enzymes are hydrolytic in action, being used for the degradation of various natural substances. Proteases remain the dominant enzyme type, because of their extensive use in the detergent and dairy industries. Various carbohydrases, primarily amylases and cellulases, used in industries such as the starch, textile, detergent and baking industries, represent the second largest group. Food enzymes can be obtained from different sources, but microbial sources are the most promising and potential one. On the other hand, the implementation of recombinant DNA technology and process engineering makes these microbial sources more comfortable, flexible, and appropriate for industrial purposes at the commercial level. Food processing is a transformation practice in the beverage and food industry to make the raw foodstuff of animal and plant origin suitable for consumption. On the basis of their difference from one another, the processed food materials can be distributed into highly processed foods, minimally processed foods, and processed foods. The processing methods are performed through a wide range of biological and chemical agents such as enzymes. The utilization of biological agents in food processing dates back to 6000 BCE and includes bread making, beer brewing, cheese making, and wine making. Currently, these products are mainly recognized in the biotechnology industry.

There is table below that indicates the enzymes used in food industry.

Table: Enzymes Used In Food Industry.

Enzymes Used In Food Industry	
Industry	enzyme class
Dairy	Protease, Lipase, Lactase, Transglutaminase, Rennin, Catalase
Beverage	Pectinase, Xylanase, β-Glucanase, Laccase, Cellulase
Fats and oils	Lipase, Phospholipase
Other Food	Pectinase, Protease, Lipase, Amylase, Xylanase, β-Glucanase
Sanitation (Antibiofilm enzymes)	Lactonase, DNase, Savinase, α-amylase, Cellulase+Pronase, Glucose oxidase+Lactoperoxidase, Acylase I+ProteinaseK
Bioethanol from Food Crops	Amylases, Cellulase, Xylanase

Enzymes used in food processing have historically been considered non-toxic. Some characteristics arising from their chemical nature and source, such as allergenicity, activity-related toxicity, residual microbiological activity and chemical toxicity are of high concern. These attributes of concern must, however, be addressed in light of the growing complexity and sophistication of the methodologies used in the production of food-grade enzymes.

Enzymes in Dairy Industry

There are many types of microbial enzymes used in the dairy industry, such as catalase, aminopeptidase, proteases, lactoperoxidase, lipases, transglutaminase, etc. They are well known in this field and are different from coagulants as they help improve shelf life. In the dairy industry, microbial enzymes have been utilized to produce diverse products, such as yogurt, cheese, syrup, bread, etc. to enhance their quality. Milk coagulants belong to four main categories: animal rennet, plant derived coagulants, microbial coagulants, and genetically engineered cymosin. Animal rennet contains the two main enzymes, cymosin and pepsin, in a ratio that depends on the age of the animals when slaughtered. Rennet is considered a famous exogenous enzyme used in dairy processing, and has been used since 6000 BCE. Rennin has enzymatic and non-enzymatic action that causes the milk to coagulate. The milk transforms to a gel-like structure during the enzymatic activity due to the temperature and calcium ion effect.

The proteases of lactate bacteria are necessary for its growth in substrate (milk) and help dramatically in enhancing flavor of fermented milk products. The proteolytic enzyme system constitutes proteinases that first break down the protein of the milk into peptides. Peptidases then breaks down the peptides into amino acids and small peptides, then the transport system takes charge of the uptake of amino acids and small peptides. Currently, most effective procedures have been developed by which enzymes are entrapped for controlled release during ripening. Direct encapsulation using food gums or lyposomes is one of the two entrapping methhods available; the second method is encapsulation of cell free extract from bacteria or whole bacterial cells in milk-fat capsules.

Different lipases of animal or microbe origin created clear cheese, low bitterness, improved flavor, and potent malodors, whereas proteinases, in combination with lipases or/and peptidases, developed cheeses with excellent flavor and low bitterness levels. In order to accelerate the ripening of cheese content of peptidases and proteinases we can use attenuate cell-free extract or starter cells to do that. Animal lypases that are isolated from the epiglottis (throat) of calves, sheep, or goats, greatly controbute to release short chain fatty acids responsible of sharp and piquant flavors. The result is due to improved fat and moisture retention in cheese caused by the increased emulsifying propert, that is reduction of fat losses in whey and cooking water.

A further recent application is enhancing flavor and structure in butter and dairy cream. Lipases from Mucor miehei, or Aspergillus niger are used to give stronger flavors in Italian cheeses from milk before adding the rennet, by a modest lipolysis, increasing the amount of free butyric acid. The free fatty acids generated by the action of lipases on milk fat endow many dairy products, particularly soft cheeses with their specific flavour characteristics. The traditional sources of lipases for cheese flavour enhancement are animal tissues, especially pancreatic glands (bovine and porcine) and pre-gastric tissues of young ruminants (kid, lamb and calf). They form inter-molecule or intramolecule cross-linkages in proteins via e-(g-Glu)-Lys bonds which is useful for modifying the physical and functional properties of foods. Transglutaminase (TG) has recently become of great interest to food scientists for its ability in strengthening the structure of protein gels. TG catalyzes the posttranslational modification of proteins by transamidation of available glutamine residues by the formation of covalent cross-links between glutamine and lysine residues. Addition of transglutaminase to milk induces of cross-linking of caseins and whey proteins that improves the strength of milk gels. Rennet cheese with modified textural and nutritional properties and improved yield could be obtained upon transglutaminase modification but simultaneous addition of rennet and transglutaminase is recommended. Moreover, transglutaminase crosslinking and calcium reduction were investigated as ways to improve the texture and storage stability of high-protein nutrition bars formulated with milk protein concentrate and micellar casein concentrate.

Lactase (EC 3.2.1.23) accelerates the breakdown of lactose into galactose and glucose. It is used to improve the sweetness, solubility, as well as digestive agent, for milk products. Lactases are important in reducing and removing lactose in milk products for lactose-intolerant patients in order to protect them from severe diarrhea, fatal consequences, and tissue dehydration. Among various features of milk treated with lactase is the increased sweetness, hence it can avert the need for adding sugars during the production of flavored milk drinks.

Lactase is used by the producers of ice cream, yogurt and frozen desserts to enhance spade and creaminess, sweetness, tastiness, and digestibility, and to decrease sandiness because of crystallization that occurs in lactose converged preparations Commercial enzymes are derived mainly from fungi like Kluyveromyces sp. and Aspergillus sp. Although many enzymes are commercially available, new efforts are still being made using recombinant engineering techniques and new enzymes are being discovered by screening metagenome databases. The enzyme's immobilization, immobilization method, and carrier type can also affect these optima rates.

Enzymes in other Food Industry

Industrial applications of enzymes include food (baking, dairy products, starch conversion) and beverage processing (beer, wine, fruit and vegetable juices), animal feed, textiles, pulp and paper, detergents, biosensors, cosmetics, health care and nutrition, wastewater treatment, pharmaceuticals and chemical manufacture and, more recently, biofuels such as biodiesel and bio-ethanol. Quality parameters like meat tenderness is influenced by a number of factors, which may be pre-slaughter or post slaughter in nature. This toughness may be resolved by the enzymes present in the muscle cell or from exogenous sources to obtain a tender meat. The meat industry's use of enzymes improves the manufacturing process and upgrades poor quality meat. Meat has been found to be a source of several bioactive peptides, and researchers have reported different health-promoting applications of these peptides. They are obtained after strategic use of enzymes on the meat as substrate. Bioactive peptides exhibiting antihypertensive, antioxidant, and antimicrobial effects have been observed in the meat protein hydrolysates.

In addition, the enzymes can be used for binding small meat pieces to improve their market value. This also enhances the efficiency of carcass utilization. The protein cross-linking enzymes, like transglutaminases (TGase), have been used extensively in several food products to improve the texture. Much of the early interest in enzymology was developed by scientists like Pasteur, Payen and Persoz, who were associated with food, wine and beer industries. Although not in isolated form, enzymes have been used traditionally for dairy, baking, brewing, and winemaking for centuries. Enzymes are needed for cheese production and a wide variety of other dairy goods. For example, their application keeps bread soft and fresh longer, leads to crispy crusts, increases dough volume and can compensate for variations in flour and malt quality. Additionally, enzymes are used to lower alcohol concentration and calories in beer. In winemaking, the sulphur content can be reduced, clarity and wine colour can be maintained, flavours can be enhanced and the filterability can be improved with enzymes.

In the production of baked goods, enzymes can be added individually or in complex mixtures at a very low level that may act in a synergistic way. Baking comprises the use of enzymes from three different sources:

- The endogenous enzymes in flour.
- Enzymes associated with the metabolic activity of the dominant microorganisms.
- Exogenous enzymes added in the dough.

In the baking industry, there is a rising focus on lipolytic enzymes. Because the enzymes degrade polar wheat lipids to produce emulsifying lipids, recent findings suggest that phospholipases can be used to supplement or substitute traditional emulsifiers.

The main constituent of bread is starch. As the starch crystallizes, bread becomes hard and unpleasant to eat with age. To avoid wasting bread and to extend shelf life, the addition of lipase and amylase enzymes in bread making reduces the crystallization of starch in the bread. The savings were chiefly determined by avoided grain production and bread transportation. The use of cathepsins in meat tenderization is doubtful, as there is dearth of evidence that they are released from lysosomes during the postmortem storage of meat. Moreover, cathepsins have been found to have an effect on myofibrillar proteins like myosin, actin, and α-actinin, and during normal aging of muscle, only a small amount of these proteins is affected. Many types of cathepsins have been identified, and out of them cathepsins B (EC 3.4.22.1) and L (EC 3.4.22.15) are said to be the key factors for deterioration of muscle proteins. Effective degradation patterns of myofibrillar proteins were observed in salmon muscle tissue which was almost reproduced upon treating myofibrils with purified cathepsin L. Studies reveal that the myofibrillar proteins are least affected during postmortem storage and the softening of muscle is due to proteolytic digestion of minor cell components that link the major structural units.

Glutaminase (l-glutamine aminohydrolase, EC 3.5.1.2), an enzyme produced by starter cultures to impart flavor, is important in products like meat sausage. Incorporating the enzyme to fermented seasoning agents leads to an increase in the glutamic acid content, and enhancing the "umami" taste in foods. The enzyme is responsible for hydrolytic deamidation of l-glutamine to produce l-glutamic acid, the element indicated as a flavor enhancer, and ammonia, another component which acts as an acid neutralizer. It is a ubiquitous enzyme in bacteria and eukaryote, but its presence in archaea, thermophiles, and plants is doubtful.

Difructose anhydride (DFA) III is a non-cariogenic sweetener and non-digestible disaccharide that promotes the absorption of calcium magnesium, and other minerals in the intestine. DFA III is produced from inulin by the exo-acting inulin fructotrnasferase from Arthrobacter ureafaciencs. Since then, inulin fructotrnasferase from Arthrobacter sp. and other bacteria has been identified. However, the industrial use of DFA III was limited by a low thermal stability and expensive inulin. A great deal of effort has been made to isolate heat-stable inulin fructotrnasferase from various microorganisms and to develop a novel process using a cheap substrate. Recent reports have shown a novel inulin fructotrnasferase Arthrobacter pascens T13-2, Arthrobacter sp. L68- 1, and other Nonomuraea species, stable up to 70–80° C after 1 h of heat treatment. Fructooligosaccharides (FOS) as a prebiotic can be synthesized fromsucrose using fructosyltransferase. Inulin has a similar fructofuranosidic linkage to FOS, which is the smallest substrate for inulin fructotransferase. The utilization of sucrose as a substrate to produce DFA III, resulting in about a 10% (w/w) yield, was attempted through a coupled enzyme reaction as a novel approach.

Papain is a highly efficient enzyme causing significant degradation of myofibrillar as well as collagen proteins. Found that papain had an optimum activity at a wide range of pH levels (5.8–7.0) and temperatures (50– 57° C) specially when the substrate used was casein. Ficin (EC3.4.22.3, MW = 26 kDa) is obtained from the latex of Ficus glabrata, Ficus anthelmintica etc., which contains about 10 proteases with the half-life of 1.5 h at 60°C. In fact, the ficin is most commonly obtained from the fig fruit and reported to have a meat tenderizing effect. Actinidin, or actinidain (EC 3.4.22.14), is obtained from the kiwi fruit (Actinidai deliciosa). The enzyme varies greatly among different cultivars of fruit and it ranges from non-detectable to 10.7 mg/mL juice. The actinidin was stable at a pH range of 7–10; however, it had an optimal activity at 58–62°C. Actinidin hydrolyzes both myofibrillar proteins and connective tissue proteins but appears to have higher proteolytic activity toward collagen. Transglutaminase (TGase) is used for the improvement of textural characteristics of food products. The enzyme is calciumdependent and catalyzes acyl transfer reactions with the ε-amino group lysine, where γ-carboxamide groups of glutamine act as acyl donors and lysyl residues act as acyl acceptors. The reaction involves the intermolecular linking between the glutamine and lysine component of proteins by creating a complex network. Hence, in raw meat products, it provides an enzymatic way of producing uniform shapes and sizes. The enzyme results in an effect on soluble proteins in such a manner that a gel network is produced and combines pieces of meat together. A large number of fat clearing enzymatic lipases are produced on an industrial scale. Most of the commercial lipases produced are utilized for flavour development in dairy products and processing of other foods, such as meat, vegetables, fruit, baked foods, milk product and beer. The function of phospholipase in egg yolk treatment is to hydrolyze egg lecithin, iso-lecithin, which improves the emulsifying capacity and heat stability. The egg yolk thus produced can be useful in the processing of custard, mayonnaise, baby foods, and dressings and in dough preparation.

Extremozymes

In addition to enzymes from conventional sources, the bioprocessing of extremozymes (enzymes produced by organisms living under extreme conditions) has the potential to expand the application scope of enzymes and therefore further increase the market value of enzymes. Extremozymes have potential application in food for a number of reasons: (a) extremozymes are hardy enzymes that can survive other-than-usual food processing conditions, (b) extremozymes are more suitable for substrates whose solubility is enhanced only under extreme conditions, (c) extremozymes may allow in situ catalysis during food processing, for example, breaking down acrylamide during the baking of food, (d) extremozymes are more suitable for use in foods that require aging under extreme conditions (high salt, low temperature, etc.), and (e) the use of extremozymes helps control microbial contamination by microorganisms that grow under normal conditions (Akanbi, 2019: 796). Extremozymes have great economic potential in many industrial processes (e.g. agriculture, food, feed and drinks, detergents, textile, leather, pulp and paper). Extremozymes are enzymes that can function under extreme conditions of temperature, pressure, pH, and alkalinity. They are derived from extremophilic microorganisms (extremophiles) that are metabolically active and grow under extreme conditions.

Extremophiles thrive in ecological niches such as deep-sea hydrothermal vents, hot springs, solfataric fields, shallow marine boiling water, heated sea floor volcanoes, hot lakes, coal and copper mines, and some environments contaminated with nuclear waste. With an average pressure of 38 MPa, the world's oceans are home to piezophilic (formerly called barophilic) microorganisms, including a variety of thermophilic and hyperthermophilic archaeal strains. Given the use of high-pressure conditions in various industries (e.g. the food industry), the availability of extremozymes capable of catalysis at high pressures would offer a novel biotechnological alternative to currently employed processes. Extremozymes may therefore be useful for (a) the production of glucose from cellulose in wood chips; (b) the release of bioactive peptides from keratinous materials such as horns, hoofs, and feathers; and (c) the in situ degradation of polyacrylamide in foods during baking, etc.

Antibiofilm Enzymes

The persistence and expansion of foodborne pathogens and their associated diseases are a prime concern for public health. It has been reported that gastrointestinal, neurological, gynecological, and immunological symptoms are some of the effects frequently caused by foodborne illnesses, often resulting in disability and mortality. If we take a look at the cleaning and disinfection aspects, enzymes have a duty here. Diverse microorganisms are able to grow on food matrixes and along food industry infrastructures. This growth may give rise to biofilms. Biofilms are polymeric mixtures (polysaccharides, lipids, and nucleic acids) produced by microorganisms, which are supported on biotic and abiotic surfaces. There are several favourable conditions for the formation of biofilms, such as the presence of moisture, nutrients, and inocula of microorganisms from the raw material. These conditions provide a persistent source of contamination, causing the deterioration of food as well as creating an important route for cross contamination during production processes. In some scenarios, biofilm formation can play an important and beneficial role, for example, plant growth promotion, the biodegradation of environmental pollutants, and controlling the microbial balance in living organisms. However, in others, such as clinical settings and some industrial processes, it can cause significant problems. Biofilms can become resistant to the chemical and physical treatments applied during cleaning and sanitizing procedures in the food industry. Therefore, the use of enzyme treatments to break down EPS in biofilms is a possible alternative when standard cleaning agents do not give satisfactory results in removing biofilms. The extracellular matrix is mainly composed of polysaccharides, such as cellulose, proteins or exogenous DNA. This matrix can be fixed to hard surfaces (food industry equipment, transport, dispensing and storage surfaces, soil, etc) or to biological structures (vegetables, meat, bones, fruits, etc.). The extracellular matrix has a structural role, which is responsible for the strong persistence of these biofilms in the food industry. It generates complex gradients with respect to nutrients and oxygen diffusion, contains extracellular enzymes used for nutritional purposes, allows for the transfer of cell communication molecules, and protects the embedded cells against toxic compounds.

In summary, biofilm formation confers many advantages to the microbial cells in a food industry environment, such as physical resistance (against desiccation), mechanical resistance (against liquid streams in pipelines) and chemical protection (against chemicals, antimicrobials and disinfectants used in the industry). One strategy that has proved effective for the removal of biofilms from industrial systems has been the use of enzymes, for example the treatment of biofilms formed in food areas. Enzymes and detergents have also been used as synergists to improve disinfectant efficacy. The combination of surfactants/proteolytic enzymes increased the wettability of biofilms formed by a thermophilic Bacillus species, thus enhancing cleaning efficiency also reported the synergistic action of phenolic antimicrobials and surfactants in combination with enzymes. The specificity of the mode of action of each enzyme makes this technique complex. It also means that it is extremely difficult to identify enzymes that are effective against all the different types of biofilms.

Food industry biofilms constitute a serious economic and health issue. On the one hand, the existence of biofilms along food manufacturing surfaces can lead to financial losses as a result of corrosion on the metal surfaces by some bacteria, requiring the replacement of these parts. Furthermore, some bacterial species, such as Pseudomonas spp. and Bacillus spp., secrete many different proteolytic and lipolytic enzymes that can generate unpleasant odors (rancid) and tastes (bitter). In these instances, the affected manufacturing batches must be removed and destroyed. These disruptions may not only represent a significant financial loss for the company, but may also potentially damage their brand with respect to their competitors.

Enzymes for Wine Fermentation

Enzymes are used in modern wine technology for various biotransformation reactions from pre-fermentation through fermentation, post-fermentation and wine aging. Over the last decades, commercial enzyme preparations have gained increasing popularity in the wine industry. They offer many advantages such as accelerated settling and clarification processes, increased juice yield, and improved color extraction.

Table: Enzymes used for winemaking and their function.

Application/Process	Enzymatic Activity	Aim
Enhancement of filtration/ clarification of must	Pectinolytic enzymes	Degradation of viscosity (pectin)
Mash fermentation/heating (red wine)	Pectinase with side activities (cellulases, hemicellulases)	Hydrolysis of plant cell wall polysaccharides. Improvement of skin maceration and color extraction of grapes, quality, stability, filtration of wines
Late phase of fermentation (white wine)	Glycosidases	Improvement of aroma by splitting sugar residues from odorless precursors
Young wine	Glucanases	Lysis of yeast cell walls, release of mannoproteins
Contaminated juice	Glucanases	Lysis of microbial exopolysacharides to improve clarification
Wine	Urease	Hydrolysis of yeast derived urea, preventing formation of ethyl carbamate
Must, wine	Lysozyme from hen egg	Control of bacterial growth
Must, wine	Proteases	Wine stabilization by prevention of protein haze; Reduction of bentonite demand

Technical enzyme preparations are usually obtained from fungi, which are cultured under optimal conditions on substrates to facilitate their preparation and purification. In contrast to grape-derived enzymes, which are often inactive under wine conditions (low pH, presence of ethanol, phenolic compounds, sulphite etc.), fungal enzymes are resistant. The production of oenological enzymes for use in the European Union is regulated by the International Organization of Vine and Wine (OIV), which has ruled that Aspergillus niger and Trichoderma sp. may be used as source organisms (i.e., have GRAS, "generally regarded as safe" status). Selected strains from A. niger are fermented under aerobic conditions in optimized growth media for production of pectinases, hemicellulases and glycosidases, Trichoderma species are used for production of glucanases and Lactobacillus fermentum for urease.

Current commercial enzyme preparations are usually cocktails of different activities, such as glucosidases, glucanases, pectinases and proteases. The search for enzymes with improved and more specific characteristics will continue. In this respect, the study and exploration of the high endogenous enzyme potential of wine and grape-associated microorganisms will assist the wine industry to meet prospective technical and consumer challenges. In contrast to filamentous fungi, yeasts with beneficial enzymatic endowment, could be directly used as starter cultures, without application of enzyme preparations.

Table: Microbial enzymes with relevance for winemaking.

Enzyme	Remarks
Fungi (Botrytis cinerea)	
Glycosidases	Influence aromatic potential of infected grapes by release of volatile aroma compounds
Laccases	Broad specificity to phenolic compounds, cause oxidation and browning
Pectinases	Depolymerizing enzymes, cause degradation of plant cell walls and grape rotting
Cellulases	Multi-component complexes: endo-, exoglucanases and cellobiases; synergistic working, degrade plant cell walls
Lipases	Degrade lipids (e.g., in cell membranes)
Esterases	Involved in ester formation and degradation
Proteases	Aspartic proteases occur early in fungal infection, determine rate and extent of rotting caused by pectinases
Yeasts	Some yeasts produce β-glucosidases which are not repressed by glucose and are resistant to ethanol and low pH; positive influence on wine flavor
Glucosidases	
Glucanases	Occur extracellular, cell wall associated and intracellular, accelerate autolysis process and release of mannoproteins
Proteases	Acidic endoproteases accelerate autolysis process and degradation of grape proteins
Pectinases	Degrade pectin in grape cell walls
Lactic acid bacteria	Convert malic acid to lactic acid
Malolactic enzymes	
Esterases	Involved in ester formation and degradation
Glycosidases	Deliberate flavor compounds
Lipases	Degrade lipids

Lichenases, Glucanases, Cellulases, Xylanases	Degradation of polysaccharides
Proteases	Hydrolysis of proteins
Tannases	Hydrolysis of tannins (polymeric phenolic compounds)
Laccases	Oxidation of phenolic compounds

In the following paragraphs we give an overview on widely used enzyme preparations for wine fermentation and a special focus on wine-associated microorganisms as alternative enzyme sources.

Pectinases

The grape cell wall consists of cellulose microfibrils linked together by a matrix of xyloglucan, mannan, xylan (hemicellulose) and pectin, all of which is stabilized by a protein network. The high viscosity of pectin, which is dissolved after berry crushing, impedes juice extraction, clarification and filtration. In addition, pectin prevents diffusion of phenolic and aroma compounds into the must during wine fermentation.

Commercial Pectinases

The complete degradation of pectin needs cooperation of several enzymes to break the complex molecule into small fragments. They include different enzymatic activities:

- Polygalacturonase (homogalacturonan-hydrolase) (PG): hydrolytic depolymerization of the polygalacturonic acid chain. One can differentiate enzymes that cleave either single galacturonic acid units from the chain end (exo-activity, exoPG, EC 3.2.1.67), or in the middle of the chain (endo-activity, endoPG, EC 3.2.1.15).
- Pectinlyase/pectate lyase (EC 4.2.2.2 and 4.2.2.9): nonhydrolytic cleavage of the polygalacturonic acid chain.
- Pectinesterase (EC 3.1.1.11): hydrolytic cleavage of methanol from the d-galacturonic acid chain, causing drastic viscosity reduction in the liquid portion of the mash and better must flow.
- Acetylesterase (EC 3.1.1.6): cleaves acetyl residues from d-galacturonic acid with release of acetic acid. By this way the interfering acetyl residues at the connecting points of the side chains of the "hairy regions" are removed which facilitates further enzymatic degradation.

Most commercial preparations are derived from fungal sources and are more or less well-defined enzyme mixtures. The application of bulk enzyme preparations is advantageous as it fulfills several functions. Examples are liquefaction enzymes, which contain cellulases and hemicellulases in addition to pectinases.

Table: Examples of commercial pectinase preparations used for winemaking modified from.

Supplier	Enzyme Preparation	Purpose of Application
AEB, South Africa	Pectocel L	Improvement of clarification, filtration and product yield
	Endozym Pectoflot	Improvement of clarification, filtration and product yield
	Endozym Contact Pelliculaire	Enhancement of extraction and color stabilization
	Endozym Rouge	Enhancement of extraction and color stabilization

	Endozyme Active	Improvement of clarification, filtration and product yield
Begerow, Germany	Siha Panzym Extract G	Enhanced extraction and release of color and aroma
	Siha Panzym Clair Rapide G	Improvement of clarification, filtration and product yield
	Siha Panzym Fino G (β-Glucanase)	Improvement of clarification, filtration and sensory
	Siha Panzym Arome G (β-Glucosidase)	Enhanced aroma development
Darleon, South Africa	Influence	Improvement of clarification, filtration and product yield
	Enzym' Color Plus	Enhancement of extraction and color stabilization
DSM, Switzerland	Rapidase Filtration	Improvement of clarification, filtration and product yield
	Rapidase Vino Super	Improvement of clarification, filtration and product yield
Enartis, Italy	Uvazym 1000S	Clarification of white juices—facilitation of fining and filtration
	Progress Quick	Must flotation
	Uvazym couleur	Enhanced extraction during short macerations
Erbslöh, Germany	Trenolin bukett DF (β-Glycosidase)	Enhanced aroma development—Improvement of clarification
	Trenolin Super DF	Improvement of clarification, filtration and product yield
	Trenolin Flot DF	Must flotation
	Trenolin 4000 DF	Enhancement of sugar yield
	Trenolin Filtro DF (β-Glucanase)	Improvement of clarification and filtration; Hydrolysis of Botrytis cinerea exopolysaccharide slime
	Trenolin Bukett DF	Enhance of color and aroma release from red grapes
Laffort, France	Lafazym press	Enhanced color and tannin extraction—Facilitation of fining and filtration
	Lafazym CL	Improvement of clarification, filtration and product yield
	Lafase 60	Improvement of clarification, filtration and product yield
	Lafase HE	Enhancement of extraction and color stabilization
Lallemand, France	Lallemand EX	Enhancement of extraction and color stabilization
	Lallemand OE	Enhancement of extraction and color stabilization
Novo Nordisk, Denmark	Novoclair FCE	Improvement of clarification, filtration and product yield
	Vinozym EC	Enhancement of extraction and color stabilization
	Glucanex (β-Glucanase)	Improvement of clarification and filtration; Hydrolysis of Botrytis cinerea exopolysaccharide slime
	Ultrazym	Improvement of clarification and filtration
	Pectinex Superpress	Improvement of clarification and filtration
Valley Research, USA	Crystalzyme	Rapid clarification–Color improvement–Increased complexity–Process efficiency

Commercially available pectinase preparations contain the active enzymes (2–5%) and additives (sugars, inorganic salts, preservatives) which stabilize and standardize the specificities of the products. Factors that generally inhibit proteins will reduce effectiveness of the enzymes. These include juice clarification with bentonite, which adsorbs and deposits the proteins. Alcohol levels above 17% (v/v) and SO2 levels above 500 mg/L inhibit pectinases. Tannin-rich wines show reduced enzyme activity as phenolic polymers react with the proteins and render them useless.

Effect on Juice Extraction, Clarification and Filtration

The pulp of the grape varieties is rich in pectin compounds. The incomplete hydrolysis of these molecules by the endogenous enzymes can therefore cause problems in processing. If pectinases are applied to the pulp prior to pressing, they can improve juice and color yield. For the clarification of musts after pressing, pectinase-based enzyme preparations are recommended. Its pectin methyl esterase and endogalacturonic activities cause hydrolysis of the pectin chains and facilitate the drainage of juice from the pomace with an increased yield of a free-flowing juice with lower viscosity. In addition, it causes cloud particles to aggregate into larger units that deposit as sediment. The acceleration of the clarification process also produces more compact lees. When applied to the pulp before pressing, it increases juice yield and color yield.

Effect on Color Extraction

Anthocyanidins are the red grape pigments, which mainly occur in the grape skin. The chemical structure, commonly referred to as "flavylium cation", is characterized by two benzene rings linked by an unsaturated cationic heterocycle. Normally, the dye molecule is linked to a glucose monomer, which improves water solubility and stability. Pelargondin, cyanidin, delphinidin, peonidin, petundin and malvidin are the main variants identified in grapes and wine.

Flavonols are light yellow pigments found in the skins of both red and white grapes. These are mainly the glycosylated forms of kaempferol, quercetin and myricetin. In red wine concentrations are in the range of 100 mg/L, in white wines 1–3 mg/L.

Under natural conditions, solubilization of phenolic compounds from grapes is facilitated by increased ethanol concentrations in the course of alcoholic fermentation. However, the extraction is incomplete as the grape skin forms a physical barrier against the diffusion of anthocyanins, tannins and flavors from the cells. Therefore, various oenological techniques have been developed that result in wines that have good visual characteristics and are as stable as possible. Especially wines made by pectinase treatment showed higher concentrations of anthocyanins and total phenols, as well as greater color intensity and optical clarity compared to untreated control wines.

Immobilization

Immobilization is a commonly used strategy to conserve the desirable properties of enzymes for biotechnological applications. In addition to improved stability, immobilization offers a number of advantages, such as reusability, ease of product separation, and better control of catalysis. Various methods have been described for the immobilization of pectinases, such as inclusion in alginate, physical adsorption to anionic resins, and covalent bonding to supports such as porous glass and nylon. A pectinase from Aspergillus niger immobilized on chitosan-coated carriers retained 100% of its original activity after several cycles of reuse.

Yeast Pectinases

S. cerevisiae strains, despite their genetic ability to secrete an endo-polygalacturonase, usually show no or only weak pectinase activity. In contrast, many so-called "wild" yeasts have been identified to be pectinase producers.

Grape fermentations at low temperatures (15–20 °C) are believed to protect the volatile compounds, thereby improving the aromatic profile of the wines. Therefore, cold-active enzymes are required for both extraction and clarification. Psychrophilic yeasts are natural sources for such biocatalysts. The pectinolytic enzymes of Cystofilobasidium capitatum and Rhodotorula mucilaginosa are effective under oenological pH (3.5) and temperature conditions (6.0 °C and 12 °C). Also, pectinases from several A. pullulans strains remain active at wine-relevant concentrations of glucose, ethanol or SO_2, bearing the potential as processing aids for low-temperature wine fermentations.

Lipases

Lipids in wine originate directly from the grape berry and by autolysis of wine yeast. It has been reported that lipid composition undergoes considerable changes during wine fermentation. Authentic lipases (E.C. 3.1.1.3) are mainly active at the oil–water interface of emulsified substrates with long fatty acid chains. Triglycerides are cleaved rendering glycerol and fatty acids. In contrast, carboxylic ester hydrolases (see below) hydrolyze soluble esters with relatively short fatty acid chains. However, the transition between both activities appears somewhat fluid.

Lipolytic activities have been detected only in few wine-relevant Lactobacillus strains, but in different genera of yeasts isolated from natural environments. In theory, lipases could be used for winemaking for decomposition of lipoid cell membranes, thereby improving color extraction from red grape berries.

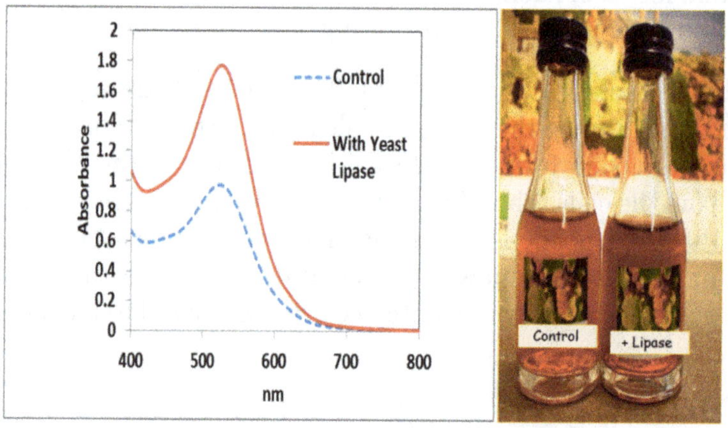

Color extraction from red grapes after a model fermentation without and in presence of a yeast lipase (Claus, unpublished data).

Glucanases

Commercial Glucanases

Polysaccharides in must and wine originate from the grape berries (cellulose, hemicellulose, pectins) and from cell walls of yeasts during growth and autolysis (beta-glucans, chitin). Several strains of lactic acid bacteria (especially Pediococcus spp.) and the grape fungus Botrytis cinerea produce viscous capsular or extracellular polysaccharides impairing wine filtration. The colloidal polysaccharides cannot be removed from wine by flocculants, adsorbents or filtration. Thus, commercial products with glucanase activities e.g., those from Trichoderma sp., and Taleromyces versatilis are useful to reduce viscosity of musts and wine caused by microbial contaminations.

Two types of glucanases are relevant for wine fermentation: (i) exo-β-1,3-glucanases split β-glucan chains by sequentially cleaving glucose residues from the non-reducing end and releasing glucose as the sole hydrolysis product and, (ii) endo-β-1,3-glucanases catalyse the intramolecular hydrolysis of β-glucans with release of oligosaccharides.

Some yeast cell constituents, in particular the wall, can exert a significant impact on the technological and sensory properties of wine. The cell wall consists of β-glucans (~60%), mannoproteins (~40%) and chitin (~2%). In particular, the mannoprotein fraction has attracted increasing interest in wine fermentation to stabilize tartaric acid and protein, improve mouthfeel and reduce astringency. The use of mannoproteins of the cell wall and chitin as binding elements for the removal of undesirable compounds such as ochratoxin A and toxic heavy metals has been proposed. The mannoprotein fractions present in the yeast cell walls are highly variable between species and even strains, providing opportunities for the development of alternative fining products that could replace conventional proteinaceous animal preparations with allergenic potential.

Commercial enzyme preparations facilitate the release of these components. As a prerequisite they have to be effective under acid wine conditions, moderate temperatures and in presence of ethanol.

Microbial Glucanases

A major microbial source of polysaccharide-degrading exoenzymes are non-Saccharomyces yeasts belonging to the genera Kloeckera, Candida, Debaryomyces, Rhodotorula, Pichia, Zygosaccharomyces, Hanseniapora, Kluyveromyces and Wickerhamomyces. Glucanolytic enzyme activities were also detected in wine-relevant lactic acid bacteria.

Table: Glucanases of non-Saccharomyces yeasts with possible use for winemaking modified from.

Species	Specificity	Substrate	MW(kDa)
Candida albicans	Endo-β-1,3-	L, OL, P	49
	Exo-β-1,3-	L	107
Candida hellenica	nd	G	nd
Candida lambica	nd	G	nd
Candida pulcherrima	nd	G, Li	nd
Candida stellata	nd	G, Li	nd
Candida utilis	Endo-β-1-3-	L, PNPG	20
	Exo-β-1,3-1,6-	L, P, PNPG	20
	Endo-β-1,3-	L, OL	21
Kloeckera apiculata	nd	G, Li	nd
Kluyveromyces phaseolosporus	Endo-β-1,3-(I)	L, OL	180
	Exo-β-1,3-(II)	L, OL	45
	Exo-β-1,3-1,6-(III)	L, P	18.5
	Exo-β-1,3-1,6-(IV)	L, Ol, P	8.7
Pichia polymorpha	Endo-β-1,3-(I)	L, OL	47

	Exo-β-1,3-1,6-(II)	L, OL, P, PNPG	40
	Exo-β-1,3-(III)	L, PNPG	30
Schizosaccharomyces pombe	Endo-β-1,3-(I)	L, OL	160
	Endo-β-1-3-(II)	L, OL	75
Schizosaccharomyces versatilis	Endo-β-1,3-	L, OL	97
	Exo-β-1,3-1,6-	L, P, PNPG	43
Wickerhamomyces anomalus AS1	Exo-β-1,3-	L, PNPG	47.5

Glycosidases

The organoleptic properties of wine are determined by a variety of different compounds that are already present in the grape (aroma) or arise during fermentation or storage (bouquet). Acids such as tartaric acid or citric acid affect the taste, but the characteristic odor and taste is mainly due to volatile organic substances such as esters, alcohols, thiols or terpenes.

Due to their low odor threshold, particularly terpenes determine wine flavor. Like other aroma active compounds (C13 norisoprenoids, benzene derivatives, aliphatic alcohols, phenols), they are secondary metabolites mainly derived from the grape skin. Approximately 90% of these compounds do not exist in a free form, but are conjugated to mono- or disaccharides, thereby forming water-soluble and odourless complexes. The most frequently occurring aroma precursors in grape varieties such as Muscat and Riesling are the glycosidic bound terpenes linalool, nerol and geraniol. The sugar residues consist of rutinoside (rhamnose-glucose), arabinoside (arabinose-glucose) or apioside (apiose-glucose).

Enzymatic hydrolysis of sugar-conjugated precursors release very aromatic, volatile terpenes (aglycones). Usually, the terminal sugars are first cleaved off by a rhamnosidase, an arabinosidase or an apiosidase. In a second step, the terpenes are released by a β-d-glucopyranosidase. This means that the latter activity alone can release only terpene compounds bound to a single glucose residue. In addition to a stepwise reaction, some glucosidases are able to hydrolyze the glycosidic bond to the aglycone, regardless of the number of sugar residues. Nowadays commercial enzyme preparations are available that can hydrolyze the disaccharide directly from the terpene in a single step. In many bulk enzyme preparations, glycosidase activities occur as side activities along with pectinase and glucanase activities.

An important microbial source of wine-related enzymatic activities is lactic acid bacteria. Researcher studied >1000 isolates for glycosidases. The β-glucosidase activities were only found in cells, but not in the supernatants of the cultures. Four O. oeni isolates retained their enzymatic activity under the conditions of winemaking. In a similar study, cell-bound glucosidase and arabinosidase activities from O. oeni strains released high levels of monoterpenes from natural substrates under optimal conditions. The enzymes showed broad substrate specificities (release of both primary and tertiary terpene alcohols) and remained active in grape juice.

Glycosidase activities have been also detected in various non-Saccharomyces yeasts (Candida, Hanseniaspora, Pichia, Metschnikowia, Rhodotorula, Trichosporon, Wickerhamomyces). Several experiments on the technical application of yeast glycosidases to improve organoleptic quality of wines gave positive results.

Table: Aroma enhancing enzymes of non-Saccharomyces yeasts with possible use for wine fermentation modified from.

Species	Enzymatic Activities*				
	β-d-Glucosidase	α-l-Arabino-Furanosidase	α-l-Rhamnosidase	β-d-Xylosidase	Carbon-Sulfur Lyase
Aureobasidium pullulans	+	+	+		
Brettanomyces anomalus	+				
Candida guillermondii	+		+	+	
Candida molischiana	+				
Candida stellata	+		+	+	
Candida utilis				+	
Candida zemplinia					+
Debaryomyces castelli	+				
Debaryomyces hansenii	+				
Debaryomyces polymorphus	+				
Debaryomyces pseudopolymorphus	+				
Debaryomyces vannjii	+				
Hanseniaspora guillermondii	+				
Hanseniaspora osmophila	+			+	
Hanseniaspora vineae	+	+	+	+	
Hanseniaspora uvarum	+	+	+	+	
Issatschenkia terricola	+				
Kluyveromyces thermotolerans	+				+
Metschnikowia pulcherrima	+				+
Pichia angusta			+		
Picha anomala	+	+	+	+	
Pichia capsulata		+			
Pichia guilliermondii			+		
Pichia kluyveri					+
Pichia membranaefaciens	+			+	
Saccharomycodes ludwigii	+				
Schizosaccharomyces pombe	+				
Sporidiobolus pararoseus	+				
Torulasporus delbrueckii	+				+
Torulasporus asahii	+				
Wickerhamomyces anomalus	+	+		+	
Zygosaccharomyces bailii	+				

Polyphenols in red wine, such as resveratrol, have gained increasing public and scientific interest due to their supposed beneficial effects on human health. A majority of the polyphenols in nature are conjugated to sugars or organic acids, making them more hydrophilic and less bioavailable to humans. The amount of glycosylated forms of resveratrol, known as piceid or polydatin, has been found to be up to ten times higher in red wines. Since these modified forms are less bioactive, experiments with β-glucosidases from various fungal sources have been undertaken to increase the trans-resveratrol content in wines by hydrolysis of glycosylated precursors. The multifunctional glucanase WaExg2 of W. anomalus AS1 released the aglycones from the model compounds arbutin, salicin, esculin and polydatin. WaExg2 was active under typical wine conditions such as low pH (3.5–4.0), high sugar concentrations (up to 20% w/v), high ethanol concentrations (10–15% v/v), presence of sulphites and various cations. Therefore, this yeast strain could be useful in wine production for several purposes: to increase the levels of sensory and beneficial compounds by cleaving glycosylated precursors or reducing the viscosity by hydrolysis of glycan slurries. In this context Researcher Underlined that glucose- and ethanol-tolerant enzymes from Wickerhamomyces are of great interest to the wine industry.

Esterhydrolases and Synthetases

Esters (e.g., ethyl acetate, isoamyl acetate, ethyl hexanoate, ethyl octanoate, ethyl decanoate) contribute to the most desirable fruity wine flavors. They are synthesized by the grapes but are also produced by yeasts in course of alcoholic fermentation. During malolactic fermentation, significant changes in the concentration of individual esters were observed. Presence of alcohol acyltransferases (ester synthesis) and esterases (ester hydrolysis) in wine yeasts and lactic acid bacteria is well documented.

Depsides are esters of aromatic hydroxycarboxylic- or phenolic acids with each other or with other carboxylic acids of the grape, such as tartaric acid. These compounds can be hydrolyzed by cinnamoyl esterases ("depsidases"), which often appear as side activities in enzyme preparations made from A. niger. The fission products can have a negative influence on wine quality. Enzymatic deliberated phenolcarboxylic acids, such as caffeic acid or coumaric acid, can be converted by the yeast metabolism to the volatile phenol derivatives 4-vinylguajacol and 4-vinylphenol, which are unpleasant side-tastes in the wine. Therefore, commercial pectinase preparations should be free of depsidase side activities. In a recent study of 15 commercial enzyme preparations, approximately half of the samples yielded significant cinnamoyl esterase activities.

Proteinases

Proteins in must in wine are derived from the grapes, and from microbial cells (yeasts, lactic, acid bacteria) and their activities. Another important source are protein-based wine additives (e.g., lysozyme, ovalbumin, gelatin, casein) which could pose allergenic-like reactions to consumers. Most of these proteins have vanished after termination of wine fermentation and subsequent fining procedures. However so-called pathogen-related (PR) proteins (β-glucanases, chitinases, thaumatin-related proteins) can still be present. They are synthesized by the plants for defence against bacterial or fungal infections and in response to abiotic stress. Due to their compact structures, they are resistant against acid wine conditions, heat, and proteolysis.

In combination with other wine ingredients, PR proteins can cause undesirable turbidity especially during cold storage of white wines with negative economic consequences. Currently, protein removal is mainly achieved by bentonite addition, a process that can be associated with decreased wine quantity and quality. Bentonite acts essentially as a cation exchanger, and individual wine proteins adsorb to different degrees on the clay. Proteins that are negatively charged at wine pH (about 3.5) and/or are highly glycosylated as laccases of Botrytis cinerea are less bound by bentonite. Thus, new fining agents are desired to remove proteins from wine.

Proteases from Fungal and Plant Sources

Enzymatic degradation of wine proteins seems an attractive alternative to bentonite treatment as it would minimize losses of volume and aroma. As a prerequisite, suitable proteases have to be active under specific wine conditions (acid pH, presence of ethanol, sulphites, phenolics) and preferably act at low temperatures. Another challenge is the resistance of PR proteins against proteolysis due to their special molecular features like disulfide-bonds and glycosylations. Nevertheless, other grape proteins might be more susceptible, and thus proteases may help to reduce effective bentonite dosages. Currently, proteases from plants (papain, bromelain) have been tested with some promising results. A fungal protease from Aspergillus sp. (aspergilloglutamic peptidase) has already approved for Australian winemaking. The enzymatic procedure involves flash-pasteurization of grape must and is thus limited to specialized wineries. In this context, a protease of Botrytis cinerea BcAp8 has been described to hydrolyze grape chitinase at moderate temperatures.

Microbial Proteases

Microbial proteases can be an alternative or supplement to bentonite treatment for removal of unwanted wine proteins. Most Saccharomyces cerevisiae strains show no extracellular protease activity on diagnostic agar media. However, a 72 kDa extracellular pepsin-like aspartic protease was characterized from a PIR 1 strain. The enzyme was active during grape juice fermentations, although it did not affect turbidity-inducing proteins, unless the wine was incubated at 38 °C for extended time.

Proteinase A (PrA, saccharomycin; EC 3.4.23.25) is the major vacuole protease of S. cerevisiae encoded by the PEP4 gene. As result of yeast autolysis, PrA enters wine in course of alcoholic fermentation. Far more, it has been found that under stress conditions (e.g., nutrient limitations) PrA is not targeted to the vacuole, but is misdirected to the cell membrane and secreted in the medium. This would be advantageous for winemaking in view of haze reduction. Interestingly, the same situation is undesirable for beer brewery as PrA degrades proteins (e.g., lipid transfer protein 1), necessary for foam formation. Apart from PrA, S. cerevisiae expresses different cell-bound proteases, some of which are not fully characterized.

Non-Saccharomyces yeasts are important sources of extracellular enzymes including proteases. Strains of Metschnikowia pulcherrima and Wickerhamomyces anomalus secreted aspartic proteases and degraded a model protein (bovine serum albumin) during growth in grape juice. In a recent study, heterologous expressed aspartic protease MpAPr1 from M. pulcherrima was added to a Sauvignon Blanc must. It was shown that the enzyme was active during fermentation and degraded wine proteins to some extent. An alternative strategy would be to perform wine fermentations with appropriate protease-positive starter cultures. In addition to cost reductions, there are no administrative restrictions for yeast applications in must and wine, which must be taken into account with enzyme preparations.

Table: Extracellular proteases of non-Saccharomyces yeast with possible use for wine fermentation.

Species	Mode of Identification	Characterization
Candida apicola	Skim milk agar (pH 3.5), Gene Sequencing	Aspartic protease CaPR1 (39.2 kDa)

Candida stellata	Casein agar	nd
Hanseniaspora guelliermondii Hansenispora valbyenis Hanseniapora occidentalis	Casein agar and broth (pH 6.0)	nd
Hanseniaspora uvarum	Skim milk agar (pH 3.5)	nd
	Casein agar	nd
Kloeckera apiculata	Enzymatic; Inhibitor studies	Acid endopeptidase
Metschnikowia pulcherrima	Skim milk agar (pH 3.5)	nd
	Azocasein hydrolysis during fermentation of grape must	nd
	Skim milk agar (pH 3.5), Sequencing of protease gene, Purification	Aspartic protease pAPR1 (40.8 kDa)
	Casein agar	nd
	Skim milk agar (pH 4.5), Enzymatic; Inhibitor studies, LC-MS/MS	Aspartic protease
Wickerhamomyces anomalus	Skim milk agar (pH 3.5)	nd
	Skim milk agar (pH 4.5), Enzymatic; Inhibitor studies, LC-MS/MS	Aspartic protease WaAPR1 (47 kDa)

Occurrence of proteolytic activities in lactic acid bacteria is also well-documented. Growth of Oenococcus oeni depends on the presence of amino acids in the culture medium because of deficiency in corresponding synthetic pathways. This bacterium secretes several proteases which may help to gain access to rare nitrogen sources during malolactic fermentation.

Phenoloxidases

Spontaneous and enzymatic oxidations exert dramatic effects on the final phenolic composition from the grape berry to bottled wine. Once the integrity of the berries is destroyed, oxidative enzymes (phenoloxidases) and their phenolic substrates are exposed to the air, resulting in enzymatic browning.

Wine browning by Botrytis-laccase (right).

There are two classes of copper enzymes responsible for these reactions : Tyrosinase (E.C. 1.14.18.1) hydroxylates monophenols to ortho-diphenols and oxidizes the latter to ortho-quinone intermediates which easily react further to polymeric, mostly colored products. Laccase (E.C. 1.10.3.2) has no monohydroxylase activity, and oxidizes a broad spectrum of different phenols and other compounds by a radical mechanism. Tyrosinase originates from grape berries, whereas laccases in must and wine are derived from epiphytic fungi, particularly Botrytis cinerea. Phenolic compounds as caffeic acid, gallic acid, vanillic acid, ferulic acid, or especially resveratrol are known for their beneficial effects on human health. In addition to be radical scavengers, they are activators of the human's intrinsic cellular antioxidant system and have antimicrobial properties. Tyrosinase and laccase oxidize phenolic wine compounds and thus alter their antioxidant and antimicrobial properties. In this context, it is an interesting observation that gallic acid oxidation by a laccase from Trametes versicolor was higher at 30 °C than at 45 °C. Although fungal laccases are usually more active at higher temperatures, the effect can be explained by reduced oxygen solubility under the experimental conditions.

Laccase is generally not very welcome in wine, but several studies have ruled out that controlled laccase treatments could promote wine stabilization and even improve sensory properties.

Volatile phenols particular produced by Brettanomyces/Dekkera sp. yeasts are associated with a serious "Brett" taste defect in wine. Researcher used a used a laccase from T. versicolor to reduce concentrations of 4-ethylguaiacol and 4-ethylphenol in a synthetic model wine.

Biogenic amines (BA) are another class of undesirable compounds in wine. They originate from the grape berries or are formed during fermentation by activities of decarboxylase-positive microorganisms. Although more common in foods such as cheese, BA have received much attention in wine, as ethanol can enhance the negative effects on human health by inhibiting the enzymes responsible for the detoxification of these compounds.

Enzymatic degradation of BA is usually catalyzed by various classes of oxidases. Depending on the type of prosthetic group, they can be classified into FAD-dependent (E.C. 1.4.3.4) and copper-containing amine oxidases (CAOs, E.C. 1.4.3.6). The latter have been detected in various yeasts such as Kluyeromyces marxianus or Debaryomyces hansenii. These enzymes belong to the class of type 2 or "non-blue" copper proteins which convert primary amines to the corresponding aldehydes with an equimolar consumption of molecular oxygen and formation of hydrogen peroxide and ammonia.

Aromatic amines such as tyramine, phenylethylamine, tryptamine or serotonin are another class of compounds that can be oxidized by laccases. Researcher detected enzymatic activities responsible for BA degradation in lactic acid bacterial strains isolated from wine. Responsible enzymes have been isolated and purified from Lactobacillus plantarum J16 and Pediococcus acidilactici CECT 5930 strains and have been identified as intracellular laccase-like multicopper oxidases. When the L. plantarum J16 laccase was overexpressed in Escherichia coli, it oxidized some BA, mainly tyramine.

Urease

Increased amounts of urea in wine can originate from yeast activities and then converted by a chemical reaction into the carcinogenic substance urethane (ethylcarbamate). During malolactic fermentation, lactic acid bacteria can produce other precursors of ethyl carbamate, such as arginine-derived citrulline and carbamyl phosphate. Especially at higher temperatures fermented wines may contain excessive amounts of urethane. Therefore appropriate precautions should be taken to prevent the production of urethane. These include, for example, the selection of suitable starter cultures for malolactic fermentations and the reduction of arginine concentrations in the grape. Urease was introduced in 1997 by the EU as a new enzymatic wine treatment agent and can be used in exceptional cases. The enzyme splits urea into ammonia and carbon dioxide, preventing urethane formation. The commercial urease from Lactobacillus fermentum is effective on urea at doses of 50 mg/L in red wines and 25 mg/L in white wines.

Lysozyme

Yeasts, lactic acid and acetic acid bacteria have a significant influence on wine quality. Microbial growth in musts and wines is conventionally controlled by the addition of sulfur dioxide. However, presence of sulphites in alcoholic beverages, particularly in wines, can cause pseudo-allergic responses with symptoms ranging from gastrointestinal problems to anaphylactic shock. Other antimicrobials such as sorbic acid and dimethyl carbonate are primarily active against yeasts, but have limited activity against bacteria.

Lysozyme (EC 3.2.1.17) is a muramidase widely used to control microbial growth in foods such as cheese and wines. Extensive enzymatic hydrolysis of the bacterial cell wall peptidoglycan, a polymer of N-acetyl-d-glucosamine units which are β-1,4-linked to N-acetylmuramic acid, results in cell lysis and death in hypoosmotic environments. Some lysozymes can kill bacteria by stimulating autolysin activity. In addition, bactericidal mechanisms involving membrane damage without enzymatic hydrolysis of peptidoglycan has been reported for c-type lysozymes, such as hen egg white lysozyme. Gram-negative bacteria (i.e., acetic acid bacteria) are rather resistant against lysozyme, because the outer membrane acts as a barrier.

Lysozyme commercially produced from hen's egg white has been approved for winemaking by the International Organization of Vine and Wine in 2001. The amount added normally ranges between 250–500 mg/L. Four main applications and dosages are: (a) prevention of the onset of malolactic fermentation (early addition of 100–150 mg/L); (b) total inhibition of bacteria activity and malolactic fermentation (500 mg/L); (c) protection of wine during suboptimal alcoholic fermentation (250–300 mg/L); (d) stabilization of wine after malolactic fermentation (250–300 mg/L). Lysozyme can be eliminated by addition of fining agents, among which bentonite and metatartaric acid are the most efficient.

It has been reported that various Gram-positive strains of Pediococcus sp., Lactobacillus sp. and Oenococcus oeni were not efficiently hydrolyzed by hen's egg white lysozyme. Reasonable explanations are structural modifications of the peptidoglycan, like N-deacetylation and O-acetylation of the glycan chains or amidation of free carboxyl groups of amino acids in the peptide chains. As a possible alternative to hen's egg white lysozyme, exoenzymes (protease and muramidase) from Streptomyces species showed a broad bacteriolytic spectrum under winemaking conditions.

It should be mentioned that hen's egg lysozyme can display pH-dependent chitinase side activities. Under adverse conditions yeast cell walls (containing 2–4% chitin primarily in the bud scar regions) can be weakened by lysozyme with significant effects on vitality and stress response of Saccharomyces cerevisiae during wine fermentation.

Legislative Regulations

The use of enzymes in wine production in the European Union is regulated by the International

Organisation of Vine and Wine (OIV). Specified resolutions define general aspects of enzymes in winemaking, the permitted enzyme activities, mode of application and enzyme activity measurements. The USA, Canada and China have national regulations in winemaking.

Genetic engineering: Today's enzyme production is based either on special selected wildtype strains or on genetically modified organisms (GMOs). The use of GMO production strains has considerable advantages: the product yields with GMOs are much higher than with wild strains and undesirable side activities become minimized. This makes it more efficient to produce and to guarantee the purity of the enzyme products. The labelling is regulated by the resolution OIV-OENO 485-2012.

GMOs for must fermentations: Although increasing numbers of Saccharomyces yeasts have been improved by genetically engineering, only two GMOs have been allowed for winemaking in three countries. The first recombinant strain to get official approval by appropriate food safety authorities (in the USA and Canada) was the malolactic wine yeast ML01. The GMO carries the Schizosaccharomyces pombe malate permease gene (mae1) and the Oenococcus oeni malolactic gene (mleA). The second strain ECM001 expresses the urease gene constitutively to prevent formation of urethane. Whether yeast strains obtained by protoplast fusion should be considered as GMO is in legal limbo.

Enzymes in Light Industry

Light industry usually produces small consumer goods, such as food, textile, furniture, papermaking, printing, daily chemicals, sporting goods and so on. It is a raw material-oriented industry. Most light industrial products are produced for end-users rather than intermediate products for use in other industries. Compared with heavy industry, light industry usually requires less raw materials, space and electricity, and causes less pollution. With the development of enzyme engineering, enzyme plays a more and more important role in the field of light industry.

- Application of Enzymes in Animal Feed Industry: Enzyme has been used as feed additive in industry for decades. During this period, the feed enzyme industry has gone through several stages of development. The first phase was the use of enzymes to enhance nutrient digestibility. The industry then started to advocate enzyme addition to poultry diets based on non-viscous grains, such as sorghum and corn. The next phase is the application of enzymes to non-cereal grain components of the diet. The enzyme industry today is constantly searching for new areas of application.

- Application of Enzymes in Detergent Industry: The detergent industry is an important application area where industrial enzymes are utilized to boost cleaning performance. Enzymes represent environmentally sound alternatives to the use of toxic chemicals and pollutants, with reduced generation of waste and added performance benefits as the result. Enzymes have effectively promoted the development and improvement of modern household and industrial detergents. The main categories of detergent enzymes-protease, lipase, amylase and cellulase-each provide special benefits for laundry and automatic dishwasher applications.

- Application of Enzymes in Textile Industry: The enzymes used in textile field are amylase, catalase and laccase, which are mainly used for starch removal, hydrogen peroxide degradation, textile bleaching and lignin degradation. The application of enzyme in textile chemical processing is rapidly recognized by all countries in the world because of its non-toxic and environmental protection characteristics, which puts forward higher and higher requirements for textile enterprises to reduce pollution in textile production. In addition, the use of enzymes reduces processing time, saves energy and water, and improves product quality and potential process integration.
- Application of Enzymes in Leather Processing: Leather processing is one of the highly polluting industrial activities. The traditional chemical method of depilation is the main cause of pollution in leather processing industry. The extensive use of sulfides not only has adverse consequences for the environment, but also undermines the efficiency of sewage treatment plants. Therefore, the systematic use of protease instead of lime and sulfides to rationalize the depilation process has become a primary problem in leather processing. This will lead to a substantial reduction of effluent load and toxicity in addition to improvement in leather quality.
- Application of Enzymes in Pulp and Paper Industry: The application of enzymes in pulping and papermaking industry began in 1986, but their use is relatively small. However, the publication of an increasing number of papers on the subject demonstrated growing interest in the issue. Cellulase, xylanase, laccase and lipase are the most important enzymes in pulping and papermaking. In addition, the main goal of enzyme development is xylanase to promote bleaching and cellulase modification of fiber. The development trend of enzymes is mainly focused on improving the thermal stability and alkaline strength of enzymes.

Enzymes in Biotechnology

Biotechnology could be followed back to a thousand years prior, when human civilisation began to utilise yeasts to make alcohol. This might have been the first dawn of biotechnology in food creation. Alongside the improvement of natural and social sciences these days, biotechnology conveys progressively vivid implications. In the modern world, biotechnology regularly alludes to the way toward making or altering items utilizing living systems or living beings. Other than the customary fermentation, biotechnological instruments have extended to apply further developed sciences, such as genetic engineering, applied immunology, and medicinal therapies and diagnostics. Among these, an assortment of chemicals is crucial—they may show up as mallets and chisels to help achieve molecular biology experiments, or they may go about as proficient smaller scale reactors in mechanical creation. They exist all over the place and take after a wide range of capacities in living cells or in harsh conditions bearing no life. In recent times, we have seen these capacities being exploited in incredible details for better uses in biotechnology.

Enzymes can be defined as biological catalysts or assistants. They comprise different sorts of proteins that work to either dispatch or speed up the chemical reaction needed for a particular activity or nutrient. Enzymes work in a mild domain like the bodily condition of a living being, and they bolster life by synthesising and degrading materials that comprise the building blocks of the living being and by making energy. They work as profoundly selective catalysts so that they selectivity catalyse explicit reactions (reaction specificity) and specific materials (substrate specificity). They can likewise be extricated from cells and afterwards used to catalyse a wide scope of monetarily significant procedures. For instance, they have significant jobs in the creation of sweetening agentS and the alteration of anti-infection agents, they are utilised in washing powders and different cleaning items, and they assume a key job in scientific gadgets and assays that have clinical, forensic and ecological applications. Technology for using enzymes for the improvement of life is a key element of biotechnology.

Biotechnology is characterised as the exploitation of natural procedures for industrial and varied purposes, and can be comprehensively characterized as "utilising life forms or their items for business purposes." Particularly the genetic manipulation of microbes is done for the creation of anti-infection agents, hormones, and so on. It incorporates a wide scope of ventures, for example:

- Pharmaceuticals Industries.
- Bio-fuels Industries.
- Paper and Pulp Industries.
- Food and Brewing Industries.
- Agricultural Industries.

Application of Enzymes in Cell Biology

Enzymes are usually utilised in cell lysis. Mechanical cell interruption techniques do not separate the release of the ideal item from among a large group of other debasing particles and cell garbage, and they may likewise harm the protein product. Conversely, the utilisation of a lytic enzyme system, which can give biological specificity to the procedure of cell lysis and item discharge, shows an intriguing potential for controlled lysis. Cell wall lytic enzymes are important instruments for the biotechnologist, with numerous applications in medication, the food business, and horticulture, and for recouping of intracellular items from yeast or microorganisms. The decent variety of potential applications has led to the improvement of lytic compound systems with specific attributes, reasonable for fulfilling the necessities of every specific application.

Applications of Enzymes in Molecular Biology

Nucleic acids utilised for sub-atomic cloning can be of natural or engineered origin, and their length ranges from a couple to a few thousand nucleotides. Nucleic acids can be broadly controlled, so as to procure specific attributes and properties. Such controls incorporate propagation, ligation, absorption, or expansion of modifying groups, viz. phosphate or methyl groups. These alterations are catalyzed by polymerases, ligases, nucleases, phosphatases, and methylases, respectively. Following are the various applications of enzymes in biotechnology:

Novel Fibres

Engineered strands produced using renewable sources of biomass are environmentally sustainable and are getting progressively monetarily viable. Biodegradable engineered polymers incorporate novel strands, such as polyglycolic acid and polylactic acid, which are produced using natural starting materials. Not every novel fibre is manufactured; they may be naturally derived. Some general organic filaments originate from essential materials found in nature, including:

- Chitin: A kind of sugar polymer found in shellfish.
- Collagen: A kind of protein found in animal connective tissue.
- Alginate: A kind of sugar polymer found in specific microbes.

A perfect example of a manufactured biomass fibre is Polylactic Acid (PLA), which is formed by fermenting cornstarch or glucose into carboxylic acid and afterwards artificially changing it into a polymer fibre. With properties like other engineered filaments, PLA based materials are strong with a smooth feel and might be mixed with fleece or cotton.

Breakdown of Hydrogen Peroxide

At the point when cotton is bleached, a chemical called hydrogen peroxide, which can react with different colours, stays on the fabric. Catalase enzymes explicitly disintegrate hydrogen peroxide and might be utilised to expel this reactive chemical before further colouring.

Biostoning and Biopolishing: Rather than utilising rough devices like pumice stones to make a stonewashed effect or to evacuate surface fluff, cellulase enzymes might be used to adequately stonewash and clean textures without abrasively harming the fibres.

Detergents

Enzymes permit detergents to adequately clean garments and expel stains. They can evacuate certain stains, for example, those made by grass and sweat, more adequately than enzymes free cleansers. Without enzymes, a great deal of energy would be needs to make the high temperatures and vigorous shaking expected to clean garments successfully. Enzymes utilised in clothing cleansers must be reasonable, stable, and safe to utilise. At present, just protease and amylase catalysts are fused into cleansers. Lipase enzymes separate too easily in the washing machine to be exceptionally valuable in cleansers. In any case, their dependability is being considered and further created through strategies, for example, genetic screening and modification.

Polylactic Acid (PLA)

PLA has likely applications in a few territories, such as:

- Textiles: Garments, fashions, upholstery.
- Horticulture: Plant mats, tree nets, soil disintegration control items.
- Sanitation: Household wipes, diaper items.
- Medication: Expendable pieces of clothing, clinical materials.

PLA limits natural waste, as it might be completely biodegraded by microorganisms under suitable conditions into carbon dioxide and water. This is unlike the non-renewable oil assets used to make conventional engineered fibres, the supply of sustainable corn biomass expected to make PLA is relied upon to outperform demand in the foreseeable future.

De-Sizing of Cotton

Untreated cotton fibres can break easily when being woven into textures. To forestall this breakage, they are covered with a jelly like substance through a procedure called sizing. After the fibres have been woven into the fabric, the agents expected to additionally complete the material cannot adhere to the jelly covered textures. Hence, the protective sizing agents must be expelled by a procedure called de-measuring. Amylase enzymes are generally utilised in de-sizing, as they do not debilitate or influence cotton filaments, nor do they hurt the environment.

From the entirety of the above portrayal of pharmaceuticals enterprises, Bio-fuels ventures, paper and mash businesses and brewing enterprises it has been presumed that catalysts are significant in biotechnology. These both are significant for the welfare of humankind.

References

- Enzyme-technology: microbenotes.com, Retrieved 10, August 2020
- Recent-Applications-of-Enzymes-in-Food-Industry-342234868: researchgate.net, Retrieved 14, May 2020
- Top-5-applications-of-enzymes-in-biotechnology: infinitabiotech.com, Retrieved 30, January 2020

Permissions

All chapters in this book are published with permission under the Creative Commons Attribution Share Alike License or equivalent. Every chapter published in this book has been scrutinized by our experts. Their significance has been extensively debated. The topics covered herein carry significant information for a comprehensive understanding. They may even be implemented as practical applications or may be referred to as a beginning point for further studies.

We would like to thank the editorial team for lending their expertise to make the book truly unique. They have played a crucial role in the development of this book. Without their invaluable contributions this book wouldn't have been possible. They have made vital efforts to compile up to date information on the varied aspects of this subject to make this book a valuable addition to the collection of many professionals and students.

This book was conceptualized with the vision of imparting up-to-date and integrated information in this field. To ensure the same, a matchless editorial board was set up. Every individual on the board went through rigorous rounds of assessment to prove their worth. After which they invested a large part of their time researching and compiling the most relevant data for our readers.

The editorial board has been involved in producing this book since its inception. They have spent rigorous hours researching and exploring the diverse topics which have resulted in the successful publishing of this book. They have passed on their knowledge of decades through this book. To expedite this challenging task, the publisher supported the team at every step. A small team of assistant editors was also appointed to further simplify the editing procedure and attain best results for the readers.

Apart from the editorial board, the designing team has also invested a significant amount of their time in understanding the subject and creating the most relevant covers. They scrutinized every image to scout for the most suitable representation of the subject and create an appropriate cover for the book.

The publishing team has been an ardent support to the editorial, designing and production team. Their endless efforts to recruit the best for this project, has resulted in the accomplishment of this book. They are a veteran in the field of academics and their pool of knowledge is as vast as their experience in printing. Their expertise and guidance has proved useful at every step. Their uncompromising quality standards have made this book an exceptional effort. Their encouragement from time to time has been an inspiration for everyone.

The publisher and the editorial board hope that this book will prove to be a valuable piece of knowledge for students, practitioners and scholars across the globe.

Index

www.ingramcontent.com/pod-product-compliance
Lightning Source LLC
LaVergne TN
LVHW080912210926
848485LV00004B/638
* 9 7 8 1 6 3 9 8 7 4 4 5 3 *